SOME BEAUTIFUL INDIAN CLIMBERS AND SHRUBS

SOME BEAUTIFUL INDIAN CLIMBERS AND SHRUBS

BY
N. L. BOR
AND
M. B. RAIZADA

Revised Second Edition

BOMBAY NATURAL HISTORY SOCIETY

OXFORD UNIVERSITY PRESS
MUMBAI DELHI CALCUTTA MADRAS

Oxford University Press, Walton Street, Oxford OX2 6DP

Oxford, New York,
Athens, Auckland, Bangkok,
Calcutta, Cape Town, Chennai, Dar-es-Salaam,
Delhi, Florence, Hong Kong, Istanbul, Karachi,
Kuala Lumpur, Madrid, Melbourne,
Mexico City, Mumbai, Nairobi, Paris,
Singapore, Taipei, Tokyo, Toronto,
and associated companies in
Berlin, Ibadan

© Bombay Natural History Society 1982

First edition, 1954
Second edition, 1982
Reprinted, 1990
Second reprint 1999

ISBN 0 19 562163 8

All rights reserved. This book, or any parts thereof, or plates therein, may not be reproduced in any form without the written permission of the publishers.

PRINTED BY BRO. LEO AT ST. FRANCIS INDUSTRIAL TRAINING INSTITUTE, MOUNT POINSUR, BORIVLI (W), MUMBAI 400 103, PUBLISHED BY THE BOMBAY NATURAL HISTORY SOCIETY AND CO-PUBLISHED BY MANZAR KHAN, OXFORD UNIVERSITY PRESS, OXFORD HOUSE, APOLLO BUNDER, MUMBAI 400 039.

FOREWORD TO THE FIRST EDITION

The present book substantially is a reproduction of a series of papers that appeared in the Journal of the Bombay Natural History Society between the years 1939 and 1948. From the beginning the series proved very popular with our readers, many of whom expressed the desire to have it in book form. The series was from the first intended as a sort of continuation of Blatter and Millard's other series on SOME BEAUTIFUL INDIAN TREES, which was also republished in book form and has been in great demand with the public.

The book which is now being offered to the public fills a great void in the popular botanical literature of India. We do possess a number of excellent local or provincial floras, which deal with the wild or indigenous plants of India; but up to the present we have been at a loss whenever a cultivated plant has come up for identification, as most of these plants are just mentioned in our floras; with the help of this book we should be able to deal with the commoner shrubs cultivated in our gardens.

A word about the scientific names adopted in this book. The authors on purpose have selected such names as are current among gardeners, or in popular garden literature, even though such names are often incorrect from the more technical point of view. The book was primarily written for laymen, many of whom have known the plant under an older name, and find it difficult to understand why scientific names should change at all once they are given. However, to satisfy the demands of the more exacting readers, the latest and only correct botanical name has been added, generally within brackets, after the more popular one.

One more word about the authors themselves. When the series of papers was planned, Dr Bor was the Forest Botanist, Forest Research Institute, Dehra Dun, and Mr Raizada his assistant. On retirement from India after a noteworthy career of service, Dr Bor was elected Assistant Director, Royal Botanic Gardens, Kew, and Mr Raizada stepped into the post of Forest Botanist, Dehra Dun. In recognition of their outstanding services in the cause of botanical science in India, Dr Bor was presented with the Bruhl Medal of the Royal Asiatic Society of Bengal, and Mr Raizada has recently been elected a Fellow of the National Institute of Sciences of India.

We wish the book and its readers every success.

St Xavier's College,
Fort, Bombay.
December 25, 1953.

H. SANTAPAU, S.J., F.N.I.

FOREWORD TO THE SECOND EDITION

Climbers and shrubs constitute a large and important sector of ornamental horticulture. The annual flowers of course are an indispensable component of gardens, contributing as they do, brilliance and an astonishing range of colours, while trees provide the natural backdrop and are part of nature's precious gifts; some of these also are a brilliant spectacle when they flower. But coming between the annuals and the trees; it is the climbers and shrubs which provide the backbone of gardens and parks throughout the world. Mainly perennial in nature, they do not need continuous attention like the annuals and can go on for years providing not only a choice range of colours but, sometimes, fragrans also.

In the rather scanty literature on the subject, the book by Dr N. L. Bor and Dr M. B. Raizada, both of them distinguished botanists, published in 1954 has been a landmark of significance for all interested in flowers and gardens. In it they covered the very wide range of beautiful climbers and shrubs available in the subcontinent. The book was so planned as to be useful not only to the ordinary gardener or nature lover but to the botanist as well. Besides general descriptions suitable for the lay reader, detailed botanical descriptions are given, and the book is profusely illustrated with colour plates, half-tone plates and line drawings. Unfortunately, the book has gone out of print and the publishers, the well-known Bombay Natural History Society, are to be commended for the decision to bring out a fresh edition.

As the senior author of the original book, Dr Bor, is unfortunately no longer alive, the present edition has been revised by the second author, Dr Raizada. With his vast experience and access to the Herbarium and the Botanic Garden of the Forest Research Institute at Dehra Dun, Dr Raizada has brought the work fully up-to-date, revising names and making other changes necessitated by the additions to knowledge which have become available since the book was first published. He has however maintained all the features which made the book so valuable to the student, the gardener and the researcher.

There has recently been an upsurge of interest in ornamental horticulture in this country. I am confident this new edition fills a real need for a comprehensive and authoritative book on beautiful climbers and shrubs. I have no doubt that it will also be warmly welcomed by garden lovers in other countries.

B. P. PAL
F.R.S., HON. F.L.S., F.R.H.S., F.N.A.

New Delhi.
January 8th, 1982.

INDEX OF CHAPTERS

		PAGE
1.	Convolvulaceae	1
2.	Aristolochiaceae	14
3.	Bignoniaceae	32
4.	Leguminosae	50
5.	Rubiaceae	84
6.	Acanthaceae	110
7.	Solanaceae	124
8.	Verbenaceae	149
9.	Caprifoliaceae	171
10.	Plumbaginaceae	177
11.	Euphorbiaceae	182
12.	Combretaceae	196
13.	Malpighiaceae	199
14.	Apocynaceae	208
15.	Oleaceae	236
16.	Scrophulariaceae	248
17.	Malvaceae	250
18.	Passifloraceae	262
19.	Polygonaceae	289
20.	Nyctaginaceae	291
	Glossary	305
	Index	313

LIST OF COLOURED PLATES

FACING PAGE

1. *Ipomoea rubro-caerulea* Hook. 1
2. *Aristolochia ornithocephala* Hook. 14
3. *Bignonia venusta* Ker. 32
4. *Tecoma grandiflora* Delaun. 40
5. *Wisteria sinensis* DC. 52
6. *Sophora tomentosa* Linn. 56
7. *Caesalpinia pulcherrima* Sw. 62
8. *Bauhinia galpini* N. E. Brown 72
9. *Mussaenda frondosa* Linn. 89
10. *Ixora coccinea* R. Br. 93
11. *Rondeletia odorata* Jacq. 104
12. *Thunbergia grandiflora* Roxb. 110
13. *Cestrum aurantiacum* Lindl. 129
14. *Solanum seaforthianum* Andr. 139
15. *Petrea volubilis* Linn. 155
16. *Holmskioldia sanguinea* Retz. 156
17. *Lonicera sempervirens* Linn. 171
18. *Plumbago capensis* Thunb. 177
19. *Poinsettia pulcherrima* Grah. 182
20. *Jatropha pandurifolia* Andr. 191
21. *Quisqualis indica* Linn. 197
22. *Stigmaphyllon ciliatum* Juss. 203
23. *Galphimia gracilis* Bartl. 206
24. *Tabernaemontana coronaria* R. Br. 209
25. *Beaumontia grandiflora* Wall. 225
26. *Jasminum humile* Linn. 238
27. *Russelia juncea* Zucc. & *R. sarmentosa* Jacq. 250
28. *Hibiscus mutabilis* Linn. 256
29. *Passiflora racemosa* Brot. 284
30. *Antigonon leptopus* Hook. & Arn. 289
31. *Bougainvillea glabra* Choisy 291

LIST OF MONOCHROME PLATES

FACING PAGE

1. *Ipomoea palmata* Forsk. 6
2. *Ipomoea palmata* Forsk. 7
3. *Aristolochia grandiflora* var. *sturtevantii* 18
4. *Aristolochia ornithocephala* Hook. 20
5. *Aristolochia ornithocephala* Hook. 21
6. *Bignonia venusta* Ker. 34
7. *Bignonia speciosa* Grah. 38
8. *Bignonia speciosa* Grah. 39
9. *Tecoma stans* H.B.K. 42
10. *Tecoma stans* H.B.K. 43
11. *Tecoma radicans* Juss. 46
12. *Tecoma radicans* Juss. 47
13. *Wisteria sinensis* DC. 51
14. *Sophora griffithii* Stocks 54
15. *Sophora griffithii* Stocks 55
16. *Sophora tomentosa* Linn. 58
17. *Sophora secundiflora* DC. 59
18. *Caesalpinia sepiaria* Roxb. 66
19. *Caesalpinia sepiaria* Roxb. 67
20. *Caesalpinia sappan* Linn. 70
21. *Caesalpinia pulcherrima* Sw. 71
22. *Bauhinia acuminata* Linn. 74
23. *Bauhinia galpini* Br. 75
24. *Bauhinia corymbosa* Roxb. 78
25. *Mussaenda luteola* Del. 88
26. *Mussaenda frondosa* Linn. .. .. between 88 and 89
27. *Mussaenda frondosa* Linn. .. .. between 88 and 89
28. *Ixora coccinea* Linn. 102
29. *Ixora parviflora* Vahl. 103
30. *Rondeletia odorata* Jacq. 106
31. *Rondeletia odorata* Jacq. 107
32. *Hamelia patens* Jacq. 108
33. *Hamelia patens* Jacq. .. .. between 108 and 109
34. *Hamiltonia suaveolens* Roxb. .. between 108 and 109
35. *Hamiltonia suaveolens* Roxb. 109
36. *Thunbergia fragrans* Roxb. 114
37. *Thunbergia fragrans* Roxb. 115
38. *Thunbergia erecta* And. 118
39. *Thunbergia erecta* And. 119
40. *Thunbergia grandiflora* Roxb. 122
41. *Cestrum aurantiacum* Lindl. 126
42. *Cestrum aurantiacum* Lindl. 127
43. *Cestrum diurnum* Linn. 130
44. *Cestrum diurnum* Linn. 131
45. *Cestrum nocturnum* Linn. 132
46. *Cestrum nocturnum* Linn. .. .. between 132 and 133
47. *Cestrum parqui* L'Her. .. .. between 132 and 133
48. *Cestrum parqui* L'Her. 133
49. *Solanum wendlandii* Hook. 140

SOME BEAUTIFUL INDIAN CLIMBERS AND SHRUBS

FACING PAGE

50.	*Datura fastuosa* Linn.	142
51.	*Datura suaveolens* H. & B.	143
52.	*Stachytarpheta mutabilis* Vahl.	150
53.	*Stachytarpheta mutabilis* Vahl.	151
54.	*Petrea volubilis* Linn.	154
55.	*Petrea volubilis* Linn.	155
56.	*Lonicera sempervirens* Linn.	172
57.	*Lonicera japonica* Thunb.	173
58.	*Plumbago capensis* Thunb.	180
59.	*Plumbago capensis* Thunb.	181
60.	*Poinsettia pulcherrima* Grah.	188
61.	*Poinsettia pulcherrima* Grah.	189
62.	*Jatropha pandurifolia* Andr.	192
63.	*Jatropha pandurifolia* Andr.	between 192 and 193
64.	*Jatropha gossypifolia* Linn.	between 192 and 193
65.	*Jatropha gossypifolia* Linn.	193
66.	*Quisqualis indica* Linn.	196
67.	*Stigmaphyllon ciliatum* Juss.	204
68.	*Stigmaphyllon ciliatum* Juss.	205
69.	*Galphimia gracilis* Bartl.	206
70.	*Galphimia gracilis* Bartl.	207
71.	*Allemanda neriifolia* Hook.	214
72.	*Allemanda neriifolia* Hook.	215
73.	*Vinca rosea* Linn.	218
74.	*Strophanthus gratus* Franch.	228
75.	*Strophanthus gratus* Franch.	229
76.	*Trachelospermum fragrans* Hook. f.	236
77.	*Trachelospermum fragrans* Hook. f.	237
78.	*Jasminum pubescens* Willd.	240
79.	*Jasminum primulinum* Hemsl.	between 240 and 241
80.	*Jasminum humile* Linn.	between 240 and 241
81.	*Jasminum grandiflorum* Linn.	241
82.	*Russelia juncea* Zucc.	248
83.	*Russelia juncea* Zucc.	between 248 and 249
84.	*Russelia sarmentosa* Jacq.	between 248 and 249
85.	*Russelia sarmentosa* Jacq.	249
86.	*Hibiscus mutabilis* Linn.	258
87.	*Hibiscus schizopetalus* Hook. f.	259
88.	*Hibiscus rosa-sinensis* Linn.	260
89.	*Hibiscus rosa-sinensis* Linn.	between 260 and 261
90.	*Hibiscus syriacus* Linn.	between 260 and 261
91.	*Hibiscus syriacus* Linn.	261
92.	*Passiflora incarnata* Linn.	268
93.	*Passiflora edulis* Sims.	269
94.	*Passiflora holosericea* Linn.	270
95.	*Antigonon leptopus* Hook. & Arn.	271
96.	*Bougainvillea spectabilis* Willd.	294
97.	*Bougainvillea buttiana* Holtt. & Standl.	295
98.	*Bougainvillea spectabilis* Willd.	298
99.	*Bougainvillea glabra* Choisy	299

PLATE 1

HEAVENLY BLUE MORNING GLORY
Ipomoea rubro-caerulea Hook.
(⅝ nat. size)

Chapter 1

CONVOLVULACEAE

A very large family of over 1,000 species distributed over the whole world. It consists of annual or perennial herbs, often twining, sometimes shrubby, rarely arborescent, with simple alternate leaves. The flowers are mostly showy, solitary and axillary. The calyx sometimes consists of separate, sometimes of combined sepals. The corolla is always gamopetalous and usually of a distinctive funnel shape with 5 points. The stamens are five in number and alternate with the petals. The ovary is superior and the fruit is usually a capsule. In most of the species the flowers remain open for one day only or even for a few hours, after which the edges of the corolla roll up inwards and close the tube.

Ipomoea Linn.

The genus *Ipomoea*, which is the largest in the family *Convolvulaceae*, is native in most of the tropical and subtropical regions of the world. Many of its four hundred species have been introduced into our Indian gardens where they are cultivated for their striking, and often very beautiful, flowers. The generic name seems to be derived from two Greek words: *Ipso*, bindweed, and *homoios*, like, similar to.

The genus *Ipomoea* as known to Linnaeus, and to Hooker in the *Flora of British India*, has been the subject of much systematic work in recent years and has been greatly modified. Garden plants well known to horticulturists as species of *Ipomoea* now appear under such generic name as *Quamoclit, Mina, Calonyction, Merremia*, and so forth. Some of these genera are quite distinct and are gradually being accepted. On the other hand the nature of the extine of the pollen, whether echinulate or not, as a character for splitting genera, may appeal to a monographer, but is hardly likely to find favour with horticulturists. Hence, in this work the name as known to horticulturists has been given followed by the correct botanical name in brackets.

The majority of the species found in our gardens are herbaceous twiners; but one, which will be treated in this chapter, is a shrubby scrambler. The presence of latex vessels in the bark is not uncommon.

The flowers of *Ipomoea*, in the wide sense, may be salver-, bell-, or funnel-shaped. They are, as a rule, large and showy, with red, blue, yellow or purple corollas and are borne singly or in clusters in the axils of the alternate, often cordate, leaves.

The method of folding in the bud, a feature which is quite apparent in the fully opened flower, is characteristic of the genus. Five triangular smooth and rigid areas arising and tapering from the base of the corolla are easily discernible. In the intermediate spaces are areas of more delicate texture which, in the bud, are folded inwards so that the triangular firmer areas are exterior, and touch by their edges. The bud is twisted to the right.

The calyx lobes are five in number and may be united or free, glabrous or hairy. The stamens, too, are five and are alternate with the triangular bands referred to above. The ovary, which may be two-, three- or four-celled, is seated upon a fleshy disc which secretes nectar.

The flowers in most species are short-lived and only remain open for a short time; sometimes only a few hours. Others remain open all day, while a few open at sunset and wither at dawn. When withering the corolla rolls up and remains tightly closed till it falls.

Some species with tubular corollas are adapted to cross pollination by birds [e.g. *Ipomoea* (*Mina*) *lobata*] while others are fertilised by insects or by the wind. If cross-fertilisation does not occur when the flower is open, self-fertilisation is almost certainly accomplished when the withering corolla rolls up.

Economic uses.—The sweet potato, *Ipomoea batatas*, is grown extensively for its large, edible, fleshy-fibrous roots. *I. turpethum* furnishes a well-known purgative named Indian Jalap. The Jalap of the British Pharmacopoeia is the resin obtained from the roots of *I. purga. I. pescaprae* has some repute as a sand binder.

It is, however, as a garden plant that the genus is most highly valued. Their gorgeous flowers and the ease with which they may be propagated have contributed to their popularity. Most species are not exacting as regards soil but they love a sunny site with plenty of water. To ensure early germination it is recommended that a small notch be filed in each seed or that they be soaked in warm water for about two hours. The perennial species can be grown from seed but they can also be propagated vegetatively by cuttings, layers or division of the roots. Owing to their thick foliage and rapid growth it is advisable to grow the climbing species over trellis-work.

KEY TO THE SPECIES

Climbing species
  Corolla salver-shaped
    Stamens and style far exserted from corolla; flowers orange or scarlet .... *I. lobata*
    Stamens and style not much exserted; flowers a brilliant crimson
      Leaves ovate-cordate .... *I. coccinea*
      Leaves with filiform segments .... *I. quamoclit*
  Corolla campanulate
    Leaves deeply lobed or palmately compound
      Sepals hairy; corolla yellow .... *I. vitifolia*

Sepals glabrous
  Lobes of leaf again lobed or deeply serrate; flowers white with a purple eye .... *I. sinuata*
  Lobes of the leaf elliptic
    Sepals purplish; flowers shining rose or light purple .. *I. horsfalliae*
    Sepals greenish; flowers violet-purple with a purple tube .... *I. palmata*
  Leaves entire
    Sepals short and thick .... *I. rubro-caerulea*
    Sepals prominently acute
      Sepals acute .... *I. purpurea*
      Sepals long acuminate .... *I. learii*
Scrambling shrub; flowers pink .... *I. carnea*

Ipomoea rubro-caerulea Hook. (I. tricolor Cav.)

Morning-glory; Heavenly Blue

(*Rubrocaerulea* is derived from two Latin words meaning red and sky blue, and refers to the colour of the corolla which in the bud is pink to red, but, when open, a brilliant blue.)

Description.—A twining glabrous creeper of extensive growth, requiring a large trellis for its support. Branches rounded, herbaceous, tinged with purple. Leaves alternate, membranous, dark green, truly cordate, with a deep and broad sinus at the base, shortly but sharply acuminate, quite entire, wavy on the surface, much veined.

Peduncles axillary, hollow, longer than the petioles, 3- to 4-flowered, the pedicels thickened. Calyx five-partite, the segments small, erect and appressed, linear-subulate; corolla funnel-shaped, white in the bud but tinted with a rich red which, when the flower is fully expanded, becomes clear azure-blue or purple, with five angles and five plicae; the angles mucronate. Stamens 5; filaments unequal, inserted at the base of the tube, hairy at the base. Anthers oblong, yellow (sometimes tinged with purple). Ovary superior, oblong, 2-celled, with 2 ovules in each cell. Style filiform, stigma 2-lobed. Fruit a capsule, usually 4-valved containing 4 seeds.

Flowers.—September-November. *Fruits.*—Cold season.

Native country.—This plant is a native of Mexico, but is now extensively cultivated in all tropical countries of the world.

Though a perennial it becomes exhausted after one season and, therefore, can be only cultivated successfully as an annual. It blossoms at the beginning of the cold season, opening its large clear blue flowers in countless numbers early in the morning and presenting then as gorgeous a sight as anyone could possibly wish for. The flowers fade in the after part of the day, turning first to a reddish or purplish tinge.

Gardening.—It is essential that the seed be sown as early as July in order that the plants may grow to perfection by the beginning of the cold weather; they do not require a rich soil, but a change of locality each year is desirable. Its rapid growth and dense foliage make it specially valuable for the rapid covering of arbours, verandas, walls, and for screening unsightly objects. It is also valuable for cut flowers.

Ipomoea rubro-caerulea Hook. var. *alba* Hort. is a variety with white flowers but not nearly so beautiful.

Ipomoea quamoclit Linn. [Quamoclit pinnata (Linn.) Bojer]
The Cypress-Vine or Indian Pink

Description.—A beautiful annual twiner. Stem smooth, slender, twining to a height of about 20 ft. Leaves short-petioled or sessile, pinnately divided into many filiform segments. Peduncles few-flowered, commonly much longer than the petioles. Corolla 1-1.5 in. long, scarlet, the tube narrowly salver-shaped, inflated above; the limb nearly flat, 5-lobed. Stamens exserted.

Flowers.—Rainy season. *Fruits.*—October-December.

Distribution.—It is a native of tropical America and is now widely spread in the warmer parts of the world.

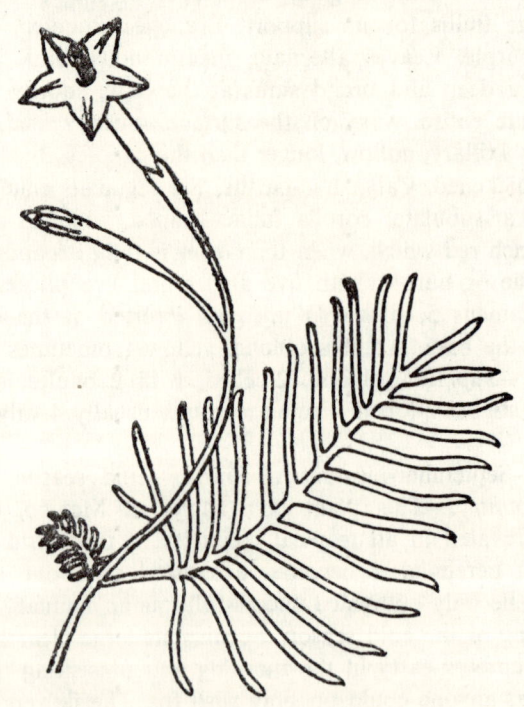

Fig. 1.—*Ipomoea quamoclit* Linn. ×1/1

Gardening.—Raised from seed usually sown in early spring. It is a very elegant species and makes rapid growth from June, and covers a trellis very rapidly.

Ipomoea quamoclit L. var. *alba* Hort. has white flowers, but, is seldom seen in cultivation in this country.

Medicinal uses.—The Hindus consider it to have cooling properties. The pounded leaves are applied for haemorrhoids, while a preparation of the juice with hot 'ghee' is administered internally. In Bombay, the leaves are used as a 'lep' (plaster) for carbuncles.

Ipomoea horsfalliae Hook.

(Dedicated to Mrs. Charles Horsfall, at whose home it was first raised in England.)

Description.—A more or less woody perennial tall glabrous twiner. Leaves more or less circular or orbicular-ovate in outline, palmate, with

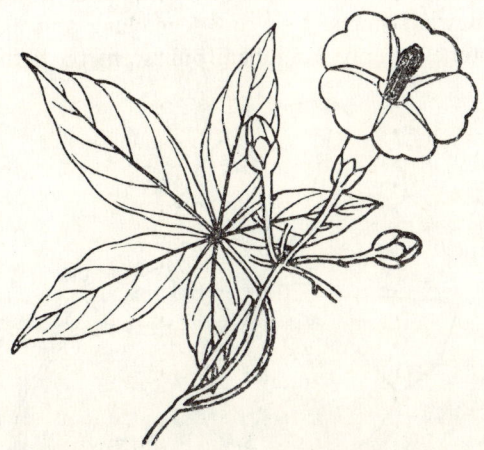

Fig. 2.—*Ipomoea horsfalliae* Hook. $\times \frac{1}{2}$

5, or rarely 7, leaflets which are obovate, oblanceolate or elliptic, 3-4 in. long, acuminate, entire, rather thick in texture, margin slightly crisped or waved. Peduncles axillary, about as long as, or longer than, the petioles, bearing a forked cyme of many flowers; pedicels thickened upward, smooth. Calyx of 5 equal, rather broad, purplish black, imbricated lobes. Corolla narrowly funnel-shaped, 2-2.5 in. long; the limb with prominent plaits and becoming revolute, of a deep rich and glossy rose, equally dark within and without. Stamens 5, shortly exserted. Ovary globose, surrounded by a large fleshy ring. Stigma capitate, two-lobed, hairy.

Flowers.—September-December. The plant does not appear to set seed in India.

Distribution.—Native of the hotter parts of America and perhaps of the Old World; now commonly cultivated in all tropical parts of the earth.

Gardening.—As this species rarely produces viable seeds it is usually propagated from root cuttings which strike with great difficulty. It is very suitable for pergolas in the open.

Ipomoea horsfalliae Hook. var. *briggsi* Hort. is a free flowering plant with magenta-crimson flowers.

Ipomoea learii Pax. (I. congesta R. Br.)
Blue Dawn-flower

(Named in honour of Mr. Lear, who is chiefly known as collector of Sri Lanka plants. The seeds from which plants were first raised in England were received from him. The plant, however, is not a native of Sri Lanka but of tropical S. America.)

Description.—A perennial climber, somewhat woody at the base, with finely pubescent stems. Leaves 3-5 in. long and as broad, mostly entire or sometimes deeply 3-5-lobed, pubescent on both surfaces, base

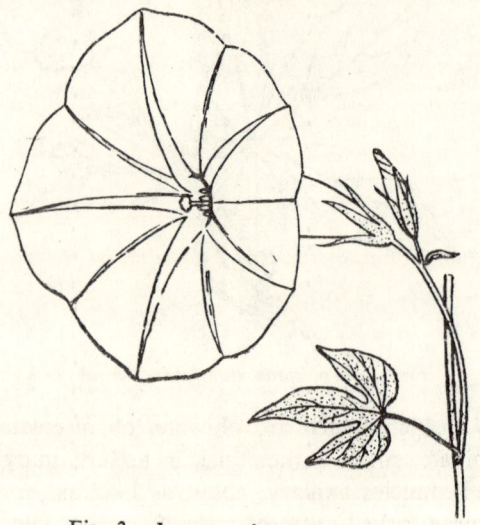

Fig. 3.—*Ipomoea learii* Pax. × ½

cordate, apex acute; petiole slender, nearly as long as the blade. Flowers in capitate clusters of 5 to many flowers, 3-4 in. long and as much across, bright violet-blue when fresh, with 5 narrow, pinkish bands. Sepals about 0.7 in. long, linear-lanceolate, hairy pubescent. Corolla very beautiful, campanulate, suddenly widened at the mouth.

Flowers.—August-October. Seeds ripen in October-December.

Distribution.—Indigenous to S. America, but now widely cultivated in the tropics of both hemispheres.

Gardening.—It is propagated by seed sown at the break of the rains. It is a magnificent species and a most useful plant for covering

PLATE 1

THE RAILWAY CREEPER
Ipomoea palmata Forsk.

M. N. Bakshi

PLATE 2

M. N. Bakshi

THE RAILWAY CREEPER
Ipomoea palmata Forsk.

waste places, embankments and the like. It is not usually grown on a trellis.

Ipomoea palmata Forsk. [I. cairica (Linn.) Sweet]
The Railway Creeper

(*Palmata* is Latin for shaped like a hand.)

Description.—A perennial climbing shrub. Stem glabrous and more or less warty. Leaves 1-3 in. in diameter, cut nearly to the base into 5-7 elliptic or lanceolate entire lobes. Flowers 2-3 in. across, purple, in axillary, often 3-flowered cymes. Sepals 0.25 in. long. Sepals 0.25 in. long, broadly ovate. Corolla campanulate, mauve or purple with a

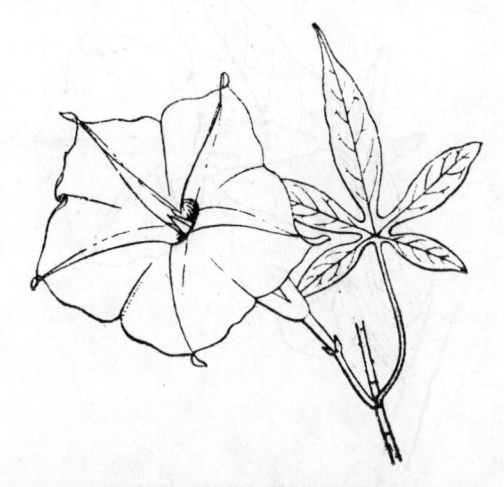

Fig. 4.—*Ipomoea palmata* Forsk. × ½

deeper tinge in the throat. Fruit a capsule 0.5 in. long, ovoid, glabrous; seeds pubescent with woolly margins.

Flowers and fruits.—Practically all the year round.

Distribution.—Indigenous or naturalized in most tropical countries.

Gardening.—Easily grown from cuttings of the branches. The small tubercles on half mature stems, on coming in contact with soil easily strike root. It is exceedingly common on trellises in gardens and fences on railway platforms (hence the name 'Railway Creeper').

Ipomoea carnea Jacq. (I. fistulosa Mart. ex Choisy)

(*Carneus* means flesh-coloured and refers to the colour of the corolla.)

Description.—A stout straggly shrub with milky juice, suberect, at times climbing to about 30 ft. Leaves about 4 in. long, 2 in. broad, ovate or ovate-lanceolate, long acuminate, base cordate, entire, soft and somewhat fleshy; petiole 2.5 in. long. Flowers large, rose or light mauve,

2.5 in. long, 2 in. across, in axillary cymes. Peduncles 2-4 in. long. Sepals 0.2 in. long, broadly ovate. Corolla campanulate, narrowly cylindrical for about 0.2 in. at the base, suddenly widened at the top. Fruit a capsule, 0.5 in. long, glabrous. Seeds silky.

Flowers.—All the year except the coldest months. *Fruits* in November-December in N. India.

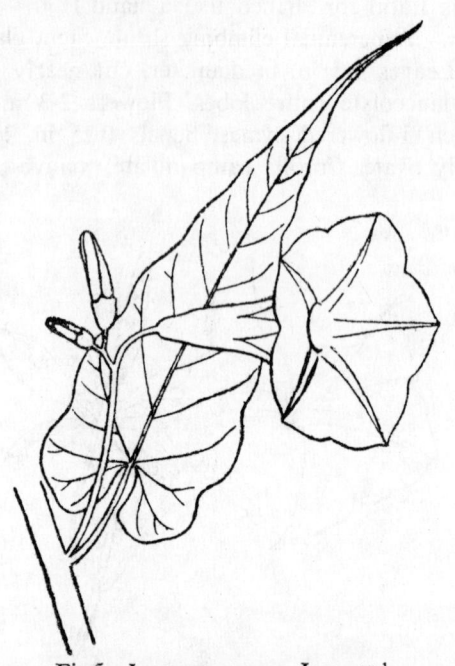

Fig.5—*Ipomoea carnea* Jacq. ×½

Distribution.—Indigenous to S. America, but now commonly cultivated in the plains especially at Railway stations.

Gardening.—It is easily grown from cuttings and is inexacting as regards site and soil.

Ipomoea vitifolia Sweet

(The specific name refers to the leaves shaped like those of a vine.)

Description.—An extensive, perennial, nearly evergreen twiner, hairy all over. Leaves hairy, suborbicular or broadly ovate in outline, cut one-third to one-half the way down into ovate or triangular, acute or acuminate lobes which are often strongly toothed, palmately-nerved with 5-7 basal nerves; petiole 0.5-4 in. long. Flowers sulphur-coloured, very pretty, in peduncled, hairy cymes of one to seven blooms. Sepals 0.5 in. long, hirsute, hardened; 0.7-0.9 in. in fruit. Corolla 2-2.5 in. in diameter. Fruit a capsule, 0.5 in. in diameter, irregularly breaking up, semi-transparent, usually 4-seeded. Seeds black or slate-coloured with a matt surface.

Flowers.—October-March.
Fruits.—May-June.
Distribution.—Throughout India (except the North-west), Malaya Islands.

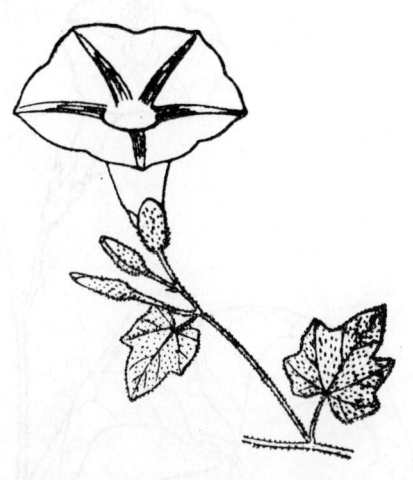

Fig. 6.—*Ipomoea vitifolia* Sweet × ½

Gardening.—It is usually propagated by stem cuttings or layers. It is a very fast grower and suitable for covering walls, trellises and pergolas.

Medicinal uses.—An infusion of the leaves is supposed to be cooling and is applied for inflammation of the eyes.

Ipomoea coccinea L. (Quamoclit coccinea Moench)

The Star Ipomoea

(*Coccineus* means crimson and refers to the flowers.)

Description.—An annual weak glabrous twiner, climbing to about 15 ft. Leaves ovate-cordate, slender, petioled, 2-4 in. in diameter, entire or angulate, acuminate; petiole 2-4 in. long. Flowers in few- to many-flowered lax cymes. Peduncle 2-6 in. long, slender. Bracts inconspicuous. Sepals 0.4 in. long, unaltered in fruit. Corolla crimson, in garden varieties often orange or yellow; tube 1 in. long, mouth 0.9 in. in diameter, limb obscurely lobed. Ovary completely 4-celled, 4-ovuled. Fruit a capsule 0.4 in. long or rather more, ovoid, smooth. Seeds furred.

Flowers.—Rainy season. *Fruits.*—October-November.

Distribution.—Probably indigenous to N. Mexico and Arizona, but now common throughout India, cultivated or semi-wild.

Gardening.—Easily raised from seed sown early in the rains. The scarlet flowers, though rather small, are produced in profusion.

Ipomoea coccinea L. var. *hederifolia* House. This form of the species has angulate, 3-lobed or even 3-5-parted, lobed leaves and usually large flowers. It is superior to the type for ornamental purposes. Rarely

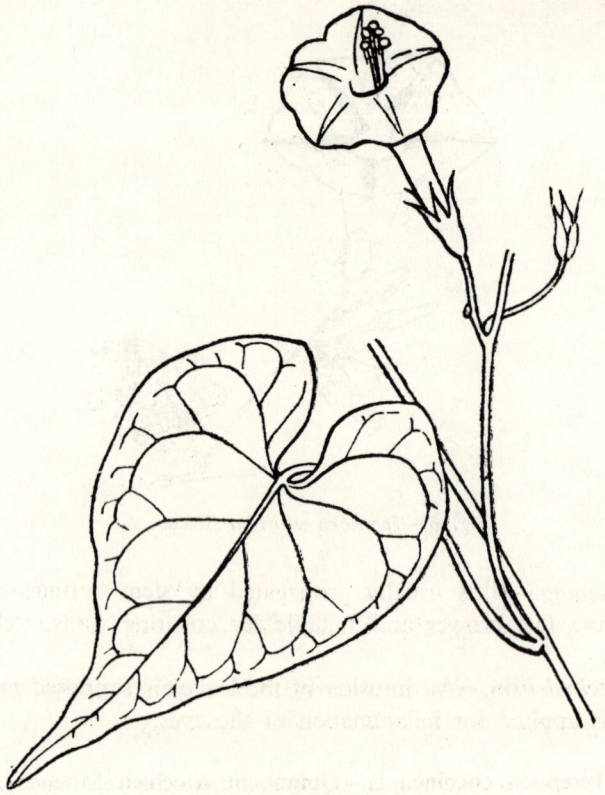

Fig. 7.—*Ipomoea coccinea* Linn. ×1/1

seen in cultivation in this country.

Medicinal uses. The root is sternutatory.

Ipomoea lobata Thell.
(Quamoclit lobata House, Mina lobata Llav. and Lex.)

(*Lobata* refers to the lobed leaves.)

Description.—An annual vigorous and quick growing creeper not attaining a very great length. Leaves cordate, with a broad sinus, 3-lobed, the middle lobe longest and narrowed below. Inflorescence a spiral cyme; peduncles stout below, several-flowered. Flowers 0.5-0.9 in. wide, when open rich crimson or orange, soon fading to pale yellow. Sepals tipped with short awn-like points. Corolla salver-shaped, stout and tubular below, abruptly widening into a cylindric or bag-shaped limb with 5 small, acute lobes. Stamens long exserted. Fruit with false partitions between the seeds.

Flowers.—September-December (occasionally in **January**). *Fruits.*—November-January.

Distribution.—A native of Mexico, but now commonly cultivated in all tropical countries.

Fig. 8.—*Ipomoea lobata* Thell. × ½

Gardening.—Propagated by seeds sown in early June either in the ground or in pots. It is best suited for growing on small arches and on posts. It is a free-flowering climber and deservedly popular.

Ipomoea purpurea (Linn.) Roth
The Common Morning Glory

(*Purpurea* refers to the colour of the flowers.)

Description.—An annual hairy climber with entire ovate-cordate shortly acuminate pubescent 3-5 in. long leaves. Flowers 1-5 on axillary peduncles, funnel-shaped, 2-3 in. long, varying in colour from white to pale-blue or purple. Calyx-lobes less than half the length of the tube, acute but not elongated and somewhat exceeding the capsule in fruit.

It resembles *I. hederacea* (another garden species) at first sight but can always be distinguished from it by its smaller and entire leaves and by the sepals not having ligulate tips.

Flowers.—During the rains. *Fruits.*—October-November.

Distribution.—It is said to be a native of tropical America but is certainly found wild in India. It is extensively cultivated in all warm countries.

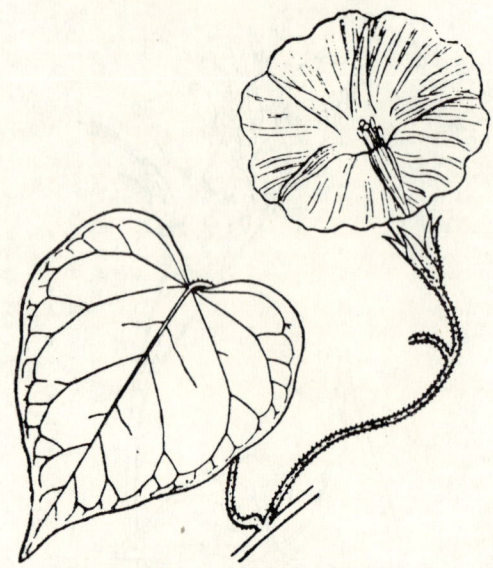

Fig. 9.—*Ipomoea purpurea* Roth × ½

Gardening.—Easily raised from seed sown in April. One of the most popular of garden annuals. It is found in many races in cultivation, varying in colour of the flowers, with variegated foliage, and sometimes with double flowers.

Ipomoea sinuata Ortega

(*Sinuata* refers to the margins of the lobes of the leaf.)

Description.—A perennial extensive twining shrub covered with long yellowish hairs. Leaves 2-4 in. in diameter, smooth, sinuately cut nearly to the base into about 7 lanceolate pinnatifid segments which are toothed. Peduncles 1-2-flowered, longer than the petioles. Flowers about 1 in. across, white, with purple centres. Sepals 0.8 in. long, glabrous, elliptic-oblong. Corolla bell-shaped. Fruit a capsule 0.5 in. in diameter, glabrous.

Flowers.—October-November. *Fruits.*—November-December.

Distribution.—It is a native of tropical America. Occasionally grown in gardens in all warm countries.

Gardening.—Raised from seed sown in early rains, or by layers of the stem, or division of the roots. It is worth growing for its delicate foliage alone and deserves more popularity. It prefers a moist situation.

The flowers usually open widely in Dehra Dhun for 2-3 hours at midday as in Texas, where it is known as the 'Noon Flower'.

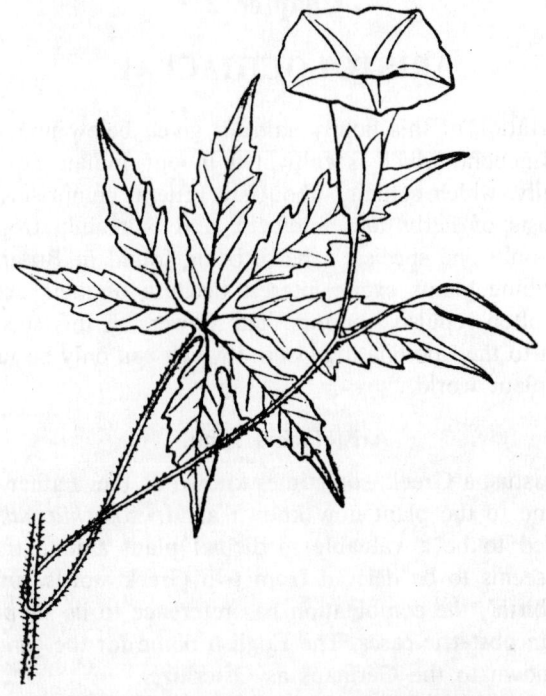

Fig. 10.—*Ipomoea sinuata* Ortega × ½

Medicinal uses.—'The leaves have an odour of oil of bitter almonds and are used in the preparation of the French Liquor known by that name' (Nadkarni). The sap contains hydrocyanic acid.

Chapter 2

ARISTOLOCHIACEAE

The derivation of this family name is given below under the name of one of its genera, which is cultivated in our Indian gardens.

The family, which contains about six genera, comprising some 200 species, consists of herbs and climbers, and is mainly tropical in its distribution—only one species, a herb, being found in Britain.

The climbing plants excite interest because of their extraordinary flowers with often repulsive odours. The flowers of this species exhibit an adaptation to their pollinating visitors which can only be rarely paralleled in the plant world.

Aristolochia Linn.

Theophrastus, a Greek, sometimes known as the 'Father of Botany' gave this name to the plant now known as *Aristolochia pallida*, which was considered to be a valuable medicinal plant among the ancients. *Aristolochia* seems to be derived from two Greek words, *aristos*, 'best' and *locheia* 'birth'; the combination has reference to its supposed medicinal virtues in obstetric cases. The English name for the genus is Birthwort. It is known to the Germans as Osterluzei.

This genus comprises about 180 species, of which the very great majority are inhabitants of the tropics, especially in South America, while only a few are to be found wild in temperate climates. The species are, however, largely cultivated in the tropics outside their native homes, and are favourite hot house plants in the colder climates of the world.

The species are grown as ornamental plants chiefly on account of their remarkable, often fantastic, flowers, but owing to the unpleasant odours which the flowers emit they are usually placed in the background. In systematic botany the genus is grouped with certain other families which have extraordinary flowers and form a well-defined cohort. Many of the species thus grouped together possess highly coloured foetid flowers, some are insectivorous and others parasites. The families represent reduced and degraded forms and include some of the monstrosities of the plant world.

The species of this genus are almost always climbing plants arising from a perennial rhizome or root. The leaves are simple, alternate and petioled. In many species these organs possess secretory cells, which, in the case of a leaf held up to the light, are visible as translucent dots. The flowers are arranged in various ways and exhibit the oddest shapes.

PLATE 2

BIRD'S-HEAD BIRTHWORT
Aristolochia ornithocephala Hook.
(½ nat. size)

The flowers of some are small, while others are gigantic blooms swollen into a trumpet with a tail up to 4 feet long. Others again, like that figured in our plate, have a large expanded broad lip and a beak.

The colours of the flowers are those usually associated with the so-called indol group of scents, though to use the word 'scent' in connection with flowers of *Aristolochia* is apt to give a wrong impression. The fact of the matter is that most of the flowers of this genus emit extremely offensive odours which apparently arise from the decomposition of albuminoid compounds developed in the perianth. Colours evolved in combination with such scents are those which one instinctively connects with decomposing carcasses. Maroon or purple or livid spots, violet streaks and red-brown veins on a greenish or fawn coloured background, are the prevailing colour schemes. The general effect, however, is bizarre and eccentric, rather than beautiful.

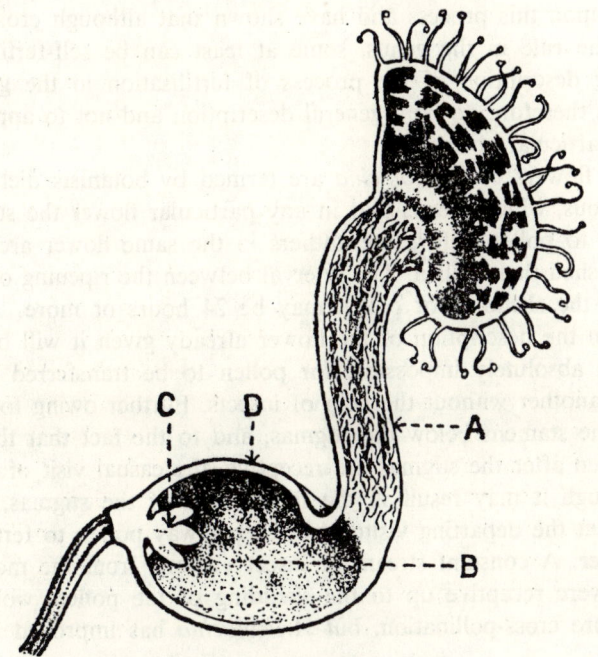

Fig. 11.—*Aristolochia ciliata* Hook.
A.—Tube with downwardly directed motile hairs.
B.—Grazing ground for insects.
C.—Gynostemium.
D.—Bulb.

It may be mentioned here that certain species which are entirely odourless to the human sense of smell, have been proved to emit scents which are attractive to insects.

The perianth is single, that is, it cannot be divided into calyx and corolla. In many respects it is similar to the spathe of the Arum Lilies, in that it consists of three distinct portions. The terminal segment, or lip, may be trumpet-shaped, or two-lipped, two-winged or expanded in various other ways. The central portion is tube-like with a wide or narrow orifice, the inner surface of which is usually covered with hinged hairs. The lower or basal portion is swollen into a bulb-like structure.

At the bottom of this bulb or pouch will be found a thick stylar column on the top of which are displayed the broad stigmas with downwardly-bent tips. A ring of linear anthers is fixed to the stylar column below the stigmas; the whole structure being called a gynostemium.

The devices adopted by plants to ensure cross-pollination are manifold and curious, but in the case of the genus *Aristolochia* they are so remarkable that they can only be described as astounding.

The researches of Petch at Peradeniya have thrown a good deal of light upon this process and have shown that although cross-fertilisation is the rule in the genus, some at least can be self-fertilised. The following description of the process of fertilisation in the genus may be taken, therefore, to be a general description and not to apply strictly to any particular species.

The flowers of *Aristolochia* are termed by botanists dichogamous-protogynous, which means that in any particular flower the stigmas are receptive to pollen before the anthers in the same flower are ripe and ready to shed their pollen. The interval between the ripening of the stigmas and the shedding of pollen may be 24 hours or more.

From the description of the flower already given it will be realised that it is absolutely impossible for pollen to be transferred from one plant to another without the help of insects. Further owing to the position of the stamens below the stigmas, and to the fact that the anthers shed pollen after the stigmas are receptive, the casual visit of an insect, even though it may result in the fertilisation of the stigmas, does not ensure that the departing visitor will carry away pollen to fertilise another flower. A constant stream of insect visitors, from the moment the stigmas were receptive up to the shedding of the pollen, would probably ensure cross-pollination, but *Aristolochia* has improved upon this solution.

The problem before the flower is threefold: (1) insect visitors have to be attracted, (2) they must carry pollen from another plant, and (3) they must carry pollen away.

As regards the first part of the problem the species develop very highly coloured and conspicuous flowers which emit odours irresistible to insects. This combination of colour and odour makes success doubly sure.

As regards the second and third parts of the problem it will be observed that the solution of the third condition in respect of one flower means the solution of the second proviso in so far as another flower is concerned.

Aristolochia gets over the difficulty of the interval between the time the stigmas are receptive and time the pollen is shed by developing a mechanism whereby an insect visitor is imprisoned. It accomplishes this unusual feat in the following manner.

When the stigmas are ready for fertilisation the lip expands into a perch or alighting platform. A visitor, usually a dipterous insect, arrives covered with pollen from an older flower and alights on the perch. Attracted by the mephitic odours from within it crawls down the tube. The interior of the tube is clothed with downwardly directed motile hairs which offer no impediment to the entrance of the insect. In some cases it has been observed that these motile hairs actually pitchfork the visitor down into the tube. Generally speaking, however, such 'strong-arm work' is not necessary and the insect makes its way into the bulb without compulsion. Once inside, it cannot escape because the hairs of the tube form a very efficient barrier.

In any case, the insect does not particularly want to escape for not only does the bulb provide 'lodging', it also provides 'board' in the shape of special areas of nutritious cells. Moreover, most bulbs have a window in the wall, formed of thinner layers of cells, and it is at this translucent spot that the prisoners congregate. The window is cunningly situated so that when the anthers dehisce the insects are drenched with pollen.

Any pollen brought by the visitors will be transferred to the receptive stigmas. Once the transference takes place the tips turn upwards and press the stigmatic surfaces together in order to prevent self-fertilisation. The anthers develop rapidly and dehisce, covering the insects with pollen.

The hairs in the tube now begin to wither and shrivel up and the prison doors are thrown wide open. The insect, bearing its load of pollen, crawls out and, undaunted by its previous experience, flies to a flower which has just opened and repeats its performance. After the emergence of the insect, the fertilised flower closes and the perianth withers.

Obviously the first flower to open will not be fertilised because it cannot be visited by pollen bearing insects. If fertilisation does not take place the flower goes through the same cycle of changes; the stigmas wither and turn upwards, the anthers dehisce and shed their pollen, the imprisoned insects are released and the problem of cross-fertilisation is solved.

The various organs of these climbers, especially the rhizome, have long been known to possess medicinal properties. Pliny and Dioscorides described several species and recommended their use for various complaints. Most of the older herbals contain references to the therapeutic properties of the roots and leaves.

The roots of *A. indica* Linn. and *A. bracteata* Retz. (*A. bracteolata* Lamk.) are sold in the bazaars of India. The former has a great reputation in this country as an antidote to snake-bite. It is recommended that the juice of the fresh leaves be applied immediately to the bites of snakes and scorpions. The root is also used in intermittent fevers as an emmenagogue and tonic and is recommended for all kinds of intestinal disorders. Decoctions of the roots of several species are reputed to have anthelmintic properties. An *Aristolochia* of South America is said to be used as an arrow poison, probably in conjunction with the juice of other plants.

Key to the Species

Flowers tailed
  Flowers very large, with large trumpet-shaped lip; leaves
    not lobed .... *A. grandiflora*
  Flowers medium sized; leaves three-lobed .... *A. macroura*
Flowers not tailed
  Corolla expanded into a large kidney-shaped lip with
    purple reticulations .... *A. ornithocephala*
  Corolla not expanded into a flat lip
    Leaves acute at the tip, not orbicular or heart-shaped
      Leaves ovate-lanceolate with a sinus at the base ...... *A. roxburghiana*
      Leaves oblong or fiddle-shaped; base cuneate .... *A. indica*
    Leaves rounded at the tip, orbicular or heart-shaped
      Flowers 5 in. long and over with a conspicuous beak
        and narrow lip .... *A. ringens*
      Flowers without a conspicuous lip and beak. Much
        less than 5 in. long
        Leaves orbicular
          Stems hairy. Flowers expanded into two finger-like
            processes which are covered with stalked glands .... *A. ridicula*
          Stems glabrous. Flowers not expanded as above;
            margin furnished with fleshy narrow processes ...... *A. ciliata*
        Leaves cordate-obtuse
          Whole plant hairy .... *A. tomentosa*
          Whole plant glabrous
            Flowers inconspicuous; lip dark purple about
              1-1.5 in. long .... *A. bracteata*
            Flowers conspicuous; lip broadly expanded up to
              3 in. long, variegated with yellow and purple ...... *A. elegans*

Aristolochia grandiflora Sw.

Pelican-flower or Poison Hog-meat

(*Grandiflora* means large-flowered.)

Description.—A shrubby climber. Leaves alternate on rather long petioles, cordate-acuminate, softly hairy. Flowers solitary on leaf-opposed peduncles; when in bud resembling a pelican's head in shape. There

PLATE 3

THE PELICAN FLOWER
Aristolochia grandiflora Sw. var. *sturtevantii*

is a leafy bract at the top of the peduncle at the point where it joins the ovary; stalk and ovary hang vertically. The bulb is roughly hexagonal in section with reinforced angles, somewhat ovate in shape, about 4 in. long, slightly narrowed into a tube which bends sharply upwards. The posterior wall of the tube expands into a hood which is pressed back upon the bulb. The interior wall forms the lip, while the hood already formed bends backwards and expands into a large limb or face which may be 1 ft across, roughly cordate in shape, slightly concave, narrowing to a thin flattened appendage up to 4 ft long. The exterior of the flower is greenish white or pale yellowish white in colour covered with a network of pale purple-red veins. The flanges of the bulb are strongly marked with purple veins. The mouth and inner hood are deep reddish purple in colour, while the face is greenish or yellowish white, streaked with broad, radial, somewhat mottled, reddish purple

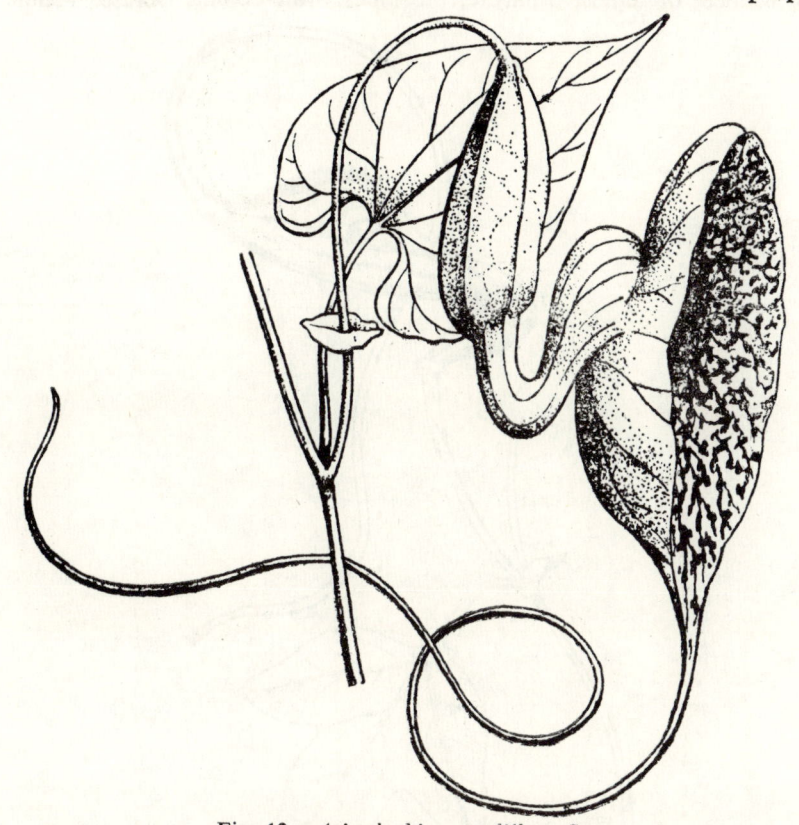

Fig. 12.—*Aristolochia grandiflora* Sw.

bands with anastomosing branches. The odour of the flower is so offensive as almost to beggar description and for this reason it is best planted away from human habitations. Fruit a strongly-ribbed 6-valved capsule, 3 in. long, oblong in shape.

Flowers.—October-November. *Fruits.*—Cold season.

Distribution.—A native of Guatemala; but now cultivated in all warm countries in the open, and under glass in temperate regions of the world.

Gardening.—A very extensive climber suitable for growing over large structures but best kept in the background on account of its offensive odour. Can be raised from seed but also grows well from cuttings of well-matured wood. Var. *sturtevantii* is the most commonly cultivated variety of the plant.

Aristolochia macroura Gomez
Livid-flowered Birthwort

Description.—A perennial climber. Leaves reniform, cordate, deeply 3-lobed, or almost 3-partite, the lobes ovate-oblong, obtuse. Petiole

Fig. 13.—*Aristolochia macroura* Gomez

1.5-2 in. long, rounded; stipules leaflike, large, somewhat cordate, acute,

PLATE 4

THE BIRD'S-HEAD BIRTHWORT
Aristolochia ornithocephala Hook.

M. N. Bakshi

PLATE 5

THE BIRD'S-HEAD BIRTHWORT
Aristolochia ornithocephala Hook.

M. N. *Bakshi*

wavy. Pedicels solitary, axillary, much curved upwards. Bulb ovate, 6-ribbed, narrowed at the apex into the tube which is curved back upon the bulb. The tube is narrow below and gradually widens into a mouth. The posterior portion of the tube ends abruptly in a truncated waved lip; the anterior wall expands suddenly into a broadly cordate lip which is abruptly attenuated into a slender twisted tail which may reach 18 in. in length. Tube and bulb dingy brownish green in colour; ribs purplish; the intervening spaces covered with a coarse purplish network of veins. Lip and tail of a rich blackish brown or deep purple colour. Fruit a 6-valved capsule.

Flowers.—Rainy season. *Fruits.*—November.

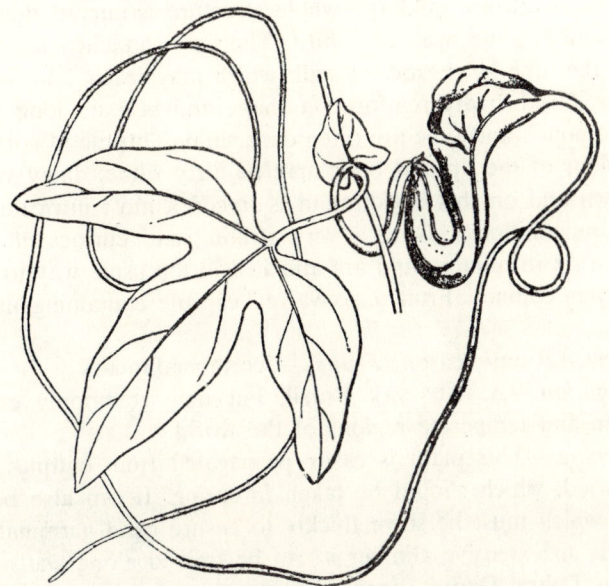

Fig. 14.—*Aristolochia macroura* Gomez
(Showing deeply partite leaves and very long tail)

Distribution.—A native of Brazil, now cultivated in gardens throughout the world on account of its striking flowers.

Gardening.—The length of the tail and the smaller upper lip in relation to the tube of the corolla is characteristic of this species. It bears its peculiar, fascinating flowers in great profusion during the rains. Easily raised from seed sown in early spring.

Aristolochia ornithocephala Hook. (**Aristolochia brasiliensis**

Mart. et Zuce.)

The Birdshead Birthwort

(The specific name is derived from two Greek words, *ornithos* bird, *kephale* head, and refers to the peculiar shape of the perianth which resembles the head of a bird.)

Description.—An extensive climbing shrub with large alternate long-petioled cordate reniform (kidney-shaped) leaves which have a broad and deep sinus at the base. At the junction of petiole and stem may be observed the foliaceous glaucous embracing stipules. The flowers are solitary and are borne on peduncles 8-10 in. long. The perianth is single, corolla-like and can be divided into three parts. The lower globose part or pouch contains the stylar column, stigmas and stamens. The tube (2.9 in. long) is suddenly inflated in the upper quarter into the so-called bird's head. Attached to the head are two expansions which may be termed beak and lip. The beak which is uppermost in the flower is lanceolate in shape but the margins are rolled inwards on the hairy inner surface, and the whole structure is curved downwards, greatly resembling the beak of a bird. The lip is attached to the lower margin of the tube by a grooved stalk which may reach 2 in. in length. The lip itself is broadly reniform in shape and is 6 in. long by 4 in. broad. The pouch and tube are often dark shining purple in colour. The ground colour of the lip and beak may be dirty white, dingy yellow or greyish fawn and on this background is traced a most intricate pattern of purple reticulations, giving a very striking and curious effect. The stamens are six in number and are attached in the usual way to the six-toothed stylar column. Fruit a six-valved capsule containing numerous flat seeds.

Flowers.—Rainy season. *Fruits.*—December-January.

Distribution.—A native of Brazil, but now commonly cultivated in all warm and temperate regions of the world.

Gardening.—This plant is easily propagated from cuttings of well matured wood, which should be taken in spring. It can also be raised from seed which must be sown thickly to ensure good germination. As this plant is an extensive climber it can be trained along walls or over trellises. In Dehra Dun it does not emit an offensive odour and so may be used to cover the walls of dwelling houses.

Aristolochia roxburghiana Klotz. (A. tagala Cham.)

Description.—A stout lofty climber with grooved glabrous stems. Leaves 3-10 in. long lanceolate to lanceolate-ovate; upper 2-3 in., lower 3-5 in. broad, broadest at the cordate base, usually with a deep and narrow sinus, reticulately veined, glabrous; petiole twinning. Flowers numerous, in loose slightly hairy racemes, mostly abortive, usually only one fertile; bracts small, oblong. Perianth 2-2.5 in. long, pale green in colour, base globose, tube curved, mouth oblique; lip straight, linear, obtuse, villous, as long as the tube. Capsule 1-2.5 in. long including the stalk, 6-valved, pear-shaped, globose or oblong-ellipsoid, transversely ridged, glabrous, splitting into six portions, each of which is held suspended on a thread-like portion of the peduncle which also splits into six parts. Seeds broadly ovate-deltoid with a deep membranous wing, flattened, usually tuberculate on one face, 0.2-0.4 in. broad.

Flowers.—April-May. *Fruits.*—May-June.

Distribution.—Indigenous to India and Burma, extending to Malaysia.

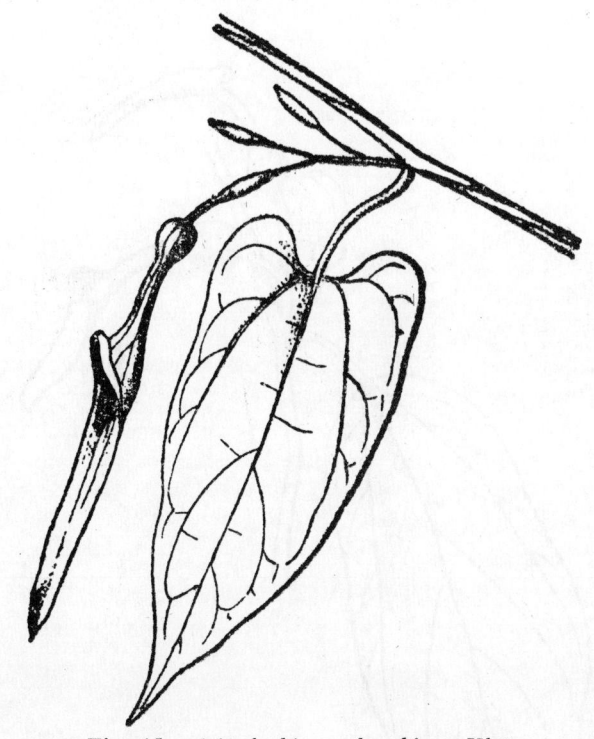

Fig. 15.—*Aristolochia roxburghiana* Klotz.

Gardening.—This evergreen twiner has small and inconspicuous flowers but its heavy and rapid growth of foliage renders it suitable for growing over unsightly buildings. It is not particular in its demands as regards site or situation and is easily raised from seed sown in March or at the break of the rains.

Aristolochia indica L.

Description.—A glabrous shrubby or herbaceous perennial arising from a woody rootstock. Branches long, slender, grooved, glabrous. Leaves variable, fiddle-shaped to linear, 2-4 in. long, 1-2 in. broad, 3- to 5-nerved at the base, somewhat cordate, acuminate, glabrous with a slightly undulate margin.

Flowers in few-flowered axillary racemes; bracts small, ovate, acuminate. Pedicels long, thickened above. Perianth 0.7-1.5 in. long, with a glabrous inflated and lobed base, which is suddenly narrowed into a cylindric tube about 0.5 in. long. The tube expands above and terminates in a horizontal funnel-shaped mouth, which is produced on one edge into a narrow, linear-oblong, obtuse lip with revolute margins. Both bulb and tube are pale green on the outer surface. The rim of the mouth

Fig. 16.—*Aristolochia indica* L.

is dark purple in colour. The upper surface of the lip is a livid brown and the lower surface dark purple. The purple areas on the mouth and lip are clothed with purple, hinged hairs. Inside the bulb are well-defined food areas. Anthers 6, stigmatic lobes 6. Capsule 1.5-2 in. long, subglobose or broadly oblong, 6-valved. Seeds flat, ovate, winged. The capsule splits up into six portions each of which is suspended from a filament-like part of the peduncle which also splits into six sections.

Flowers.—June-October. *Fruits.*—November-March, but the parachute-like old capsules may be found up to the end of the hot weather.

Distribution.—Throughout the low hills and plains of India from Nepal and lower Bengal to Chittagong, and in the Deccan Peninsula from the Konkan southwards. Common in Ceylon up to 8,000 ft.

Gardening.—This deciduous herbaceous or perennial twiner sometimes dies back to the root. It has inconspicuous, small flowers and is not at all showy. According to Roxburgh the root is nauseously bitter. The plant is supposed by the natives to ward off snakes and to be an antidote for snake-bite. This belief is probably founded on the resemblance which the flower bears to the head of a serpent and is interesting, in as much as several species of this genus enjoy a similar reputation in both North and South America, the Philippine Islands and elsewhere. It is usually raised from seed. The flowers of this plant have no appreciable smell unless bruised.

Aristolochia ringens Vahl

(*Ringens* means snarling in Latin and refers to the gaping flowers.)

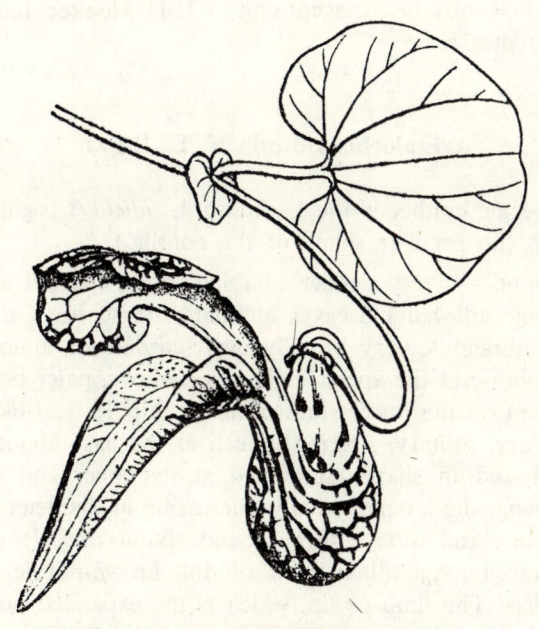

Fig. 17.—*Aristolochia ringens* Vahl

Description.—A tall, glabrous slender twiner. Leaves petioled, round reniform, pale green above, glaucous beneath, digitately 5-7-

nerved at the base. Stipules foliaceous, very deeply two-lobed, kidney-shaped. Peduncle slender, four times as long as the petiole. Flowers 7-10 in. long, green, marked with dark purple. Perianth with an obovoid ventricose sack 2.5 in. long, woolly inside; tube ascending obliquely from the sack, round, dividing into two very long lips. Upper lip (actually the lower for the flower hangs upside down) oblong-lanceolate, obtuse, concave, recurved, hairy inside from below the middle or hairy up to the edge; lower shorter, spathulate; claw long with recurved margins; blade broadly ovate or orbicular or almost kidney-shaped. Stamens 6, equidistant on the almost sessile stylar column. Ovary slender, twisted, grooved, expanded at the apex into a dark purple, callous disk.

Flowers.—November-December. *Fruits.*—Cold season.

Distribution.—A native of Brazil, but now in cultivation throughout the tropics.

Gardening.—Like other members of this genus it can easily be raised from seeds sown in spring or by cuttings or root suckers. It generally prefers a cool sheltered situation at Dehra Dun. It is suitable for growing over trellis-work.

The roots of this striking plant are esteemed as an antidote for snake bites in New Granada, where it is known as 'Guaco', from which place it was first introduced, according to J. D. Hooker, into the Royal Botanic Gardens, Kew.

Aristolochia ridicula N. E. Brown

(The specific epithet is the Latin word, *ridiculus*, signifying 'droll' and refers to the peculiar shape of the corolla.)

Description.—A very slender climbing shrub, covered all over with spreading long stiff hairs. Leaves alternate, borne upon a petiole 1-2 in. long, membranous, very broadly ovate-cordate or almost round in shape, very obtuse at the apex, bright green above, paler beneath; veins very prominent on the lower surface, hairy. Stipules leaf-like, recurved. Flowers axillary, solitary, striking. Pouch at the base about 1 in. long, obliquely-ellipsoid in shape contracted at its upper end into a tube about 1 in. long, the latter gradually increasing in diameter from below upwards. Pouch and tube with a ground colour of pale yellow upon which are traced longitudinal veins of dull brown-purple with anatomosing veinlets. The limb or lip, which is the expanded portion of the tube, is recurved at the margins and is extended dorsally into two widely spreading linear obtuse lobes 1 in. long by 0.5 in. broad, bright yellow, spotted with reddish purple, in colour; lobes sparsely covered with stalked, club-shaped glandular hairs. The mouth and throat of the tube are filled with deflexed white hairs. Stylar column short, with six obtuse stigmatic lobes, and six linear anthers.

Flowers.—Rainy season. *Fruits.*—Cold season.

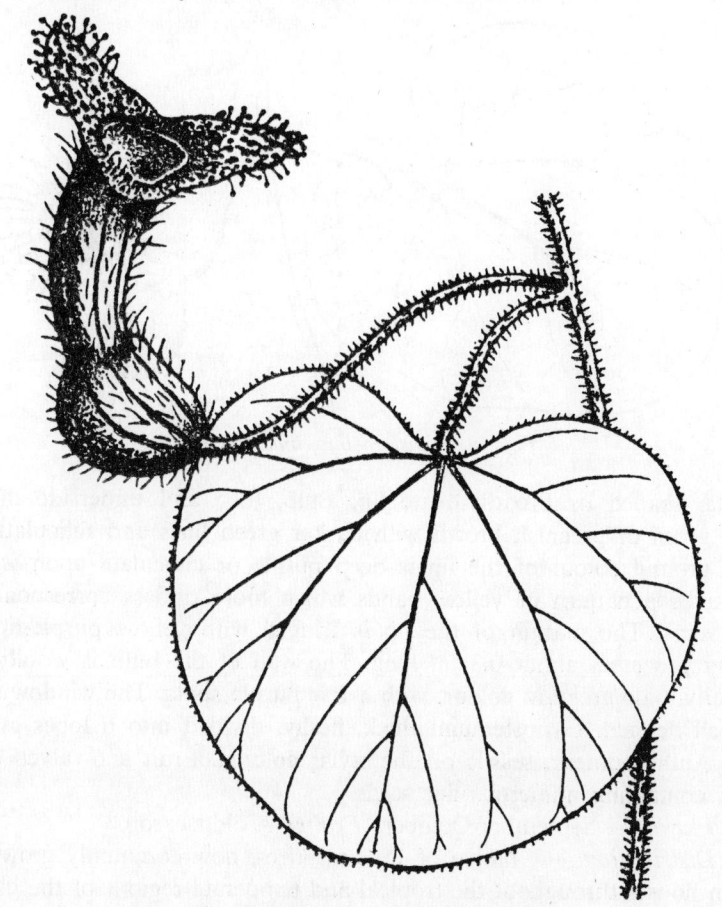

Fig. 18.—*Aristolochia ridicula* N.E. Brown

Distribution.—This grotesque looking plant is a native of Brazil, but is now cultivated throughout the world.

Gardening.—Like its allies it is suitable for culture in cool and protected places or in conservatories. Easily raised from seed, suckers or cuttings.

Aristolochia ciliata Hook. (A. fimbriata Cham.)

Fringe-flowered Aristolochia

(*Ciliata* refers to the margins of the flowers.)

Description.—A weak, slender, straggling plant. Leaves cordate-reniform, very obtuse, with a deep sinus at the base, pale beneath petioles 1-1.5 in. long. Flowers solitary on peduncles which are shorter than the petioles. Bulb pear-shaped, constricted into the tube which is bent back upon it. The tube suddenly expands at the mouth into a

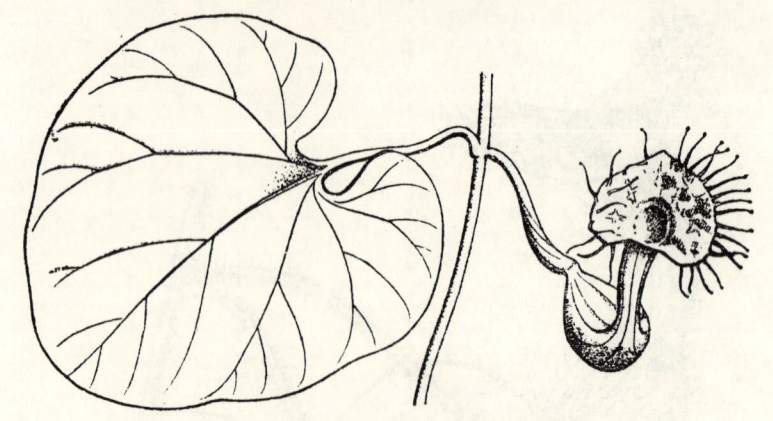

Fig. 19.—*Aristolochia ciliata* Hook.

kidney-shaped or broadly-ovate lip. Bulb, tube and underside of lip pale green or greenish brown with darker green lines and reticulations. The ground colour of the lip is deep purple or chocolate upon which is traced a pattern of yellow bands which more or less correspond to the veins. The margin of the lip is fringed with yellow purple-tipped fleshy processes about 0.3 in. long. The wall of the bulb is woolly internally, pale green in colour, with a few purple spots. The window area is well defined. Gynostemium thick, fleshy, divided into 6 lobes at the top. Anthers linear, sessile on the stylar column. Fruit a 6-valved capsule, containing numerous flat seeds.

Flowers.—September-October. *Fruits.*—Cold season.

Distribution.—A native of Buenos Aires, now commonly grown in green houses throughout the tropical and temperate regions of the globe.

Gardening.—The fringed, peculiarly shaped and coloured flowers render this species worthy of cultivation in green houses and cool, shady situations, particularly as the flowers themselves have no particularly offensive odour, although the whole plant has a rank smell. It is easily raised from seeds sown in April.

Aristolochia tomentosa Sims

The Hairy Birthwort

(*Tomentosa* means hairy.)

Description.—A tall woody climber, hairy all over. Leaves leathery in texture, rounded, cordate at the base, tomentose, 4-7 in. long, and as much across. Flowers solitary or 2 or 3 together in the axil of a leaf.

Bulb pear-shaped, constricted into a narrow tube which is bent back upon the bulb. The tube expands above into a lip which is extended into three reflexed triangular obtuse lobes. The bulb, tube and lower surface of the lobes are yellow or yellowish green. The lobes them-

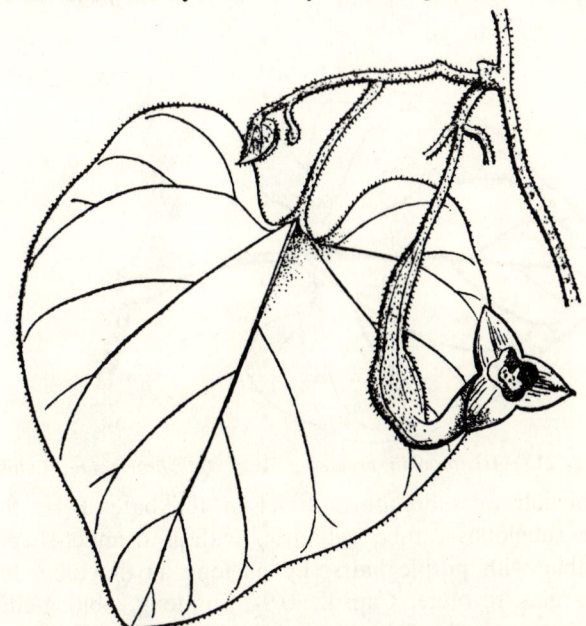

Fig. 20.—*Aristolochia tomentosa* Sims

selves are greenish purple with a dark brown centre. Bracts absent. Stamens six, short, adnate to the stylar column. Fruit a 6-valved capsule; seeds numerous, flat.

Flowers.—May-June. *Fruits.*—Cold season.

Distribution.—Indigenous to North America (California to Illinois), now commonly grown in gardens throughout the tropical and subtropical regions of the world.

Gardening.—This plant is quite hardy in India, and will grow to a great height if properly supported. It can be grown with great ease over trellises and arbours, and its profuse foliage renders it particularly useful in hiding unsightly objects. The flowers are rather inconspicuous amid the mass of luxuriant foliage but they are rather curious in appearance and the plant is well worth cultivating in gardens. The root and bark are said to possess an aromatic flavour. The plant is said to have been first sent to England from Philadelphia in 1763.

Aristolochia bracteata Retz. (A. bracteolata Lamk.)

The Bracteated Birthwort

Description.—A slender perennial herb; stems 12-18 in. long, weak, prostrate, branched, glabrous, striate. Leaves 1.5-3 in. long and as much across, reniform or broadly ovate, usually widely and shallowly cordate at the base, glaucous beneath, finely reticulately veined, glabrous; petiole 0.5-1.2 in. long. Flowers solitary; pedicels with a large

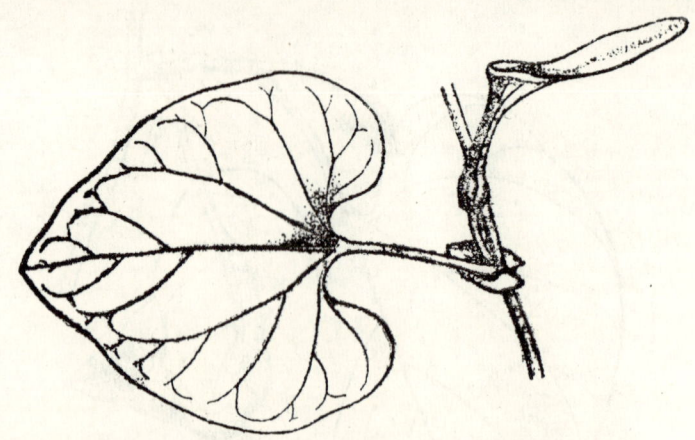

Fig. 21.—*Aristolochia bracteata* Retz. (*A. bracteolata* Lamk.)

sessile orbicular or subreniform bract at the base. Perianth 1-1.9 in. long, base subglobose; tube cylindric, with a trumpet-shaped mouth, villous within with purple hairs; lip as long as the tube, linear, dark purple, margins revolute. Capsule 0.9-1 in. long, oblong-ellipsoid, 12-ribbed, glabrous. Seed 0.3 in. long, deltoid with a slightly cordate base.

Flowers.—At the end of the rainy season.

Distribution.—Upper Gangetic Plain, Bengal, Western Peninsula, Sri Lanka, extending to Arabia and Tropical Africa.

Gardening.—This species has very small, inconspicuous flowers and is not worthy of cultivation but is valued for its medicinal properties which are extensively prescribed in Ayurveda. Decoctions of the plant are purgative, anthelmintic, and are said to be useful in the treatment of 'kapha', fevers and painful joints. The juice is applied to sores to kill maggots (hence the vernacular name *kera-mar*).

According to Kirtikar and Basu the extremely bitter taste of the plant persists for a long time, chiefly in the throat. Mhaskar and Caius repudiate its supposed medicinal virtues as an antidote to snake venom.

Commonly raised by seed sown in early spring.

Aristolochia elegans Mast. (A. littoralis Parodi)

Calico Flower

(*Elegans* means graceful in Latin.)

Description.—A very slender glabrous climber with pendulous branches and foliage. Leaves 2-3 in. long, and as much broad, borne on long (1-2.5 in.) petioles, very broadly ovate-cordate in shape, with a wide sinus and rounded lobes at the base, obtuse or rounded at the tip; bright green on the upper surface, somewhat glaucous beneath, stipules leaf-like, curved. Flowers solitary, seated on long pedicels, Perianth-pouch pale yellow-green in colour, swollen, contracted at the the sack. Tube somewhat funnel-shaped at the top, expanding abruptly into a nearly circular shallow cup, cordate at the base and about 3 in.

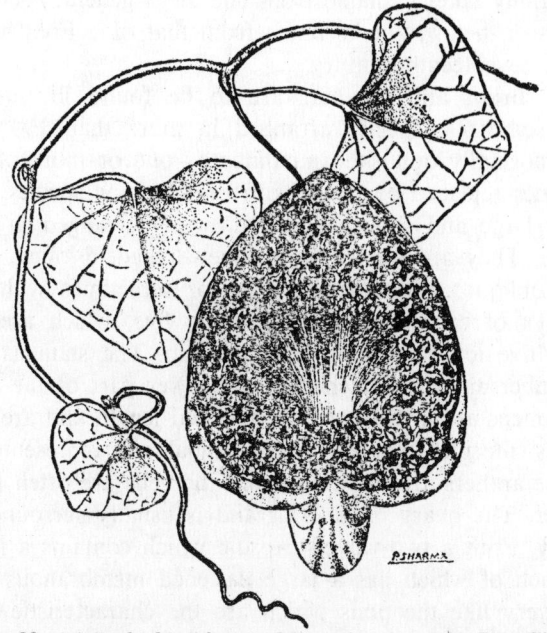

Fig. 22.—*Aristolochia elegans* Mast. (*A. littoralis* Parodi)

diameter. Outside the cup is whitish in colour covered with a network of red-purple veins; inside the ground colour is a deep rich purple-brown becoming yellowish green and velvety at the mouth and white towards the margins, where the purple flows into the white in irregular patches. Stylar column short, cylindrical, expanded upwards into six oblong-obtuse lobes; anthers six. Fruit a capsule with innumerable flat seeds.

Flowers.—Rains. *Fruits.*—Cold season.

Distribution.—This plant is a native of Rio de Janeiro in Brazil, but is now commonly cultivated throughout the tropical and temperate regions of the globe.

Gardening.—It is a rather small-flowered graceful species and a most desirable climber for a green house. The hanging basket-like fruits are very attractive. Easily raised from seed which is freely produced under cultivation and usually flowers in the very first year. It can also be propagated by cuttings, layers or suckers. It is entirely devoid of the unpleasant odour which is characteristic of the flowers of this genus. In Florida it reproduces itself freely.

Chapter 3

BIGNONIACEAE

This family takes its name from one of its genera, *Bignonia*, which, as will be seen below, got its name from that of a Frenchman, prominent in the seventeenth century.

Trees, shrubs and climbers are to be found in this very large family of some 600 species arranged in more than 100 genera. The leaves are normally opposite and pinnate—one or more of the leaflets are sometimes replaced by tendrils in the climbing species. The flowers are usually large and showy, and are mostly arranged in terminal inflorescences. They are funnel or trumpet-shaped, 5-lobed, usually irregular and oblique, sometimes fragrant or sometimes with a repulsive odour. Calyx of combined sepals, usually five, which are represented usually by five lobes at the top of the rim. The stamens are four or five in number and are attached to the lower part of the corolla; four of these stamens are in two pairs of unequal length and are thus termed didynamous, the fifth is very often vestigial and represented by a staminode; the anthers are two-celled and the cells are often placed above one another. The ovary is superior and is usually surrounded by ring-like nectary. Fruit a two-valved capsule which contains a large number of seeds each of which has a large flattened membranous wing. These fruits are very like the pods which are the characteristic fruits of the sweet-pea family, but they can be easily distinguished from them by the central partition and the winged seeds.

Most of the species of this family are found in America, mainly in South America. A good many of these exotics are grown in India and they add a great deal of beauty not only to our gardens, but also to our avenues and public places. One has only to mention the Jacaranda (*J. filicifolia*), the Indian Cork Tree (*Millingtonia hortensis*), the African Sausage Tree (*Kigelia aethiopica*), the Tulip Tree (*Spathodea campanulata*), to show how much we owe to this family for beauty of tree form, foliage and flower.

It is however the climbers that are so striking in our gardens. Nothing can possibly equal the profusion and exquisite colour of the flowers of the Golden Shower, which are produced in succession for several weeks during the cold weather.

Many of the species of Bignoniaceae are night flowering. The fragrant white flowers of *Millingtonia*, one of our native trees, must certainly be cross-pollinated by long-tongued moths. A good many of our climbers do not set seed in this country, possibly because the

GOLDEN SHOWER
Bignonia venusta Ker.
(⅝ nat. size)

insects which visit the flowers in their native land are not to be found in this climate.

The genera as treated in this chapter can be separated by the following key:

> Leaves with tendrils .... *Bignonia*
> Leaves without tendrils .... *Tecoma*

Bignonia Linn.

The genus *Bignonia* was erected in honour of Abbe Jean Paul Bignon (1662-1743), court librarian to Louis XIV of France. Systematic botanists in the past have held very divergent views regarding the delimitation of this genus and modern research has resulted in the genus, as conceived by Bentham and Hooker with its 150 species, being broken up and the majority of its species transferred to other genera. We would warn our readers that the species, hitherto known as *Bignonia*, found in our Indian gardens and treated in this chapter, have all, with the exception of one, been placed in other genera of the *Bignoniaceae*.

The species of the genus, as known to horticulturists, have been in high favour for many years on account of the beauty and profusion of their flowers. They are mainly extensive climbers and require generous space to be seen at their best.

The wood of the stem is very peculiar. Young stems do not show any anomalous structure, but in old stems the wood is cleft by wedge-shaped radially projecting masses of bast, which are regularly arranged in multiples of four. Those with four wedges are called 'Cross Vines' in South America and are regarded with superstitious awe. Some of these climbers grow to an immense size, and these alternating wedges of bast and wood enable them to withstand twisting and bending without fracture.

Climbing is accomplished in various ways—by twining round a support, by the possession of tendrils, by the rotation of the petioles, or by means of aerial roots. The species treated in this chapter make use of all these methods, either singly or in various combinations. The possession of tendrils is a very common feature.

Tendrils are always modified leaves or leaflets. In the well-known *Bignonia venusta* Ker. [*Pyrostegia venusta* (Ker.-Gawl.) Miers], the foliage seems at first sight to consist of a pair of opposite leaves upon a common peduncle the apex of which is continued and ends in three filiform branches. The correct interpretation of this arrangement is that the leaf is compound and consists of three leaflets, the terminal of which is modified as a branched tendril. The tendril after attachment to a support contracts spi⌊ and not only raises the stem but also permits a certain elasticity in the attachment of the liane to its support and thereby minimises the danger of damage from storms. The tips of the tendrils after attachment often become transformed into small disks.

The leaves of the species may be simple or compound and are exstipulate, though the development of small leaves in the axils of true leaves often simulate foliaceous stipules.

The flowers are arranged in terminal or axillary pairs or racemes or in corymbose fascicles. The calyx is of combined sepals with obscure lobes or teeth. The corolla is large and showy, zygomorphic and usually bell- or trumpet-shaped, less often cylindrical, rarely 2-lipped, 5-lobed. The stamens are five in number but the posterior one is usually rudimentary and represented by a small stipe. The other four are paired, two with long filaments, and two with short; a condition which is termed didynamous. The anther cells are usually divergent at the base.

The ovary is seated upon a disk which may be cupular or platter-shaped, and is two-celled. The ovules are numerous and are attached to axile placentas. The style is simple with two flattened stigmatic lobes. The fruit is usually a two-valved capsule.

In species of this genus, whose life history has been studied, it is found that the stigmas and pollen become mature at different times and that the stigma-lobes close before the pollen from the same flower can fertilise them. Cross-fertilisation seems to be the rule in the genus and in the case of those species imported into India from abroad the agent of pollination seems to be absent in this country, for very few of our exotic species set fruit.

KEY TO THE SPECIES

Tendrils 3-partite
  Corolla cylindrical .... B. venusta
  Corolla swollen upwards
    Tendrils clinging by disks .... B. capreolata
    Tendrils clinging by claws or hooks .... B. unguis-cati

Tendrils simple
  Disk present; flowers rose with darker stripes; leaves shining .... B. speciosa
  Disk absent; flowers pale mauve, purplish with a white
    or yellowish throat; leaves dull .... B. magnifica

Bignonia venusta Ker. [Pyrostegia venusta (Ker.-Gawl.) Miers]
Golden Shower

(*Venusta* is Latin for charming, beautiful.)

Description.—An evergreen plant climbing extensively by means of tendrils. Stems stout, angled. Leaves compound, consisting of two leaflets with a terminal branched tendril, or sometimes of three leaflets. Leaflets ovate or ovate-oblong, bluntly and shortly acuminate, wedge-

The Golden Shower
Bignonia venusta Ker.

M. N. Bakshi

shaped at the base, 2-4 in. long, glabrous on the upper surface, shortly and sparsely hairy beneath.

Flowers very showy, arranged in corymbose cymes or racemes, drooping. Calyx small, campanulate in shape, with five, very small, hairy teeth. Corolla tubular, 3 in. long, gradually expanding to the mouth where it ends in five linear lobes which are valvate in the bud (i.e. they touch by their margins), the point of junction being very evident as each lobe has white villous margins. On opening the lobes curve backwards and form two lips, the upper of which is 2-, the lower 3-lobed. Stamens four, in pairs, with filaments of different lengths. The longer pair is well exserted from the tube, the shorter reaches the base of the lobes. Ovary linear, seated on a fleshy disk. Style very long, almost as long as the stamens. This plant does not produce fruit in this country.

Flowers.—Cold season.

Fig. 23.—*Bignonia venusta* Ker.

Distribution.—Indigenous to Brazil, but now a very common cultivated plant in all tropical countries.

Gardening.—Probably no plant in the world presents so gorgeous an appearance as *Bignonia venusta* when in full bloom during Jan.-Feb. The plants are not fastidious as to soil but a good fibrous loam, to which one-third cow or sheep manure has been added, suits them admirably. Propagation is effected by cuttings of the wood taken in late spring and inserted in sand preferably under a bell jar. It is suitable for pergolas in the open, porches, verandas and the like.

Bignonia capreolata Linn.

Trumpet Flower; Cross Vine; Quarter Vine

(*Capreolata* is a Latin word meaning 'winding' or 'twining'.)

Description.—An extensively climbing glabrous species. The older stems in section exhibit a perfect cross of four wedge-shaped insertions of bast in the wood. The leaves are opposite and compound, consisting of an ovate or oblong, acuminate and subcordate pair of stalked entire leaflets and a compound tendril which clings by small disks. Stipules are absent but accessory leaves or leaflets in some of the axils simulate foliaceous stipules.

The flowers are borne on pedicels which are arranged in fascicles of 2-5 on axillary spurs. The calyx is membranous, green, 0.2 in. long. The corolla is tubular with a stout limb, 2 in. long, yellowish red in colour, rather lighter within.

Fig. 24.—*Bignonia capreolata* Linn.

Flowers.—March-April. Does not set fruit in Dehra Dun.
Distribution.—This climber is indigenous in North America, but

is now extensively grown as an ornamental plant in the tropics of the Old World.

Gardening.—This handsome vine is very suitable for covering walls, embankments and the like. Propagated by cuttings or layers of half matured wood.

Bignonia unguis-cati Linn.

Cat's Claw (in Latin *unguis cati.*)

Description.—An extensively climbing slender evergreen species. The leaves are opposite and compound, exstipulate. Leaflets 2, lanceolate or ovate-acuminate, cordate, 3 in. or less long; the terminal leaflet is represented by a three-partite claw-like tendril, the arms of which do not form disks after attachment. The flowers, which are a beautiful clear yellow in colour, with deeper yellow lines in the throat, are borne in pairs seated on slender peduncles in the axils of the leaves. Calyx 0.5 in. long, obtusely 5-lobed, membranous, bowl-shaped, with conspicuous veins. Corolla with a short tube and broadly ventricose limb with

Fig. 25.—*Bignonia unguis-cati* Linn.

spreading lobes, 2-4 in. across and 2-2.5 in. long. Ovary linear, 0.25 in. long, seated on a fleshy disk. Style curved, 1 in. long.

Flowers.—April. Occasionally a second flush in August. *Fruit.*—July.

Distribution.—This climber is a frequently cultivated plant in India. It is a native of Argentina.

Gardening.—An extensive climber reaching the tops of the tallest trees and forming masses of foliage and yellow flowers in pendent bunches. It sows itself freely in Dehra Dun and also probably elsewhere. Easily raised from seed or by layers or cuttings.

Bignonia speciosa R. Grah. [Clytostoma callistegioides (Cham.) Bur.]

(*Speciosa* is Latin for handsome.)

Description.—An evergreen shrub, extensively climbing by means of tendrils. Leaves opposite, compound, consisting of two opposite stalked leaflets and a terminal unbranched tendril. Leaflets about 3 in. long, but may be longer, elliptic-oblong or ovate-acuminate in shape, glabrous, shining, reticulated below; margins undulate; base subcordate, rounded or acute.

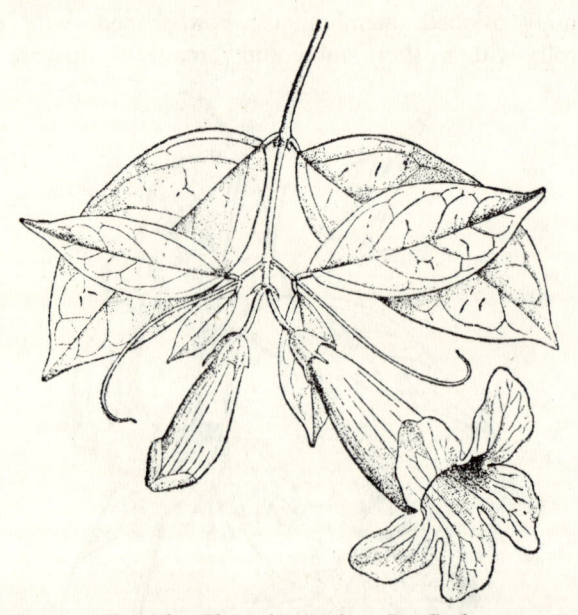

Fig. 26.—*Bignonia speciosa* R. Grah.

Flowers large and showy, borne in pairs on a terminal peduncle. Calyx obliquely campanulate, with 5 acute or subulate lobes. Corolla pubescent, about 3 in. long; limb 3 in. broad, broadly ventricose from a short yellowish tube, lilac rose or pale purple in colour, streaked inside with darker purple veins. Lobes five, broadly ovate or orbicular, the upper reflexed. Ovary seated on a fleshy disk with crenulate margins. Stamens included, didynamous; anther-cells much divaricate. Pod 2.5 in. long, shortly oblong, densely covered with short spines.

PLATE 7

THE HANDSOME BIGNONIA
Bignonia speciosa R. Grah

N. L. Bor

PLATE 8

THE HANDSOME BIGNONIA
Bignonia speciosa R. Grah.

N. L. Bor

Flowers.—March-April. *Fruits.*—Cold season.

Distribution.—Indigenous to tropical America, but now widely cultivated in the tropics of both hemispheres.

Gardening.—This is a very showy and ornamental species best suited for training over trellis work or over fences and the like. Easily raised, like most other species of the genus, by layers.

Bignonia magnifica Bull.

The species known to gardeners by the name above should really be called *Saritaea magnifica* (Sprague) Dugand, syn. *Arribidaea magnifica* Sprague. Actually the name *Bignonia magnifica* Bull. belongs to a plant which differs widely in respect of leaf and calyx and may not be a *Bignonia* at all [Chatterjee in Bull. Bot. Soc. Beng. 2: 78, 1948].

Description.—A climbing shrub with branches compressed when young, terete when old, rather rough to the touch. Leaves compound, opposite, consisting of 2 leaflets and usually a tendril. Leaflets opposite, obovate oblong, obtuse at the tip, attenuate at the base, rather dull on both surfaces, 3-5 in. long by 1.75-2.5 in. wide, membranous, glabrous, entire, with a short petiolule 0.1 in. long; nerves impressed above, pro-

Fig. 27.—*Bignonia magnifica* Bull.

minent below. Tendril issuing between the leaflets, hooked, up to 6 in. long.

Inflorescence of four-flowered axillary or terminal cymes. Flowers large, showy, almost sessile. Calyx gamosepalous, cupular-campanulate, very slightly narrowed above, olive green, truncate at the top or with a few teeth, shining within, dull without. Corolla large, up to 2.5 in. wide across the limb, purplish violet, mauve or light purple in colour, whitish or primrose in the throat with longitudinal purple veins. Tube tubular-infundibuliform, cured above the base, 2.5 in. long, 5-lobed, lobes orbicular, almost 1 in. in diameter. Stamens 4, didynamous, fifth a rudimentary stipe; filaments curved, inserted at the top of the narrow portion of the tube, swollen at the base and glandular, included; anthers widely divergent. Ovary superior, at the base of the tube; style 1-1.5 in. long, with 2 stigmatic lobes.

Flowers.—Cold season. Does not set fruit in India.

Distribution.—Indigenous in Colombia, South America; now commonly cultivated in all tropical countries in the open and under glass in temperate climes.

Gardening.—It is a free flowering handsome climber, bearing large flowers of a delicate mauve changing to rich crimson, with the throat of a light primrose colour. Suitable for growing on arches, pergolas and the like. Easily raised by layers or cuttings.

Tecoma Juss.

A genus of the *Bignoniaceae*. The name itself is derived from the Mexican term for a plant, *Tecomaxochitl*, which means 'vessel-flower' and refers to the large cup-shaped or trumpet-shaped blooms.

The genus *Tecoma*, as known to botanists, contains plants with a shrubby habit, but horticulturists in India include several scandent species, now known by other names, in this genus.

The species of *Tecoma*, in the horticultural sense, are deservedly popular in Indian gardens, not only for the large size and profusion of their gorgeously coloured flowers, but also for their extremely handsome foliage.

In the scandent species treated in this article the methods of climbing are either by the production of aerial rootlets from the nodes which clasp the support, or by simple twining round a support.

The leaves of our species are opposite, exstipulate and may be evergreen or deciduous. The leaf is compound and odd pinnate with entire or toothed leaflets.

The flowers, which are large and showy, are borne in terminal or axillary panicles or racemes. The calyx is gamosepalous, tubular, with five lobes. The corolla gamopetalous, more or less irregular, campanulate, or funnel-shaped with five lobes.

Stamens five, of which one is rudimentary, inserted on the corolla tube alternate with the lobes. The ovary is seated upon a fleshy disk

LARGE-FLOWERED TECOMA
Tecoma grandiflora Delaun.
(5/9 nat. size)

and may be one- or two-celled, with many ovules attached to parietal placentae. The style is slender and divided into two stigmatic lobes. The fruit is a capsule, septicidal or loculicidal. The seeds are furnished with papery wings.

Glandular hairs are developed on the aerial parts of several species of *Tecoma*. Parija and Samal who investigated these structures in *Tecoma capensis* came to the conclusion that they were hydathodes or water secreting organs. The importance of the hydathodes lies in their power to secrete water when transpiration is suppressed, an arrangement which prevents the hydrostatic pressure within the conducting system from becoming excessive, and hence protects the ventilating system against the danger of flooding (Haberlandt).

The same two investigators discovered glandular areas on the floral parts of *Tecoma capensis*. They proved that these glands (extrafloral nectaries) secrete glucose and cane sugar, and that they are active long before the sexual parts of the flower are mature. This being so, it seems that the large black ants which visit these nectaries in crowds, day and night, have nothing to do with the pollination. Some insects, instead of entering the mouth of the flower, to reach the floral nectaries developed at the base of the corolla, and so assisting the plant in cross-fertilisation, bore through the base of the corolla and steal the honey. It is possible and even probable that the presence of ants on the extrafloral nectaries prevent insect robbers from getting their booty in this way without working for it. Parija and Samal, however, consider that the extrafloral nectaries prevent the ants from visiting the flowers and so favour self-fertilisation. Cross-fertilisation is, however, extremely likely in view of the fact that intrafloral nectaries are developed in the brightly coloured corollas.

Tiwary has given some account of the pollination of *Tecoma radicans*. The flowers are protandrous and the anthers are found to have completely dehisced even before the flower is open. The bifid stigma of the immature ovary, with the halves closely pressed together, remains tucked away behind the filaments. The honey is secreted at the base of the corolla and is capable of being extracted by long-tongued insects. The flowers are much visited by bees which remove quantities of pollen. Susequently the style elongates and the stigmatic lobes separate. Cross-fertilisation is now possible. Tiwary concludes that cross-fertilisation does not take place because long-tongued insects do not visit the flowers, but surely the pollen gatherer can perform this service just as well. Black ants are also found on the calyx and corolla of this plant, and they may exercise the same function as in the case of *Tecoma capensis*.

The presence of ants, however, cannot prevent sunbird (*Nectarinia asiatica*) from boring a small hole at the base of the corolla and stealing the honey. They are often seen hovering about the flowers of *Tecoma stans* at Dehra Dun.

Key to the Species

Shrubs .... *T. stans*
Climbers
  Stamens exserted .... *T. capensis*
  Stamens included
    Flowers scarlet or orange; climbing by aerial rootlets
      Leaflets 9-11, pubescent beneath, at least on the midrib;
      calyx teeth short .... *T. radicans*
      Leaflets 7-9, glabrous beneath; calyx 5-lobed to middle .... *T. grandiflora*
    Flowers white, yellowish or pink, climbing without aerial rootlets
      Corolla 1.5-2 in. long .... *T. jasminoides*
      Corolla 0.5-0.75 in. long .... *T. australis*

Tecoma stans (Linn.) H. B. K. [**Stenolobium stans**]

The Yellow Elder; Trumpet-Flower; Yellow Bells

(The specific name means 'erect' in Latin.)

Description.—A large shrub sometimes reaching the size of a small tree. The leaves are opposite, compound and odd-pinnate, 4-8 in. long. Leaflets opposite, 5-11 in number, ovate, ovate-oblong to lanceolate in shape, 1-4 in. long, acuminate, sharply serrate, sessile or nearly so. The new leaves are a beautiful glossy green but the foliage looks very tired after a cold winter and dry hot weather. It is usual in older trees to see a profusion of erect shoots from the base of the stem. These shoots bear flowers.

The flowers are borne in large terminal panicles and are fragrant. Calyx cup-shaped with 5 narrow lobes, glabrous or sparsely covered with short crisp hairs, 0.2-0.3 in. long. Corolla gamopetalous, of a beautiful yellow-gold colour, contracted to a narrow tube at the base, expanding above the calyx into an irregular funnel with five wavy lobes. Stamens four fertile, one rudimentary, included inside the corolla tube. This is the usual arrangement but one often finds only three fertile stamens and one unfertile anther on a filament only half the normal length. The anthers are widely divergent from the apex and the connective is crowned with a small foliaceous limb. The ovary is seated in a cupular disk which is wavy on the margin. The style is long and filiform and divides at the apex into 2 stigmatic lobes. The capsule is linear, 6-9 in. long, compressed, green, turning brown at maturity, and contains numerous seeds, each with 2 large thin wings. The eastern and southern sides of a bush flower first and by the time the pods have replaced the flowers, the northern and western aspects flower.

PLATE 9

THE YELLOW ELDER OR TRUMPET-FLOWER
Tecoma stans H.B.K.

M. N. Bakshi

PLATE 10

THE YELLOW ELDER OR TRUMPET-FLOWER
Tecoma stans H.B.K.

N. L. Bor

BIGNONIACEAE 43

Flowers.—Practically throughout the year. The capsules are produced during the cold weather and remain hanging for a long time on the plant.

Fig. 28.—*Tecoma stans* (Linn.) H.B.K. ×$\frac{3}{4}$

Distribution.—Native of Tropical America. Very common in gardens in the plains throughout India. It is also found in the hills up to 5,000 ft.

Gardening.—This species grows exceedingly well in our gardens and flowers profusely. Owing to its rapid growth and dense foliage down to the ground, the 'yellow elder' is highly valued as a screen for unsightly objects. It ripens its seeds so abundantly that hundreds of seedlings come up self-sown around old plants. Easily raised from seeds sown in March.

Tecoma stans (L.) H. B. K. var. *incisa* Hort. is a form in which the leaflets are narrow and are cut almost to the midrib. A plant with very graceful foliage.

Tecoma capensis Lindl. [Tecomaria capensis (Thumb.) Spach]
The Cape-Honeysuckle

(The specific name indicates the origin of the plant.)

Description.—A rambling shrub about 6 ft. high. Stem and branches slender, brown, with prominent raised lenticels. Leaves opposite, compound, odd pinnate. Leaflets 7-9, 0.5-1.5 in. long, broadly ovate to almost orbicular in shape, glabrous, shortly petiolate, coarsely serrate; teeth often obtuse.

Fig. 29.—*Tecoma capensis* Lindl. × ¾

Inflorescence a peduncled terminal raceme. Flowers seated on pedicels up to 0.5 in. long. Calyx gamosepalous with five regular triangular teeth, 0.2-0.25 in. long, cup-shaped. Corolla tubular or funnel-shaped, slightly curved, gamopetalous, 2 in. long, orange or scarlet or flame-coloured, ending in four lobes the upper of which is notched. Fertile stamens four, inserted on the corolla tube and alternate with lobes. Anther-cells spreading at the apices of the filaments which are longer than the corolla tube. The connective projects as a small knob above the anther cells. Ovary oblong; seated in a cup-like disk. Style long, slender, ending in two stigmatic plates. Fruit a capsule usually 2 in., but occasionally up to 4 in. long.

Flowers.—Sept.-Oct. *Fruits.*—Cold season.

Distribution.—Indigenous to the Cape, South Africa. Cultivated rather frequently in gardens throughout India.

Gardening.—The Cape-Honeysuckle grows luxuriantly in our gardens. If the long shoots are cut back severely, the plant can easily be trained into shrub form. It is usually propagated by cuttings which strike root readily during the rains. It is rather frost-tender and must be protected from ground frosts.

Tecoma smithii W. Wats. is supposed to be a hybrid between *T. mollis* H. et B. and *T. capensis* Lindl. It is an upright bush with odd-pinnate leaves of 11-17 leaflets. Flowers in large compound panicles, bright yellow tinged with orange.

Tecoma radicans Juss. [Campsis radicans (Linn.) Seem.]
Trumpet-Vine

(The specific name means 'rooting' in Latin and refers to the aerial roots of this plant.)

Description.—An extensive scandent deciduous shrub, climbing by means of adventitious roots. Stems and branches cylindrical, smooth and polished, emitting roots where they touch the soil. Leaves compound, odd-pinnate, opposite; leaflets 9-11, petioled, ovate to ovate-oblong in shape, 1.5-2.5 in. long; acuminate or tailed at the apex, coarsely toothed, glossy dark green above, paler beneath and hairy at least on the midrib. This shrub is leafless during the cold weather.

Flowers in terminal clusters or panicles. Calyx tubular-campanulate, rather leathery, glandular, shortly 5-lobed, 0.5-1 in. long. Corolla narrow, tubular below, funnel-shaped or campanulate above, about 3 in. long, with 5 spreading lobes which are fimbriate on the margins, slightly 2-lipped, orange in colour with a scarlet limb. Stamens four, inserted on the corolla tube, two long and two short, included within the tube. Ovary oblong, seated in a large disk. Style filiform with two stigmatic plates at the apex. Fruit an elongated angled capsule, 3-5 in. long curved and beaked at the apex. Seeds many, compressed, with two large, translucent wings.

Flowers.—Practically throughout the year except the cold weather. *Fruits.*—Cold season.

Distribution.—Indigenous to North America. Commonly cultivated in gardens throughout India.

Gardening.—This species is particularly adapted for covering walls and rocky embankments, as it will climb with aerial rootlets and clings

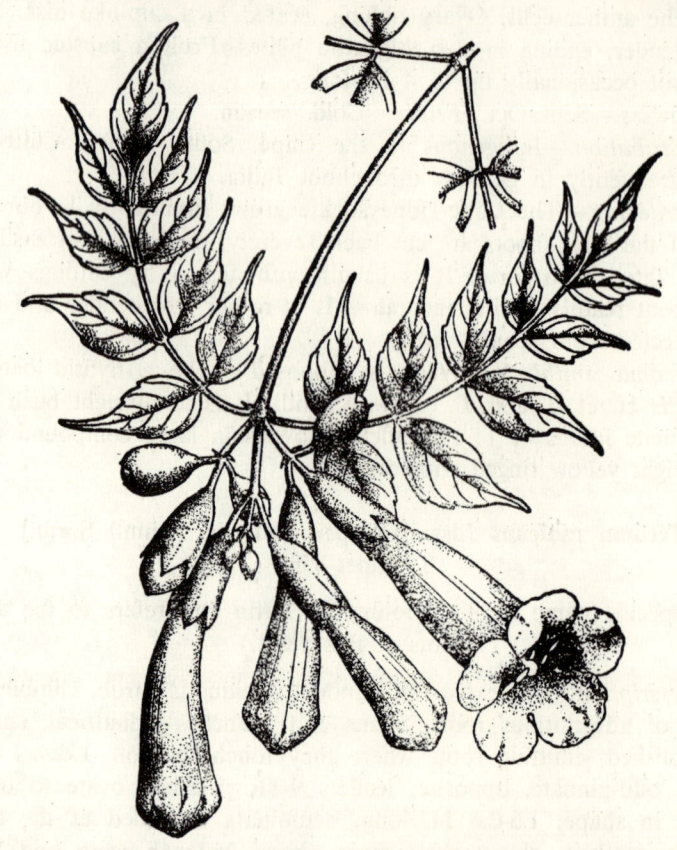

Fig. 30.—*Tecoma radicans* Juss. ×¾

firmly to its support. It is an excellent plant for covering bare trunks of trees. It is constantly in bloom with a profusion of drooping corymbs or orange-scarlet tubular flowers. Usually propagated by cuttings or root-suckers.

Tecoma grandiflora Delaun. [**Campsis grandiflora** (Thunb.) K. Schum.]

Large-flowered Tecoma; Chinese Trumpet Creeper

(*Grandiflora* means large-flowered in Latin.)

Description.—An extensive climber with few or no aerial roots. Stems somewhat angled, smooth and glabrous. Leaves opposite, com-

PLATE 11

M. N. Bakshi
TRUMPET-VINE
Tecoma radicans Juss.

PLATE 12

TRUMPET-VINE

M. N. Bakshi

pound, odd pinnate. Leaflets petioled, often unequal-sided at the base, 1-2 in. long, ovate, ovate-oblong or lanceolate in shape, acuminate, coarsely toothed, perfectly glabrous on both surfaces.

Flowers in terminal pendulous racemose panicles. Calyx campanulate, divided almost to the middle into five lanceolate lobes which end in a subulate point. Corolla up to 2 in. long, shorter and broader than that of *T. radicans,* scarlet or orange in colour, sometimes showing marked scarlet striations in the throat, terminating in 5 spreading lobes. Stamens four, two long, two short not exserted from the corolla. Anther-cells divaricate. Ovary seated in a disk. Style filiform. Fruit a capsule, not beaked at the apex.

Flowers.—Hot and rainy season. *Fruits.*—Nov.-Dec.

Distribution.—Indigenous to China and Japan, now widely cultivated in the tropical and subtropical parts of the world.

Gardening.—This graceful climber is very beautiful when in full bloom during the hot and rainy season and is an ornament to any garden. It sheds its leaves in the cold weather, when it is advisable to cut it well back and remove the numerous suckers which it sends up all round from the roots. Easily propagated by rootsuckers, cuttings and seed.

Tecoma jasminoides Lindl. [**Pandorea jasminoides** (Lindl.) K. Schum.]

The Bower Plant of Australia

(The specific name means jasmine-like and probably refers to the foliage.)

Description.—A climbing shrub. Stems terete. Leaves opposite, compound, odd pinnate. Leaflets 5-9, narrowly elliptic, ovate or lanceolate in shape, sessile, blunt at the tips, quite entire, dark green in colour, glabrous, 1-1.5 in. long.

Flowers in axillary or terminal many-flowered corymbose panicles. Calyx 0.25 in. long, almost truncate or obscurely lobed, tubular in shape. Corolla tubular, campanulate expanding into 5, almost orbicular, wavy lobes, white or rosy pink in colour, dark pink or almost purple in the hairy throat, 2 in. wide at the mouth and as long. Stamens four, two long two short; anther cells spreading so that they come to be on the same line. Ovary seated on a ring-like disk. Style filiform, 2-lobed.

Flowers.—Hot season. Does not fruit in Dehra Dun.

Distribution.—Indigenous to Australia. Commonly cultivated in gardens throughout this country.

Gardening.—This 'Vine' has rather handsome evergreen foliage

Fig. 31.—*Tecoma jasminoides* Lindl. × ¾

and beautiful large pink flowers. It is quite frost hardy, but requires a rich soil and a sunny site. Propagated by cuttings during the rains.

Tecoma australis R. Br. [**Pandorea pandorana** (Andrews) Van Steenis]

Wonga-Wonga Vine

(The specific epithet refers to the country or origin of this plant.)

Description.—A very extensive climber. Stems cylindrical, striate. Leaves opposite, compound, odd pinnate, exstipulate. Leaflets 3-9, opposite, ovate, elliptic or ovate-lanceolate, almost sessile, perfectly smooth and glabrous, entire or wavy on the margin, glossy green.

Flowers heavily scented borne in axillary or terminal racemose panicles. Calyx very small, campanulate, regularly 5-lobed, glabrous or slightly hairy. Corolla 0.5-0.75 in. long, funnel-shaped with five short reflexed lobes, yellowish white in colour, spotted with purple in the throat. Stamens four, included. Ovary seated in a ring-shaped disk. Style filiform, exserted.

Flowers.—March-April. Does not fruit in Dehra Dun.

Distribution.—Indigenous to Australia. Cultivated in gardens throughout India.

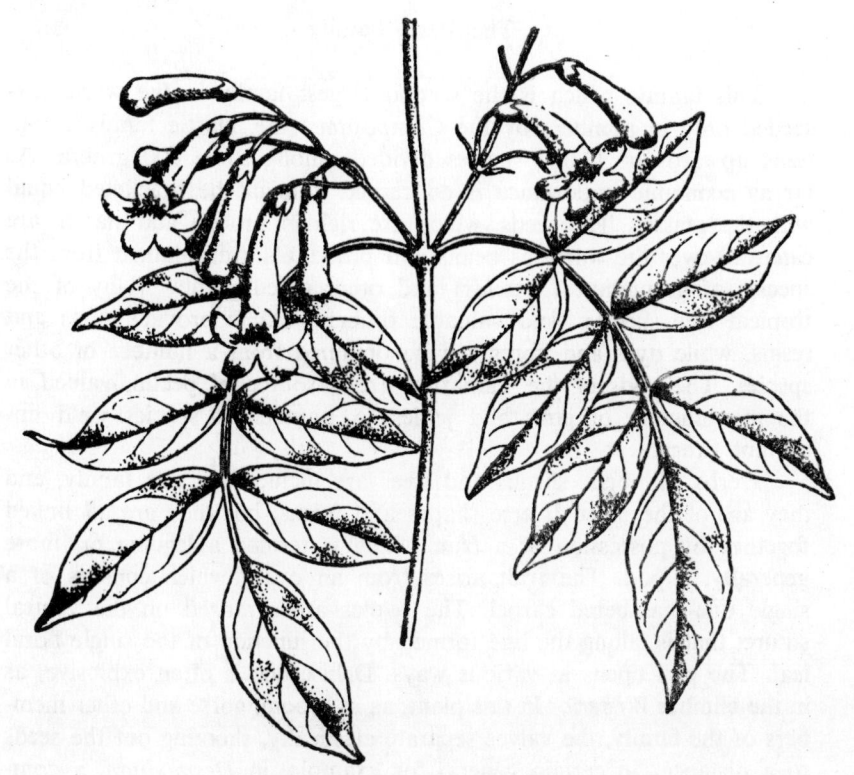

Fig. 32.—*Tecoma australis* R. Br.

Gardening.—This vine has beautiful dark green glossy foliage and comparatively small flowers, but is worth cultivating for its foliage alone. It prefers a soil rich in humus and, if liberally fertilized, it will grow very vigorously. It does not produce fruit in this country, and is propagated by cuttings.

Chapter 4
LEGUMINOSAE
The Bean Family

This family, which is the second largest family in the world, exceeded only in numbers by the *Compositae* (the Thistle family), contains upwards of 12,000 species divided among some 600 genera. As far as economic importance is concerned it might be bracketed equal with the grasses. The seeds, which are rich in protein and starch, are eaten all over the world as beans and peas. Oils are obtained from the species of groundnut (*Arachis*) and other allied plants. Many of the tropical tree species yield valuable timbers, others provide gums and resins, while dyes and fibres can be obtained from a number of other species. The pods of the Tamarind are a source of pectin, valued in the manufacture of jams and jellies. A number of species yield important drugs.

Herbs, climbers, shrubs and trees are included in this family, and they are of the most diverse shapes and forms, but they are all linked together by possession of a fruit which is termed a legume or, more generally, a pod. The fruit arises from an ovary which consists of a single one-chambered carpel. The ovules are arranged on the ventral suture, that is, along the line formed by the junction of the single floral leaf. The pod opens in various ways. Dehiscence is often explosive, as in the climber *Wisteria*. In this plant, as in broom, gorse and other members of the family, the valves separate elastically, shooting out the seeds to a distance. In certain genera, for example, in *Desmodium*, a common genus in this country, the pod breaks up into sections each containing a single seed, and these are provided with hairs, hooks, or other devices, which enable them to be distributed through the agency of animals.

The pods show great variety in form and size. In size they may vary from colossal pods one yard long and several inches wide, down to mere specks a fraction of an inch long. While in general they are more or less flattened, even paper thin, they may also be cylindrical, globular or pyriform in shape.

The other characters are—leaves alternate, compound and stipulate. The flowers may be irregular or regular; petals free or united. Stamens ten, rarely free and more usually joined together in various ways, most often the ten joined together, or nine joined together and one free. It seems as if evolution had, after taking 10 stamens, unicarpellate ovary and pod as fruit, as a basis, tried out every possible com-

PLATE 13

THE CHINESE WISTERIA

M. N. Bakshi

bination of characters and produced a family of the most diverse forms, but a very natural one after all.

Insects play an important part in the pollination of the flowers of the family. In almost every case there is some characteristic of the flower which appears to exist solely to be attractive to the pollinating insect. In the sweet pea section of the family, nectar is secreted inside at the base of the stamens and it is only accessible to insects which possess a long proboscis.

The seeds of many of the species have exceeding tough coats which apparently enable the seeds to exist for several, and even for many years in the soil, before germinating. The process of germination can be hastened in various ways, such as by allowing strong or weak sulphuric acid to act upon the seed coat, by dropping the seeds into boiling water, or even by rubbing the seed coat with a sharp file.

Despite the naturalness of the family, it can be divided into three distinct subfamilies, sections which are often given full family rank by some botanists.

In one of these the flowers are like those of the sweet pea and the family is called the *Papilionaceae*, from the fancied resemblance of the flower to a butterfly. The other two families have more or less regular flowers. They are, *Caesalpiniaceae*, with showy flowers with spreading petals, and the *Mimosaceae* with small flowers in heads with a corolla tube. The so-called garden mimosas (really *Acacias*) belong to the latter family.

Members of the family are familiar plants in our gardens. The common shrubby and climbing plants belong to the *Papilionaceae* and *Caesalpiniaceae*. The species treated in this chapter are easily separated as follows:

Flowers like those of the sweet pea
   Climbers with mauve flowers .... *Wisteria*
   Shrubs or small trees .... *Sophora*
Flowers more or less regular with spreading petals
   Prickly climbers, or shrubs; leaves pinnate .... *Caesalpinia*
   Unarmed climbers; leaves of two leaflets, joined on inner
      margin .... *Bauhinia*

Wisteria Nutt.

(This genus was called after Caspar Wistar (1761-1818), a famous Philadelphian anatomist. The name was spelt *Wisteria* by Nuttall the author of the genus, and this spelling holds good.)

This is a genus of the *Leguminosae*, sweet pea section. The genus contains woody veins often reaching a great age. The leaves are alternate, compound, odd-pinnate, with 9-12 leaflets. The flowers are in long drooping racemes and may be white, lilac, purple or blue in colour.

The calyx is bell-shaped, 5-toothed. Corolla papilionaceous. Stamens 9+1. Fruit a pod, twisted when mature.

The pods of this climber explode when the seeds are ripe and the seeds are flung out. Ridley mentions the results of two observers who experimented with *Wisteria* in order to see to what distance the seed would be ejected. Haesteal stated that on dehiscence of the pod, the seeds flew 10 feet across a room and struck the window violently. From this he calculated that the seeds would have travelled at least 16 feet. Zabriski says that the seeds were thrown with great force against a window 16 feet away and he judged that this flight would have been 30 feet. The explosion appears to be effected by a layer of strongly thickened elongate cells which runs transversely across the valve and which winds into a spiral when drying.

In the *Gardener's Chronicle* for 1940 appears this account of a *Wisteria*:

'The specimen was planted somewhere about forty years ago; it covers the face of a building twenty feet high and eighteen feet wide, and rambles into a Yew tree. About four years ago, rats were found to be working at the base, which was about two feet in circumference, and while trying to evict these rodents, I found the old stem was completely rotten at eighteen inches from the ground. I looked at the plant in amazement, as it was still growing and in full vigour. When casting around to discover what had happened to enable it to keep on growing, I found that a branch nine feet up had gone round the corner of the building and *joined* on to another plant eighteen feet away at the other corner, twisting itself round a stem and making a natural graft. This means that the flow of sap must now be reversed, and flows down to the base of the old stem. I have never seen or heard of anything like this, and therefore thought these notes might be of interest.'

Wisteria sinensis (Sims.) DC.

Chinese Wisteria; Blue Rain; Blue Acacia

(*Sinensis* means of Chinese origin, and refers to the natural home of this plant.)

Description.—A sprawling deciduous vine with a twisted trunk reaching 6-9 in. in diameter, and many long whip-like branches. Bark brown, often covered with brown felt. Leaves compound, alternate, odd-pinnate, deciduous. Rhachis up to 12 in. long, swollen at the base, grooved on the upper surface and slightly silky, terete and striate. Leaflets 7-11, densely silky when young, when old, sparsely covered with silk on the upper surface, often very hairy below, margins ciliate, ovate-acuminate, ovate-lanceolate or elliptic in shape, 2-3 in. long, somewhat oblique, cuneate at the base, acute-acuminate and sometimes even caudate at the tip, ciliate on the margins; petiolule short, 0.1 in.

PLATE 5

CHINESE WISTERIA; BLUE RAIN; BLUE ACACIA
Wisteria sinensis (Sims.) DC.
(nat. size)

long, densely hairy, articulate on the rhachis; stipellae subulate, 0.2 in. long.

Flowers arranged in terminal lax racemes up to 13 in. long which appear with the young leaves, pedicellate; rhachis striate-hairy; pedicel 0.75 in. long, hairy; bracts obsolete. Calyx bell-shaped, densely covered with short appressed hairs, 5-toothed; one tooth almost subulate, the other triangular-acute, larger than the lower. Corolla papilionaceous; standard orbicular-emarginate, or broader than long, clawed, with 2 appendages at the top of the claw, violet-blue or blue in colour. Wing petals slightly shorter than the standard, obovate-oblong, violet-blue; tip rounded; base excised with a cusp; claw slender. Keel-petals shorter than the wings, obovate-oblong, almost hatchet-shaped, violet-blue, base narrow, rounded. Stamens 9+1; the upper stamen free, the remaining nine joined by thin filaments; anthers small. Ovary elongate, pubescent; style curved at right angles to the ovary. Fruit a pod, up to 5 in. long, oblanceolate in shape, flattened, narrowed from the upper third to the base, top acute or cuspidate; valves thinly woody, clothed externally with a dense brown velvety pubescence, internally with a thin puberulous white corky layer, twisted after dehiscence. Seeds 0.25-0.5 in. in diameter, orbicular, flattened, dark brown smooth.

Flowers.—Hot season. *Fruit.*—Rainy season.

Distribution.—A native of China now widely cultivated in the hills and plains of India.

Gardening.—An attractive, large climbing shrub with handsome foliage and long racemes of large pale purple or violet flowers. It is very suitable for covering porches, arbours and the like. According to Bailey it is 'the noblest of the woody vines for temperate regions.' This vine succeeds well in Northern India but in the climate of Calcutta it thrives indifferently and at high elevations in the hills it never flowers profusely. Propagated usually by layers as the seeds, though they germinate with ease, do not reproduce the horticultural forms.

Wisteria sinensis (Sims.) DC. var. *alba*, a white flowered variety, has recently been imported but it is not so beautiful

Medicinal and Economic Uses.—A resinous substance and a glucoside have been isloated from the bark of this plant. Both these substances are said to be poisonous but apparently nothing is known of their chemical constitution.

Sophora Linn.

(The generic name is derived from the Arabic name, Sophera, for *Cassia sophora* Linn.)

Sophora belongs to the sweet-pea family (*Papilionaceae*) which is so well known as to obviate the necessity of a description. The genus itself comprises trees and shrubs with alternate stipulate odd-pinnate leaves. The flowers are arranged in racemes or panicles and are yellow or violet-purple in colour. The calyx is gamosepalous, oblique, broadly campanulate in shape, surmounted at the top by short deltoid teeth. The corolla consists of five clawed petals. The upper petal is broad, erect, and is termed the standard or vexillum; the two side petals are narrower and are called the wings or alae; the 2 lower petals, loosely connate by their lower margins, are the narrowest and form the so-called keel or carina. The stamens are 10 in number, free or obscurely joined at the very base; the anthers are versatile. The ovary is shortly stalked, with many ovules on the ventral suture. The pod is constricted between the seeds and is termed moniliform, that is, somewhat like the beads on a necklace.

The species of this genus are valued for their pretty often fragrant flowers, and for their handsome foliage and curious pods. They are inhabitants of the warmer countries of the world but mainly at high altitudes, and therefore some of them have been introduced with success into Europe. *Sophora japonica* L. was introduced into France by Jussieu in 1747, and is common in that country and in Germany. It has been found that it is capable of withstanding the smoke-vitiated atmosphere of towns, especially in the neighbourhood of railway stations.

The brightly coloured and fragrant flowers indicate that these plants are cross-pollinated through the agency of insects.

As has already been said, the leaves of *Sophora* are odd-pinnate, that is, they are composed of an odd number of leaflets arranged on either side of a central rhachis. The leaflets can be distinguished from ordinary leaves by the fact that they do not possess a bud in the axil of the stalklet or pulvinus. During the night some of these species arrange their leaves in the sleep-position. Towards nightfall the pulvini move and the leaflets drop downwards and their under surfaces are pressed together. This is one of the ways in which a plant avoids loss of heat and water through excessive transpiration.

Several Sophoras are well known for their poisonous properties. This is due to the presence of an alkaloid which is variously known as sophorine or cytisine. It is a very dangerous drug causing convulsions and death. The seeds and roots of *Sophora tomentosa* Linn. contain cytisine, and are used as remedies in small doses for cholera and diarrhoea.

The unopened flower buds and pods of several species give a yellow dye; when used in conjunction with the indigo plant (*Indigofera flaccidifolia*) the resultant colour is given. In Baden, Germany, the unopened flower buds are used to colour Easter eggs. In Japan this yellow dye is said to be used exclusively to dye the garments of the Royal House.

PLATE 14

GRIFFITH'S SOPHORA
Sophora griffithii Stocks.

M. N. Bakshi

PLATE 15

GRIFFITH'S SOPHORA
Sophora griffithii Stocks.

M. N. Bakshi

The wood, being extremely hard, is of use in joinery and parquetry.

KEY TO THE SPECIES

Flowers yellow
    Leaflets up to 0.5 in. long .... *S. griffithii*
    Leaflets 1 in. long or over .... *S. tomentosa*

Flowers blue, violet or white
    Leaflets leathery, over 1 in. long .... *S. secundiflora*
    Leaflets soft, up to 0.5 in. long .... *S. viciifolia*

Sophora griffithii Stocks

[The specific epithet commemorates the name of William Griffith (1811-1845), one of that noble band who did so much for Indian Botany in the early nineteenth century.]

Description.—A small erect deciduous shrub. Young twigs densely hoary. Stipules minute, villous, often persistent after the fall of rhachis. Leaves alternate, compound, odd-pinnate, 4-8 in. long. Leaflets 21-41, opposite or alternate, 0.25-0.5 in. long, ovate, oblong or obovate in shape, thick, rigid, glabrous above when mature, densely appressed silvery velvety beneath.

Fig. 33.—*Sophora griffithii* Stocks × ½

Inflorescence of pedunculate racemes from old leaf scars, 1.5-5 in. long; peduncle usually leaf bearing. Flowers 0.1 in. long, yellow, appearing with the young leaves; pedicels silvery hoary, 0.1-0.2 in. long, in the axils of hoary buds. Calyx gamosepalous, very oblique, 0.2 in. long, densely appressed-hoary, ending in 5 shorter triangular teeth. Petals 5, yellow; standard longer than the keel, markedly veined, almost orbicular in shape, emarginate at the top, ending below in a long strap-shaped claw; wings obliquely oblong-obtuse, clawed below, toothed on either side above the claw; keel of two petals loosely connate along the lower margins; upper margin toothed above the claw. Stamens ten; filaments free to the base. Ovary hirsute with appressed white hairs; style slender, incurved; stigma capitate. Pod up to 4 in. long, with four winged ridges, constricted between the ellipsoid seeds, hairy.

Distribution.—Salt Range and Trans-Indus hills, Baluchistan, Afghanistan and Persia. Occasionally cultivated in gardens in India.

Gardening.—A small shrub with velvety silvery grey leaves and large bright yellow flowers appearing shortly before or with the leaves. Usually propagated by seed.

Sophora tomentosa Linn.

(*Tomentosus* means hairy in Latin.)

Description.—A large shrub or small tree, the whole plant being covered with a soft grey velvety pubescence. Leaves alternate, odd-pinnate, 6-10 in. long, with 15-19 leaflets; rhachis terete, tumid at the base; leaflets often alternate on the rhachis, shortly stalked, 1-1.5 in. long, broadly elliptic in shape, rounded or obtuse at both ends, dull grey-green in colour, thinly downy above, the lower surface covered with a dense pubescence; margins somewhat reflexed.

Inflorescence a terminal raceme up to 6 in. long. Individual flowers pedicellate, often crowded; pedicels densely silky, bracteate, 0.25-0.3 in. long, jointed just below the calyx. Calyx tube broadly campanulate, almost truncate with 5 very small obscure lobes, minutely silky, oblique, 0.25-0.4 in. long. Corolla yellow, of 5 petals. Standard long-clawed, 0.6-0.75 in. long; blade orbicular or elliptic, slightly emarginate. Wings oblong-obtuse, narrow, clawed. Keel of two narrow clawed petals. Stamens 10; filaments free to the base or nearly so. Ovary shortly stalked, linear, tapering at the apex, densely hairy; style straight; stigma capitate. Pod long-stalked, issuing from the persistent calyx, moniliform, the joints being separated by narrow necks as long as themselves, pointed, covered with a brown velvety down; seeds 1-8 in each pod, 0.35 in. in diameter, nearly globular, pale brown.

Flowers.—Rainy season with a second flush in the cold weather.
Fruits.—Cold season.

PLATE 6

Sophora tomentosa Linn.
(nat. size)

Distribution.—Cosmopolitan in the tropics on the sea coast. Widely cultivated in gardens throughout the plains of India.

Gardening.—The shining dark green handsome foliage of this, shrub, contrasting with its bright yellow flowers, render it a very ornamental plant. Unfortunately in Dehra Dun it is liable to be infested by a caterpillar which, at times, strips it of all its leaves and beauty. Easily raised from seed during the rains. It is apt to be damaged by frost in northern India.

Medicinal and economic uses.—The seeds and roots of this plant contain the alkaloid cytisine and are considered medicinal in Malaya. Small doses of it are used as remedies in cases of cholera and diarrhoea. According to Heyne, pounded leaves are applied on wounds caused by certain poisonous fish.

Burkill says that the timber is hard and heavy, but is not much used.

Sophora secundiflora DC.

(*Secundiflora* is a Latin word meaning flowering to one side, from the fact that the flowers are arranged in one-sided racemes.)

Description.—An evergreen shrub or a small tree, which may reach 35 ft in height, with smooth bark and young shoots covered with an appressed silvery pubescence. Buds ovoid, densely covered with short hairs. Leaves compound, alternate, odd-pinnate, 4-6 in. long, with 7-9 leaflets; rhachis glabrous or hairy, grooved on the upper surface; leaflets opposite, dark green and shining, rather leathery, evergreen, oblanceolate or obovate in shape, with swollen pulvini, 1.25 in. long, emarginate at the apex; nervation reticulate, very prominent below.

Inflorescence of one-sided, terminal racemes, 2-3 in. long. Individual flowers very fragrant, about 1 in. long, pedicelled; pedicels 0.3 in. long, supported by a bract and bearing two small bracteoles halfway up, covered with silvery appressed pubescence. Calyx gamosepalous, 0.3 in. long, including the lobes, brown in colour, covered with appressed pubescence, lobes 5, the upper broadly obtuse, the lower four deltoid in shape. Petals five; standard oblong, 0.5 in. long, 0.5 in. wide, cordate at the apex, contracted into a strap-shaped claw below; wings and keel obliquely oblong, rounded at the tip, sagittate at the base, with a strap-shaped claw. The petals are violet-blue or pale blue in colour, and the standard has a few darker spots near the base. Stamens 10, equal; filaments slightly connate at the base, inserted on the hypanthium. Ovary narrowly elliptic-oblong, stipitate, covered with coarse hairs, 0.3 in. long; style slender, curved, glabrous; stigma small, capitate. Pod 1-7 in. long, 0.5-0.75 in. in diameter covered with white hairs, constricted between the seeds. Seeds bright scarlet.

Flowers.—March-April. *Fruits.*—Rainy season.

Distribution.—Indigenous to the Southern United States and North Mexico; now frequently cultivated in gardens throughout the plains of India.

Fig. 34.—*Sophora secundiflora* DC. × ½

Gardening.—A shrub or a small tree with short slender trunk and upright branches, forming a narrow head. Flowers fragrant, handsome, violet-blue. It is a very free flowering and ornamental plant but its growth is slow. Raised from seed sown during the rains.

Medicinal and economic uses.—The seeds of this species are poisonous due to the presence of the alkaloid cytisine or sophorine. This alkaloid if taken in big doses causes nausea, then convulsions, and finally death by asphyxia.

Sophora viciifolia Hance

(*Viciifolius* expresses in Latin the similarity of the leaves of this species to those of *Vicia* (the Vetch), another genus of the family.)

Description.—A small shrub reaching 6 ft in height, with striate

THE HAIRY SOPHORA
Sophora tomentosa Linn.

M. N. Bakshi

PLATE 17

THE ONE-SIDED SOPHORA
Sophora secundiflora D .

M. N. *Bakshi*

stems. Lateral branches covered with short white hairs, often arrested and then appearing as spines. Stipules spinescent. Leaves alternate, compound, odd-pinnate, up to 2 in. long; rhachis cylindrical covered with short appressed silvery pubescence; leaflets 11-15, almost sessile, 0.25-0.5 in. long, oblong or elliptic, obtuse, mucronulate, glabrous above, covered with white pubescence beneath; nervation, apart from the midrib, obscure.

Fig. 35.—*Sophora viciifolia* Hance × ½

Inflorescence of short axillary or terminal 6-12-flowered racemes. Individual flowers seated on hairy pedicels 0.3 in. long, bluish violet or nearly white in colour. Calyx gamosepalous, violet in colour, membranous, shortly 5-toothed, cylindrical in shape, somewhat oblique, covered with appressed pubescence. Petals 5; standard spathulate-obovate, reflexed, clawed; wings oblong-obtuse, with a tooth on one side; keel-petals clawed, oblique. Stamens 10; filaments free or slightly connate at the base. Ovary stipitate, oblong, pointed at both ends, covered with coarse hairs; style curved, glabrous, slender; stigma capitate, small. Pod 2 in. long, slender, long-beaked.

Flowers.—March-April. Has so far not fruited in Dehra Dun.

Distribution.—Indigenous to central and western China, now cultivated in various parts of India.

Gardening.—The violet calyx makes an attractive contrast to the milky white corolla of this hardy graceful shrub. It was introduced into the Royal Gardens, Kew, in 1898, where it flowered for the first time in July 1902. Easily grown from seed.

Caesalpinia Linn.

A genus of trees, shrubs and woody climbers belonging to the family *Caesalpiniaceae*. The name perpetuates the memory of Andreas Caesalpinius, 1519-1603, an Italian botanist.

The *Caesalpiniaceae*, at one time considered to be a section of the *Leguminosae*, but now accepted by most botanists as a distinct family, is a very well-defined group of plants. Its flowers are intermediate between the regular flowers of *Mimosaceae* and the extremely irregular flowers of *Papilionaceae*. The fruit, however, is the characteristic and familiar pod of the sweet pea family.

The genus *Caesalpinia* is well represented in India and several of the indigenous species as well as some exotics are cultivated in Indian gardens on account of their showy flowers and handsome foliage. The family comprises trees, shrubs and woody prickly climbers. The leaves are large and abruptly bipinnate. The flowers which are usually yellow, sometimes red, are arranged in large many-flowered racemes. The calyx consists of five segments which are imbricate in the bud. The petals are orbicular in shape with a distinct claw. Stamens ten in number, free. The ovary is sessile upon a disk and is usually few-ovuled. Pod various, sometimes covered with spines.

Key to the Species

Stamens very long, much exceeding the petals
  Stamens 3-5 in. long; sepals hairy on the margins,
    flowers yellow .... *C. gilliesii*
  Stamens less than 3 in. long; sepals glabrous, flowers red,
    red and yellow or yellow .... *C. pulcherrima*
Stamens not very long, hardly exceeding the petals
  Pod prickly .... *C. bonducella*
  Pod not prickly, dry
    Pod flat
      Large shrub or small tree .... *C. sappan*
      Climbing shrubs
        Leaflets 2-3 pairs .... *C. nuga*
        Leaflets 8-12 pairs .... *C. sepiaria*
    Pod twisted
      Climber; flowers in racemes .... *C. digyna*
      Large shrub or tree; flowers in dense panicles .... *C. coriaria*

Caesalpinia gilliesii Wall.

Bird of Paradise

(Called after Dr. Gillies who introduced it to Kew in 1829 from South America.)

Description.—An erect shrub attaining a height of 6-7 feet.

Branches round, striate, covered with small circular raised lenticels, nearly glabrous but sometimes covered with short crisped white hairs. Leaves alternate, compound, up to 1 ft. long, stipulate, bipinnate; stipules at the base of the main rhachis, ovate-acuminate in shape, ciliate or laciniate on the margins. Leaves abruptly bipinnate; pinnae opposite or alternate, up to 1.5 in. long, without stipellae; rhachis of the pinnae slightly swollen at the base. Leaflets oblong, 0.16 in. long, rounded at

Fig. 36.—*Caesalpinia gilliesii* Wall. × ½

both ends, shortly petioluled, rather thick, with a definite row of black glands inside each margin on both surfaces.

Flowers arranged in a terminal raceme; rhachis thick, woolly and covered with numerous red-stalked glands. Individual flowers pedicellate, each pedicel being supported by an ovate-acuminate deciduous bract which is hairy and glandular on the back and glandular-subulate-laciniate on the margins; pedicel of the open flowers 0.75 in. long. Calyx-tube short, turbinate, 0.16 in. long, glandular and hairy; calyx-limb of five equal oblong segments covered on back and margins with flat-topped, shortly stalked glands, hairy outside, glabrous inside, about 0.5 in. long. Petals five, of a beautiful yellow colour, shortly clawed, seated on the margin of the calyx-tube 1.5 in. long, obcordate in shape. Stamens 10, free, arising from the top of the calyx-tube; filaments 4-5 in. long, crimson in colour, glabrous. Anthers oblong, versatile. Ovary seated at the bottom of the calyx-tube, shortly stipitate, hairy, prolonged into a long style with a capitate stigma. Fruit a falcate leathery pod, beaked, 2.5 in. long.

Flowers.—Hot and rainy season. *Fruits.*—July-August.

Distribution.—Native of Mendoza, South America, now frequently cultivated in all tropical and subtropical parts of the world.

Gardening.—A very hardy shrub reaching 7 ft in Dehra Dun, with graceful feathery bipinnate foliage of small leaflets. It flowers profusely during the hot and rainy seasons, the flowers being of a pale yellow colour. The petals seldom expand fully but the long stamens which are crimson in colour, are conspicuous. After two seasons or so the plant is apt to decay and look unsightly; it is consequently advisable to replace it by plants from fresh seed. In Dehra Dun and elsewhere in this country the pods are frequently attacked by a borer which destroys the seeds. They should, therefore, be covered with muslin to protect them from the ravages of these Gillies's insects. Easily raised from seed sown during the rains or even earlier. According to Sir W. J. Hooker it was introduced by Dr. Gillies into the Royal Botanic Gardens, Kew, in 1829. It is popularly known as Dr. Gillies's *Poinciana,* as Hooker, when the plant was first discovered, described it as *Poinciana gilliesii*. It is a hardy plant and can be grown out of doors in England.

Caesalpinia pulcherrima (Linn.) Swartz

Dwarf Poinciana; Barbados Pride; Peacock flower

(*Pulcherrima* means most beautiful in Latin.)

Description.—A handsome shrub reaching 6 ft. in height. Branches smooth, green or glaucous, glabrous, shining, with a few prickles here and there. Leaves abruptly bipinnate, stipulate, up to 1.5 ft long; pinnae 6-12 pairs, opposite, with small stipellae at the base, up to 3 in. long; leaflets oblong, 0.5 in. long, oblique at the base, emarginate

DWARF POINCIANA
Caesalpinia pulcherrima Sw.
(½ nat. size)

at the apex with the midrib produced as a short mucro, very shortly stalked; each leaflet with a very small pair of stipels at the base of the stalk.

Flowers arranged in an erect terminal raceme. Flowers pedicelled; the pedicels of the fully opened flower being up to 4 in. long; pedicels supported at the base by a rapidly deciduous lanceolate-acuminate bract 0.1 in. long, which can be seen at the apex of the raceme surrounding the young flowers. Calyx-tube turbinate, 0.2 in. long, glabrous; limb 4-partite, one sepal larger than the others, hooded, glabrous, coloured red or orange in the bud. Petals about 0.75 in. long, distinctly clawed; limb orbicular in shape, often lobed on the margins, variously coloured in red and gold, very often claw and centre of limb crimson, red or golden red, with a narrow margin of gold. Stamens 10, free, seated on the margin of the calyx-tube; filaments long, red, rather thick at the base where they are covered with white hairs, 2.25 in. long, tapering to the versatile anthers. Ovary seated on a short gynophore arising from the base of the calyx-tube, glabrous, compressed, terminating in a long yellowish red style. Pod nearly straight, narrow and thin, 2-3 in. long.

Flowers.—Hot and rainy seasons. *Fruits.*—Cold season.

Distribution.—Native country uncertain, perhaps South America. Extensively cultivated throughout the tropics.

Gardening.—A hardy drought resistant showy shrub up to 6 ft or so in height. It flowers profusely during the hot and rainy season making the plant very ornamental. It should be cut back closely in the cold season, as it is apt to grow very straggly and become unsightly. It can hardly stand the cold of the Punjab but flourishes there and in the United Provinces better in the hot and rainy seasons than in Bengal. It is better to replace old plants every now and then by fresh plants raised from seed. This shrub starts to flower as early as 8 months old. It is best suited for growing on lawns and for hedging. Easily raised from seed which it produces abundantly. According to Rheede, Hortus Malabaricus, it was in the gardens of India as long ago as 1680. It was introduced into Holland from Amboyna about the year 1670. On account of its showy flowers and attractive finely divided foliage, this plant is a great favourite in all tropical and subtropical countries. Although it will thrive in poor soil, an application of manure or chemical fertilizer may be given to advantage, causing it to make more vigorous growth and give better and larger heads of flowers.

C. pulcherrima Sw. var. *flava* Hort. A race with bright yellow flowers, not nearly so handsome as the type. It is called 'Radha chura' in this country, the red race being named 'Krishna chura'.

Medicinal and economic uses.—Burkill (Economic Products of the Malay Peninsula) remarks that this plant has a few medicinal uses in the Dutch Indies. The pounded roots are said to be useful in cases of infantile convulsions. The flowers are used as a remedy for intestinal worms, for coughs and chronic catarrh. The leaves are reputed to have a purgative action, and also to be abortificient. A decoction of the leaves is said to be useful in cases of fever.

Caesalpinia bonducella (Linn.) Fleming [**C. bonduc** (Linn.) emend. Dandy & Exell]

Fever-Nut; Physic-Nut

Description.—A prickly shrub, rambling or scandent. Branches more or less covered with short, soft hairs and armed with stout sharp straight or recurved prickles. Leaves compound, 12-18 in. long, abruptly bipinnate, with the rhachis often produced as a short mucro; rhachis stout, covered with very short brown hairs and bearing sharp recurved

Fig. 37.—*Caesalpinia bonducella* Fleming × ½

prickles; pinnae up to 16 pairs, opposite; base of the rhachis of the pinnae with recurved stipellate spines. Stipules large, foliaceous, often lobed. Leaflets 6-10 pairs, seated on very short pedicels with a pair of recurved prickles at the base, somewhat oblique at the base, oblong or elliptic, obtuse with the midrib produced as a short mucro, glabrous or sparsely hairy above, puberulous below.

Flowers arranged in axillary or terminal racemes, bracteate; bracts linear-lanceolate, tip subulate, 0.4-0.5 in. long, covered with rusty brown hairs. Flowers pedicelled; pedicel 0.2-0.3 in. long, rusty tomentose. Calyx-tube very short; lobes 5, 0.2-0.3 in. long, rusty tomentose on both surfaces. Petals 5, golden yellow or the uppermost sometimes spotted with red oblanceolate in shape. Stamens 10; filaments short, hardly exceeding the petals, covered with white hairs below. Ovary seated on a short gynophore which comes from the base of the calyx-tube. Pod broadly elliptic in shape, beaked, coriaceous, 2-3 in. long, covered all over with sharp wiry prickles.

Flowers.—August-September. *Fruits.*—Cold season.

Distribution.—Wild or naturalized throughout the tropics.

Gardening.—A scandent prickly shrub with yellow flowers which are produced during the rains. It is commonly met with in a wild state in abandoned village sites or in hedges. It makes an effective hedge-plant. Propagated easily from seed.

Medicinal and economic uses.—This plant, which bears the Sanskrit name of 'Pulikaranja', has long enjoyed a reputation for medicinal properties and was well known to Clusius and Rumphius. The leaves and seeds possess a substance known as bonducin of which the properties are not yet fully known. The substance is extremely bitter and appears to be absent from the roots. In India the seeds as well as the leaves are much used in native medicine to relieve colic, fever, hydrocele, diarrhoea and rheumatism.

Caesalpinia sappan Linn.

Sappan Wood

(The specific name comes from its local name, *sepang*, in Java.)

Description.—A thorny, shrubby tree reaching a height of about 20 ft and a diameter of 6-10 in. Leaves up to 18 in. long. stipulate alternate, with 16-24 pinnae, each 4-6 in. long. Leaflets 20-36, chartaceous, oblong, very oblique at the base, almost sessile; rounded and slightly emarginate at the apex, glabrous above, puberulous below.

Flowers in panicled racemes, 12-16 in. long; separate racemes 4-6 in. long. Calyx-tube short, bowl-shaped; lobes 0.4 in. long, 5 in number. Petals 5, orbicular, 0.3 in. long, yellow; base of upper petal pink. Stamens 10, free, arising from the lip of the calyx-tube; filaments white, woolly in the lower half. Ovary stipitate, grey velvety. Pod woody, oblong, flattened, 3-4 in. long; 1-5-2 in. wide, polished, brown, beaked.

Flowers.—Rainy season. *Fruits.*—Cold season.
Distribution.—From India throughout Malaysia.
Gardening.—A large thorny shrub, quite ornamental when laden with its large panicles of yellow flowers. Easily propagated from seed which it produces abundantly.
Medicinal and economic uses.—The wood yields a beautiful red

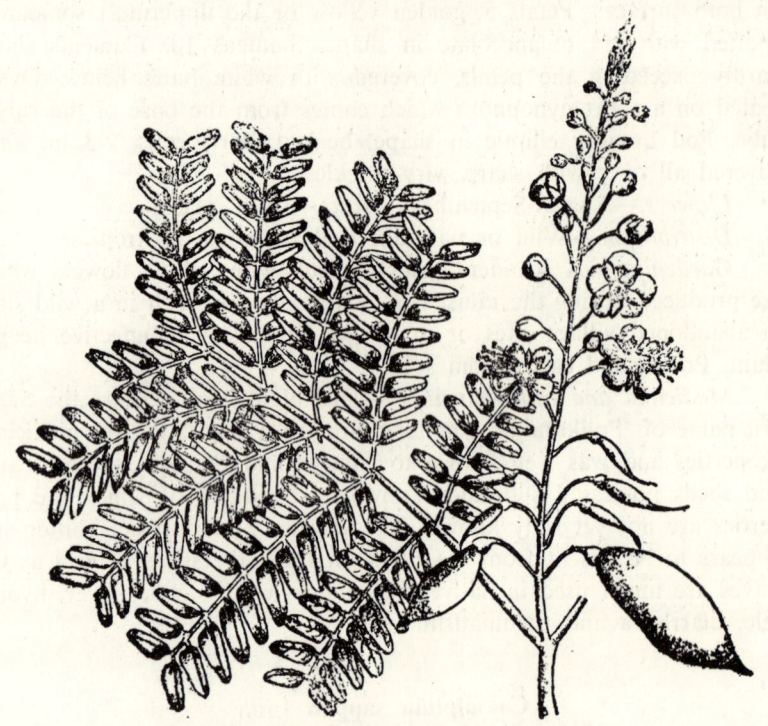

Fig. 38.—*Caesalpinia sappan* Linn. ×1/3

dye which is used to colour silk. The dye is also used for colouring starch which is scattered on the occasion of the Holi festival.

Caesalpinia nuga Ait.

Description.—A stiff wiry prickly climber with a blackish bark and few prickles. Leaves alternate with small stipules, bipinnate. 6-12 in. long, with 6-8 pinnae, often much less on flowering branches, rhachis with recurved spines. Rhachis of the pinnae with recurved stipular spines at the base. Leaflets 2-3 pairs, light green above, rather pale below, coriaceous, ovate or elliptic, acute or obtuse, 1-2.5 in. long, 0.5 to 1 in. wide.

Inflorescence a panicle of racemes; racemes racemosely arranged,

PLATE 18

M. N. Bakshi

THE MYSORE THORN
Caesalpina sepiaria Roxb.

PLATE 19

THE MYSORE THORN
Caesalpinia sepiaria Roxb.

M. N. Bakshi

up to 6 in. long. Flowers fragrant, seated on pedicels, 0.3 in. long. Calyx broadly obconic, very short (0.1 in. long), smooth, glabrous, 10-ribbed; limb consisting of 5 sepals, oblong, smooth and glabrous. Petals 5, yellow, clawed; limb orbicular crumpled; upper spotted with red; stamens ten, free; filaments hardly longer than the petals, dilated at the base and woolly below; anthers versatile. Ovary seated on an oblique gynophore, elliptic-compressed, sparsely hairy; style short. Pod turgid, beaked, indehiscent, 2 in. long.

Flowers.—May-October. *Fruits.*—Cold season.

Distribution.—Found from Bengal to the Pacific, chiefly along the coast, but also inland.

Gardening.—A vigorous prickly climber with yellow flowers which against the glossy green leaves appear very striking. Readily propagated

Fig. 39.—*Caesalpinia nuga* Ait. $\times \frac{1}{2}$

by seed which (as in all other species of this genus) should be well soaked in warm water for some hours before sowing.

Medicinal and economic uses.—The roots of this plant are stated to be diuretic by Watt. The same authority remarks that the roots as well as the roasted seeds are used externally as well as internally in diseases of the eye. In India the lac insect feeds on this plant.

Caesalpinia sepiaria Roxb. [C. decapetalia (Roth) Alston]
The Mysore-thorn

(*Sepiaria* is derived from the Latin verb *saepio*, to surround with a hedge, as this plant is often used as an impenetrable hedge.)

Description.—A branchy scrambling or scandent shrub with dark red bark; branches glabrous or covered with dense rusty or golden brown pubescence. Prickles numerous, recurved, stout. Leaves compound, bipinnate, up to 1 ft long, stipulate; stipules small, caducous. Pinnae opposite, stalked, with a pair of recurved stipular thorns at the base, 6-10 pair; leaflets rather thin, 8-12 pairs, shortly stalked, opposite, oblong, obtuse at both ends, emarginate at the apex, puberulous on the lower surface, green above, pale below.

Fig. 40.—*Caesalpinia sepiaria* Roxb. ×½

Inflorescence a terminal, erect raceme, many-flowered. Individual flowers set obliquely on pedicels 1.25 in. long; pedicels ascending. Main rhachis of the raceme and pedicels fulvous hairy. Calyx-tube a broad inverted cone, fulvous hairy, 10-ribbed, 0.15 in. long. Sepals 5, oblong, covered outside with golden hairs, glabrous inside, 0.3 in. long. Petals 5, obovate-obtuse, 0.5 in. long, yellow, becoming reflexed as the flower opens fully to expose the bases of the filaments. Stamens 10, free, arising from the lip of the calyx tube. Filaments flattened at the base and distinctly woolly, 0.4 in. long; anthers versatile. Ovary seated on an oblique gynophore arises from the base of the calyx-tube, flattened-oblong, hairy. Style short. Pod beaked, 3.5 in. long, woody, glabrous, recurved, indehiscent, with the upper suture expanded into a narrow wing.

Flowers.—March-June. *Fruits.*—Rainy season.

Distribution.—Indigenous and naturalized throughout India and Burma (ascending to 5,000 ft in Jaunsar), extending to Sri Lanka, the Malay Archipelago, China and Japan.

Gardening.—A large prickly climber. The large racemes of bright yellow flowers make a fine show. It makes an excellent hedge plant. Easily raised from seed.

Medicinal and economic uses.—This creeper, thanks to its prickly nature and very close mode of growth, is much used by Nagas as one of their village defences. The Nagas of Henima in days gone by, grew it very thickly around the village. The branches were erected on forked poles over the paths into the village during the day, while at night the poles were removed and the creeper laid on the ground forming an impenetrable barrier to any marauder. This creeper may be seen to this day near the village entrance of most Angami villages.

The bark is said to be used for tanning in South India. It is also stated that the lac insect feeds on this species.

Caesalpinia digyna Rottl.

Description.—A large scandent prickly shrub. Bark dark brown or dark red with plentiful, strong, recurved prickles. Leaves alternate, compound, bipinnate, stipulate, 6-12 in. long, with 8-12 pairs of pinnae; stipules lanceolate, small, caducous; rhachis sparsely hairy or glabrescent. Pinnae up to 2 in. long, shortly petioled at the base, with a pair of stipular thorns. Leaflets 7-10 pairs, 0.3 in. long, rounded at the top, slightly and obliquely cordate at the base, very shortly stalked; rhachis hairy.

Flowers yellow, arranged in terminal or supra-axillary racemes. Individual flowers numerous, seated on pedicels up to 0.8 in. long; bracts very small, caducous. Calyx-tube very short, almost salver-shaped, glabrescent with age; sepals five, oblong, hooded, rounded at the top, imbricate in the bud, the upper arching over the others and falling as the flower opens. Petals 5, inserted on the lip of the calyx-tube, orbicular, obovate or oblong, rounded at the apex, very, shortly clawed,

0.25 in. long. Stamens 10, free, inserted on the lip of the calyx-tube; filaments dilated at the base and very woolly. Pod fleshy, shortly stipitate, 1.5-2.5 in. long, beaked, 1-2 seeded, twisted.

Flowers.—Rainy season. *Fruits.*—Cold season.

Distribution.—Found in India, extending to the Malay Peninsula and Sri Lanka.

Gardening.—A large woody prickly climber. It flowers abundantly during the rains, the petals being yellow streaked with red. Propagation is by seeds which are very hard and must be kept in hot water overnight

Fig. 41.—*Caesalpinia digyna* Rottl. $\times \frac{1}{2}$

or filed or abraded in some other way before they will germinate.

Medicinal and economic uses.—The pods of this plant contain an excellent tanning material. The roots are said to be of use in phthisis and scrophulous affections.

Caesalpinia coriaria Willd.

The Divi-divi plant; American Sumach

(*Coriaria* is a Latin word derived from *corium,* a hide or skin, and refers to the use of the pods of this plant as a tanning material.)

Description.—A large bush or sometimes a small tree with mimosa-like foliage on unarmed stems. Leaves compound, bipinnate; pin-

PLATE 20

M. N. Bakshi

SAPPAN WOOD
Caesalpinia sappan Linn.

PLATE 21

M. N. Bakshi

PEACOCK FLOWER
Caesalpinia pulcherrima Sw.

nae impari- or pari-pinnate, up to 6 in. long. Leaflets very numerous and narrow, 0.25 in. long, linear, green above, pale below, slightly obliquely-cordate at the base, emarginate at the apex, shortly stalked; main rhachis and rhachis of the pinnae hairy.

Flowers arranged in axillary and terminal dense panicles up to 2 in. long. Individual flowers on short pedicels, bud 0.2 in. long. Calyx-tube minute; lobes 5, oblong, rounded at the top, 0.16 in. long. Petals 5, inserted on the lip of the calyx-tube, spatheate, yellow. Stamens 10,

Fig. 42.—*Caesalpinia coriaria* Willd. ×¼

free, arising from the lip of the calyx-tube, dilated at the base and hairy in the lower half; anthers versatile. Ovary stipitate, glabrous; style short, Pod twisted, thin, up to 3 in. long by 0.5 in. wide.

Flowers.—Sept.-Oct. *Fruits.*—Cold season.

Dristribution.—Native of the West Indies and Central America, now grown in gardens in this country.

Gardening.—A spreading umbrella-shaped tree with dark green foliage and delightfully scented pale yellow flowers, suitable for planting in compounds. It is easily raised from seed, but is rather sensitive to frost.

Medicinal and economic uses.—The pods of this tree contain a powerful tanning material and it is much cultivated for this purpose in South India.

Bauhinia Linn.

The genus *Bauhinia* was created in honour of the brothers Jean and Caspar Bauhin, French herbalists of the sixteenth century, in consequence of most of the species having their leaves composed of two lobes, which are either quite separate, or, more frequently, joined by a portion of their inner margins. This arrangement of the leaves was considered to be symbolic of the great services the two brothers had rendered to science.

The genus belongs to the family *Caesalpiniaceae*, which was formerly considered to be a subsection of the Leguminosae, or sweet-pea family. The flowers are, however, not papilionaceous and the *Caesalpiniaceae* is considered to be distinct from, though closely allied to, other sections of the *Leguminosae, Papilionaceae* and *Mimosaceae*, and like them has been given full family rank.

Blatter and Millard in *Some Beautiful Indian Trees* have treated two tree species, *B. variegata* and *B. purpurea*, but the genus contains a number of other species, shrubs and climbers, which are cultivated in gardens in this country on account of their showy flowers or handsome foliage.

The characters of the genus are as follows:

Trees, shrubs or climbers, the latter with tendrils. Leaves alternate, compound, usually consisting of two leaflets which are joined together by the lower part of their inner margins, simulating a 2-lobed leaf. The top of the common rib or nerve between the 2 leaflets is produced as a small spur; a vestigial structure which represents the remains of the rhachis of the compound leaf. The flowers are showy and are arranged in simple or panicled, terminal or axillary racemes. Calyx-tube sometimes long and cylindrical, sometimes short and turbinate, with the disk produced to the top; limb entire or spathaceous, or cleft into 2 or 5 teeth. Petals 5 in number, slightly unequal, usually narrowed at the base into a claw, variously coloured, ranging from red to purple, white or yellow. Stamens 10 or reduced to 5 or 3, if less than 10 with sterile filaments absent or present; filaments free, filiform. Anthers versatile, dehiscing longitudinally. Ovary seated on a stalk (gynophore); ovules usually many. The style is long or short and usually curved, ending in an oblique or terminal stigma. The fruit is a linear pod, dehiscent or indehiscent.

In many species the flowers are fragrant, and this characteristic combined with their showy appearance points to pollination through the agency of insects.

It is well known that the cotyledons of dicotyledonous seedlings, which are widely spread during the day, press their inner surfaces together during the night in order to prevent loss of heat and to protect

PLATE 8

BAUHINIA GALPINI N. E. BROWN
(93.4 per cent nat. size)

the tender first leaves. This phenomenon is particularly well seen in the seedlings of all Bauhinias. The cotyledons, moreover, seem to function as ordinary foliage leaves to some extent, as well as performing their role of reservoirs of food.

Bauhinia vahlii, a climber, often reaches a length of over 100 yards and scrambles over the tallest trees. The exceedingly tough and fibrous bark of this particular species had, and still has, a great reputation as a suitable material for making strong ropes. Before steel came into use, ropes made from this species were used to carry suspension bridges. About the middle of the 19th century the ropes for the suspension bridge over the Jumna at Kalsi were made from the bark of *B. vahlii*.

The bark of certain Indian species smoulders very slowly when set alight. This property was made use of in bygone days to construct the slow matches whereby primitive artillery were discharged. Nowa-days such torches are often carried by cattle-boys to free themselves from the attention of sandflies and other insects.

Key to the Species

Erect or prostrate shrubs
  Stems spiny .... *B. candicans*
  Stems not spiny
    Stamens 10
      Flowers usually in axillary pairs .... *B. tomentosa*
      Flowers in axillary racemes .... *B. acuminata*
    Stamens 3 .... *B. galpini*
Climbers
  Stems flattened and undulate .... *B. anguina*
  Stems cylindrical
    Leaves very large, 4 in. to 18 in. long .... *B. vahlii*
    Leaves small not longer than 1.5 in. .... *B. corymbosa*

Bauhinia candicans Benth.

(*Candicans* means hoary, covered with white hairs.)

Description.—A shrub or small tree with zigzag, very hoary, tomentose branches. Leaves compound, 2-3 in. long, ovate or somewhat oblong, shallowly cordate at the base, 2-lobed for one-third or almost to the base, with the common midrib produced as a short spur, 11-13-nerved from the base, glabrous on the upper surface, very hoary below, petiolate, stipulate; lobes obtusely rounded at the top; petioles hoary, up to 1 in. long; stipules represented by a pair of stout thorns.

Flowers few, in axillary racemes. Pedicels short. Calyx tube short, turbinate, hoary. Calyx-limb spathaceous; segments valvate and connate in the bud, shortly 5-lobed at the apex, up to 2.5 in. long, densely and shortly hairy outside. Petals 5, free, long-clawed, with a pronounced midrib, spathulate, creamy white in colour, opening at night. Stamens 10; 5 on short, 5 on long filaments; filaments bound together at the base by a short membrane which is apparently a prolongation of the disk. Ovary seated on a long gynophore, hairy, glabrescent, ending above in a long style and capitate stigma. Pod dehiscent, very leathery,

Fig. 43.—*Bauhinia candicans* Benth. ×2/3

polished, 3-4 in. long.

Flowers.—June-July. *Fruits.*—Oct.-Nov.

Distribution.—A native of Brazil, now commonly cultivated throughout the plains of India.

Gardening.—An almost evergreen shrub with large creamy white flowers which usually open in Dehra Dun during the night and last only for a day. Easily raised from seed.

Bauhinia tomentosa Linn.

St. Thomas-Tree

(*Tomentosa* is a Latin word meaning hairy and refers to hairiness of the leaves and pods.)

Description.—A handsome shrub which sometimes grows into a

M. N. Bakshi

The Sharp-leaved Bauhinia
Bauhinia acuminata Linn.

GALPIN'S BAUHINIA
Bauhinia galpini N.E. Br.

M. N. Bakshi

small tree. Leaves compound, on slender petioles up to 1 in. long, stipulate; stipules long, subulate, hairy. Leaf-blade of two connate leaflets, coriaceous, 1-2 in. long, broader than long, 7-9-nerved from the base, glabrous on the upper surface, tomentose below, 2-lobed; lobes rounded at the apex; reticulation conspicuous.

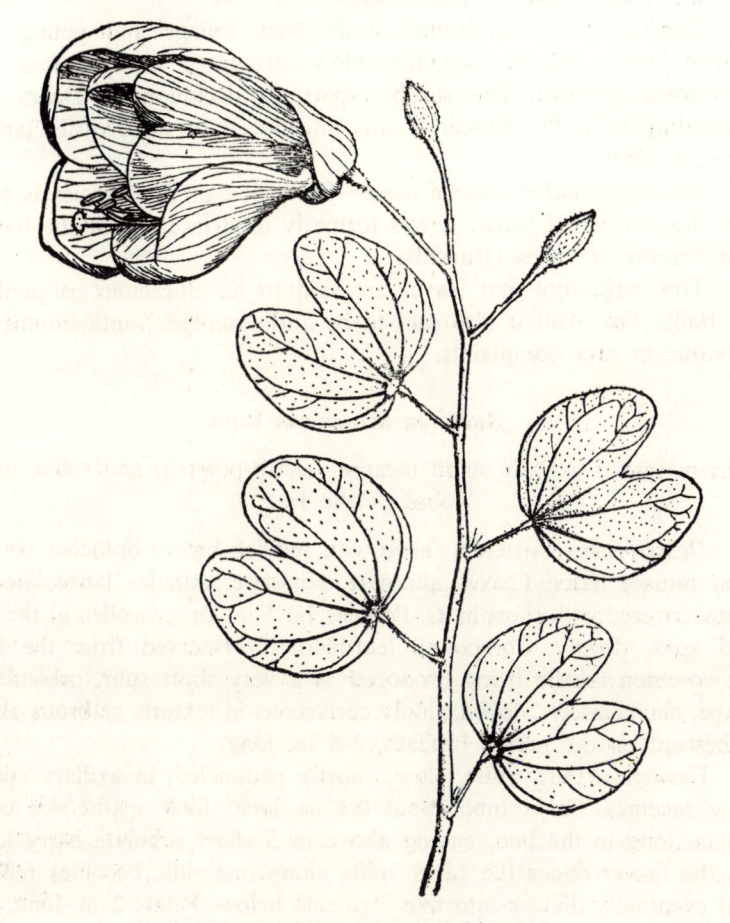

Fig. 44.—*Bauhinia tomentosa* Linn. × 1

Flowers axillary in pairs (sometimes 1 or 3) on pedicels bearing a pair of subulate persistent bracteoles. Calyx tube short, covered with appressed pubescence; limb spathe-like with 2 small teeth at the apex in bud, splitting down one side when the flower opens, about 0.5 in. long. Corolla of 5 distinct petals. Petals obovate in shape, 0.75-2 in. long, not clawed, yellow in colour, conspicuously veined, one with a red blotch on the inner surface. Stamens 10 in number arising from the lip of the calyx tube; anthers sagittate. Ovary on a gynophore; style nearly 0.75 in. long. Pod dehiscent, stalked, 4-5 in. long, 6-10-seeded, tomentose or glabrous.

Flowers.—Aug.-Oct. *Fruits.*—Dec.-Feb.

Distribution.—Throughout India, wild or cultivated, extending to China and tropical Africa. Now commonly cultivated in the tropics of both the old and the new worlds.

Gardening.—A handsome shrub. New foliage and young parts downy brown. Flowers sulphur-yellow, drooping. Easily raised from seed sown in April. This shrub is partially deciduous in Dehra Dun. According to J. D. Hooker it was introduced to the Royal Gardens, Kew, in 1860.

Medicinal and economic uses.—The wood of this species is tough and the heartwood black. It was formerly used in Java for the handles and sheaths of krises (Burkill).

The bark, root and leaves are said to be efficacious as poultices for boils. The plant is also said to be anti-dysenteric, anthelmintic and of value in liver complaints.

Bauhinia acuminata Linn.

(*Acuminata* is a Latin word meaning sharp-pointed, and refers to the lobes of the leaf.)

Description.—An erect shrub with reddish brown branches covered with minute hairs. Leaves alternate, stipulate; stipules lanceolate-subulate, covered with short hairs. Petioles 1-1.5 in. long, swollen at the base and apex, downy. Compound leaf-blade 9-11-nerved from the base, the common midrib being produced as a very short spur, orbicular in shape, acuminately 2-lobed, thinly coriaceous in texture, glabrous above, pubescent on the nerves beneath, 3-6 in. long.

Flowers a pure white, close, shortly peduncled, in axillary corymbose racemes. Calyx tube about 0.3 in. long; limb spathe-like up to 1.5 in. long in the bud, ending above in 5 short subulate hairy lobes. As the flower opens the calyx splits along one side, becomes reflexed and eventually divides into five segments below. Petals 2 in. long, 0.75 in. wide, oblong-obtuse, white, not clawed but rounded at the base, seated on the margin of the calyx tube. Stamens 10, filaments white, of various lengths; anthers yellow, versatile, hairy. Ovary on a gynophore at the base of the calyx tube. Style short, 0.5 in. long; stigma a 2-lobed disk. Pod pendent from a 0.5 in. stalk, 4-5 in. long, 0.9 in. broad, firm, glabrous, 8-12 seeded, with a rib on each side of the upper suture.

Flowers.—March-May. *Fruits.*—Cold season.

Distribution.—Indigenous to Central India, Sri Lanka, Malaya and China. Very frequently cultivated in gardens all over the country.

LEGUMINOSAE

Gardening.—This is one of the most satisfactory species of Bauhinia for cultivation in the open. It is quite frost-hardy and starts flowering when the plants are only a few months old and but a foot or so high. According to Aiton's 'Hortus Kewensis' it was introduced from

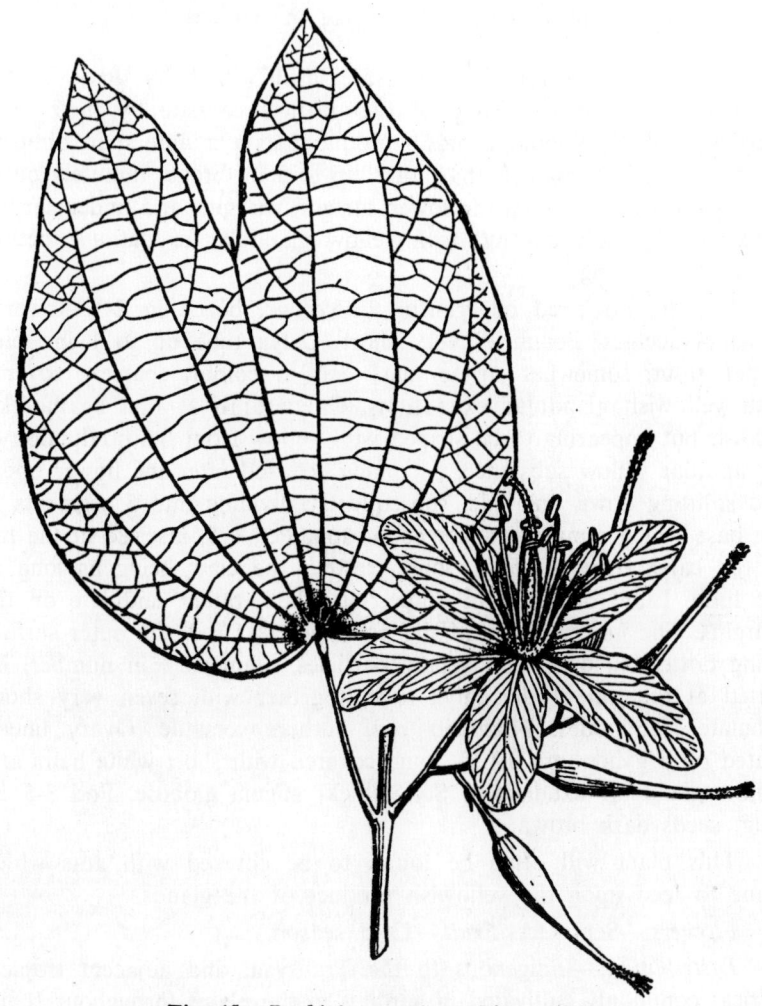

Fig. 45.—*Bauhinia acuminata* Linn. ×2/3

India into England by Dr. Francis Russell and flowered at Kew in the months of May and June. It grows readily from seed and bears its numerous, large, snow-white flowers practically all the year round but chiefly during March-May.

Bauhinia galpini N. E. Brown

Galpin's Bauhinia

(This plant was named in honour of E. E. Galpin, who was one of the earliest collectors of the species.)

Description.—A rambling prostrate shrub with hairy stems. Leaves alternate, stipulate, consisting of two leaflets connate by their inner margins, with the common midrib produced as a minute spur, obtusely 2-lobed, 7-nerved, broader than long, up to 3 in. broad, minutely pubescent with white hairs on the lower surface; margins and under surface thickly or sparsely covered with yellow glandular exudations; petiole about 0.5 in. long.

Flowers brick-red or crimson in colour, borne in 2-10-flowered axillary racemes. Peduncles very short. Calyx tube up to 1 in. long, rather stout, somewhat striate, dark red in colour, sparsely covered with yellowish glandular exudations. Calyx limb at first spathe-like, reddish but appearing to be striped with yellow from the production of a glandular yellow substance, becoming reflexed after the flower opens and splitting down one side, subsequently dividing into 5 segments at the base which remain connate at the top. Petals 5, inserted at the top of the calyx tube, 1-1.5 in. long, clawed, the claw being as long as the limb. Limb orbicular in shape, cuspidate, rather undulate on the margins. The inner surface of the petals is clear red, the outer surface being dotted all over with yellow particles. Stamens 3 in number, inserted at the top of the calyx tube, together with seven very short, subulate staminodes. Filaments red, anthers versatile. Ovary linear, seated on a gynophore 0.3 in. long, covered with short white hairs and yellow glandular exudations. Style thick; stigma globose. Pod 3-5 in. long; seeds dark brown.

This plant will often be found to be covered with ants which come to feed upon the yellowish product of the glands.

Flowers.—Sept.-Oct. *Fruit.*—Cold season.

Distribution.—Indigenous to the Transvaal and adjacent tropical Africa; commonly cultivated in gardens in the plains throughout India.

Gardening.—This fine species, which in its native country is said to be a climber, is found only as a straggling or prostrate shrub in India. It thrives better on a well-drained soil and bears its bright scarlet flowers profusely during Sept.-Oct. Propagated by seed which germinates rather sparingly. The seedlings are liable to damp off during the rains if not properly looked after. This beautiful shrub is well worth growing in spite of the initial difficulties of propagation.

PLATE 24

M. N. Bakshi

THE CORYMROSE BAUHINIA
Bauhinia corymbosa Roxb.

Bauhinia anguina Roxb. [Lasiobema scandens (Linn.) De Wit]
Snake Climber

(*Anguina* means snake-like and refers to the peculiar stems of this species.)

Description.—A woody cirrhose climber with very peculiar com-

Fig. 46.—*Bauhinia anguina* Roxb. ×2/3

pressed flattish stems, alternately concave and convex in the central portion, with stout margins. This is an adaptation to take strain. The majority of the sap vessels are in the corrugated part so that if tension is applied the strain is taken by the margins and the ascent of sap is not impaired. Leaves alternate, densely tomentose when young, glabrous when old, thin, 2-5 in. long, shortly lobed or with 2 very long acuminate lobes. On old plants the leaves are often quite entire and acuminate.

The flowers are very small for the genus and are arranged racemosely in lax, pubescent, terminal panicles. Individual flowers seated on very short pedicels. Calyx very small, about 0.05 in. long, broadly campanulate, with 5 deltoid teeth. Petals 5, oblong-lanceolate in shape, about 0.1 in. long. Fertile stamens 3. Pod thin flat, oblong or elliptic, glabrous, indehiscent, 1.5-2 in. long, 1-2-seeded.

Flowers and fruits.—Cold season.

Distribution.—Sikkim Terai, ascending to 2,000 ft, Khasi hills, Chittagong, Martaban, and Kerala.

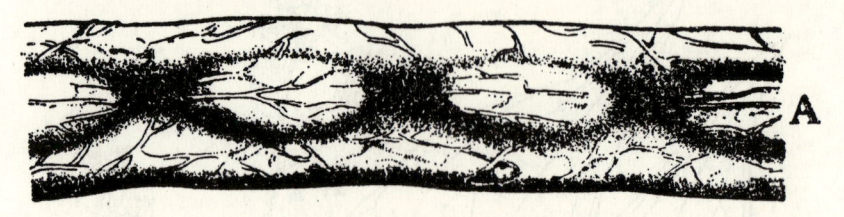

Fig. 47.—*Stem of Bauhinia anguina* Roxb.

A.—A surface view of the strap-shaped stem.

B.—A longitudinal section of the stem seen sideways to show the sinuous centre portion.

Gardening.—A large evergreen climber having a curious stem and small inconspicuous white flowers. According to Roxburgh 'the most regularly serpentine pieces of the stems and large branches are carried about by our numerous mendicants to keep off serpents'. Easily propagated by seed or layers.

Bauhinia vahlii W. & A. [Phanera vahlii (W. & A.) Benth.]

(This species was called after Martin Vahl, a Danish botanist, 1749-1804.)

Description.—A gigantic climber, with densely pubescent branches and abundant circinate tendrils, mostly leaf-opposed. Leaves compound, alternate, petiolate, deeply cordate at the base, lobed at the top, 11-15-nerved from the base, almost orbicular in shape, up to 18 in. long, sparsely hairy on the upper surface, densely ferrugineous-tomentose on the lower surface; petiole 3 in. long.

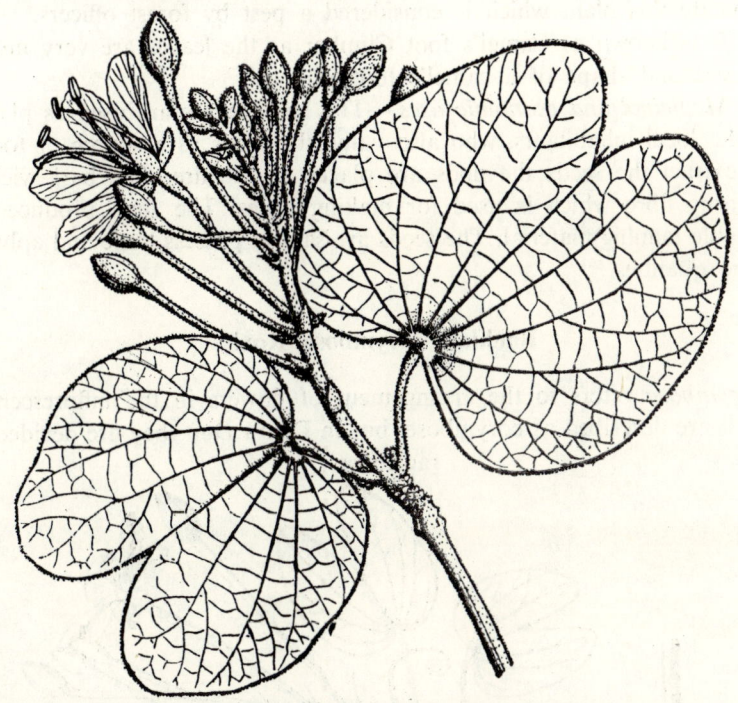

Fig. 48.—*Bauhinia vahlii* W. & A. ×2/3

Flowers numerous, arranged in long-peduncled terminal dense subcorymbose racemes. Individual flowers seated on pedicels 1-2.5 in. long, with persistent bracteoles at the base. The whole inflorescence is covered with a dense rusty tomentum. Calyx tube short, it and the limb very hairy; limb splitting into 3-5 valvate segments when the flowers open and becoming reflexed. Petals 5, white, fading to yellow, shortly clawed, broadly spathulate in shape, spreading, glabrous within, covered on the outside with rusty villi. Stamens 3, fertile; a number of staminodes may be found on the lip of the calyx tube. Ovary on a short gynophore, very densely hairy as is also the style. Pod woody, up to 12 in. long finally dehiscent, velvety, 8-12-seeded.

Flowers.—April-June. *Fruits.*—Cold season.

Distribution.—Subhimalayan tract and outer valleys ascending to 3,000 ft from the Chenab eastwards, chiefly in sal forests; Assam, Bihar, Western Peninsula.

Gardening.—This gigantic woody climber needs plenty of space for its growth. It is useful for covering unsightly embankments and the like. Easily raised from seed. It is one of the most distinctive climbers in the Indian forests. The trunk may reach a girth of 4 feet and is often deeply fluted. When cut down shoots more than 50 feet long may be produced in one season, and for this reason it is difficult to eradicate this plant which is considered a pest by forest officers.

It is known as Camel's foot Climber as the leaves are very much the size and shape of a camel's footprint.

Medicinal and economic uses.—The large leaves are used as plates by the local inhabitants, who also value the seeds as a source of food. To extract the seeds the pods are placed in the fire. The bark yields a strong fibre which is used for making ropes. The stem produces a valuable tanning material. The seeds are said to possess tonic and aphrodisiac qualities.

Bauhinia corymbosa Roxb.

(*Corymbosa* refers to the arrangement of flowers in the inflorescence which are described as corymbose, but in Dehra Dun they are decidedly racemose.)

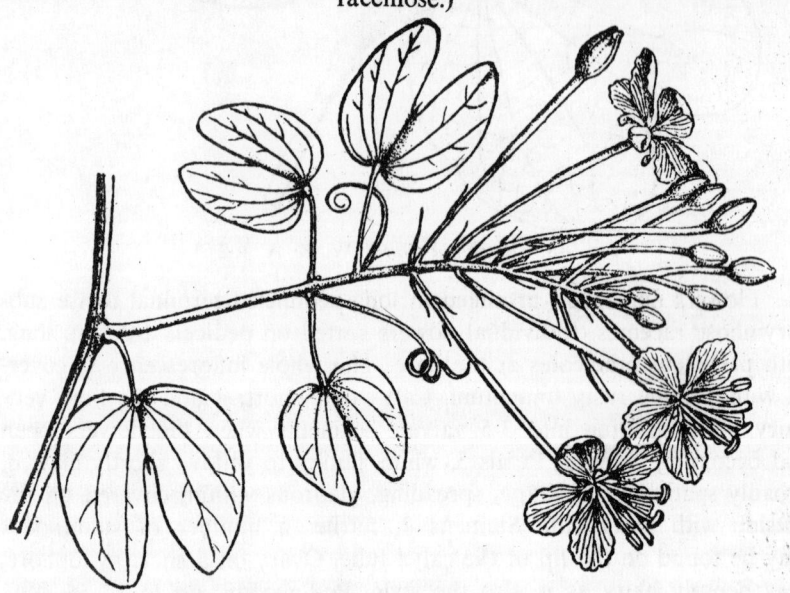

Fig. 49.—*Bauhinia corymbosa* Roxb. ×1

Description.—A woody climber, branching from the ground, with grooved branches and circinate tendrils. Leaves compound, 1-2 in. long, divided almost to the base; common midrib very short, produced into a short spur; lobes rounded. The two leaflets fold together at night. Petiole up to 1 in. long, swollen at base and apex, often covered with appressed brownish hairs.

The fragrant flowers are borne in terminal long-peduncled racemes or corymbs. The individual flowers are seated on pedicels up to 1 in. long, and are supported by bracts and bracteoles. The calyx tube is 0.75 in. long, green in colour, with 10 well-marked red ribs covered with brownish hairs. Calyx-limb short, 0.25 in. long, red, 5-lobed; lobes valvate, splitting into 5 segments, which turn downwards as the flower opens. Petals white, with pink nervation, rather crumpled, seated on the margins of the calyx tube. Fertile stamens 3 on pink filaments, posterior; staminodes 5, pink, anterior, all seated on the calyx tube. Ovary on a gynophore arising from the calyx tube, red in colour, produced into a short style and globose stigma. Pod 4-5 in. long, 0.7 in. broad, thin, smooth, dark brown.

Flowers.—April-June. *Fruits.*—Cold season.

Distribution.—Native of China. Commonly cultivated in gardens throughout India.

Gardening.—Bauhinia corymbosa is one of the most beautiful of climbing Bauhinias; and even with its rather small flowers, it is a most charming plant with exceedingly slender stems and very small, pretty, dark green shiny foliage. The rosy-white, fragrant flowers appear in great abundance during April-June. Easily raised by layers. It has long been cultivated in Indian gardens for its ornamental foliage alone.

Chapter 5

RUBIACEAE

A very large family of over 4,500 species, comprising herbs, shrubs, climbers, and large trees. A small number are epiphytes. Many beautiful shrubs belonging to this family, are cultivated in our Indian gardens for their handsome, occasionally fragrant, flowers.

The leaves are opposite or whorled, stipulate, usually entire, with pinnate nervation. The stipules are interpetiolar, that is, the two adjacent stipules are joined together across the node, forming a half sheath, simple or divided into lobes or fringed. The flowers are variously arranged, sometimes in globular heads, sometimes in corymbose cymes or panicles, axillary or terminal. Calyx often campanulate, adnate to the ovary, truncate or with 5 lobes. Corolla gamopetalous, seated on the top of the ovary, more or less tubular with spreading lobes; lobes 4-10, imbricate or valvate. Stamens as many as the lobes, seated on the corolla and alternate with the lobes. Disk present, annular or lobed; lobes as many as the ovary cells. Ovary inferior, 2- or more-celled with axile basal or apical placentation. Ovules one to many. Fruit a capsule, berry or drupe.

The family takes its name from *Rubia cordifolia*, a well-known Himalayan plant, which used to be the source of madder before the days of synthetic dyes. It is still largely used by the hill tribes to dye thread.

Apart from plants of ornamental value there are many species of *Rubiaceae* which are of economic importance. Quinine, so widely used in malarial cases, is a product of *Cinchona ledgeriana* and other species of *Cinchona*, which are indigenous in South America.

The drug, quinine, is of such importance in all tropical malarial countries that it will be of interest to detail briefly, in narrative, the history of this plant and its introduction into India. A great deal of the information contained in this account has been obtained from *Travels in Peru and India* by C. R. Markham (afterwards Sir Clements Markham), who had a great deal to do with the introduction of this genus of plants into India. His book may be recommended as a first class travel book, full of curious information apart from its worth as botanical history.

The genus *Cinchona* was erected by Linnaeus in honour of the Countess of Chinchon, of whom more later. As there has been some dispute regarding the correct spelling of this name, it will perhaps be better to settle the matter at once. Some hold that the spelling should be *Chinchona*, on the grounds that *Cinchona* implies a mark of disrespect to the Counts of Chinchon, the hereditary Alcaides of the Alcazar of Segovia.

A reference to the Species Plantarum of Linnaeus, Vol. I, page 172, shows that the famous Swede spelled the name *Cinchona*, a spelling which, following the rules of International Botanical Nomenclature, must stand. Linnaeus mentions one species, *Cinchona officinalis*, and in addition informs us that it is also called Quinquina and 'habitat in Loxa Peruviae'.

The name 'quinquina' is the Peruvian name for the tree and means 'bark of bark', a circumstance that Markham uses to refute the idea prevalent at the time, that the Peruvians had no knowledge of the febrifugal properties of what afterwards came to be known in the trade as Peruvian bark. The reasons for this belief were that this bark was not found in the wallets of the Peruvian doctors, nor did they communicate their knowledge of its virtues to their conquerors. This, however, can be easily explained by their hatred for the Spanish invader.

Whether the inhabitants of Peru knew of the febrifugal properties of the bark of quinquina or not, it is certain that the Spaniards must have known of them before 1638. In that year the wife of Luis Geronimo Fernandez de Cabrera Bobadilla y Mendoza, fourth Count of Chinchon, lay ill of intermittent fevers in the palace at Lima in Peru. 'The Corregidor of Loxa, Don Lopez de Canizares, on hearing of her illness, sent a parcel of powdered quinquina to her physician, Juan de Vega, who was also captain of the armoury, assuring him that it was a sovereign and never-failing remedy for tertiaria' (Markham). The gallant captain fortunately was able, despite his duties at the armoury, to administer the drug to the Countess, and the result was a complete cure. As we have already seen Linnaeus immortalized the lady by calling the genus *Cinchona* after her, doubtless influenced by the extraodinary belief of those days that a person of rank conferred an honour upon a drug by kindly consenting to be cured by it. The lords of Chinchon are, however, said to have made history in another direction by giving Charles I of England in 1623 a supper of 'certaine trouts of extraordinary greatnesse'.

Following the cure of this important personage the Jesuits seem to have been primarily responsible for the introduction of Peruvian bark into Europe. In 1670 it appears to have been largely used with success in Rome to cure fevers of all kinds. Apparently it was at that time known as 'Jesuits' bark and as such was naturally avoided by all good Protestants. Gradually, however, prejudice was overcome and by 1726, the bark was held in high repute in all European countries as a cure for various fevers.

The popularity of Peruvian bark naturally led to the destruction of hundreds of thousands of trees to provide for the European trade, so that in time the trees were threatened with extinction.

The Dutch apparently were the first to take steps to introduce the tree into their colonies and *Cinchona* plants were landed in Java in 1854. Sir Clements Markham led an expedition from England to the Andes in 1855, and the seeds and plants of *Cinchona* obtained by him and his assistants reached Kew in 1860; and the same year Markham landed in Bombay with 125 seedlings. These were planted in the Nilgiris but all died. Others, however, arrived from Kew and extraordinary progress was made, so that by 1862 there were something like 30,000 plants established. From these beginnings the cultivation of *Cinchona* spread all over India, and has reached enormous proportions. In 1937-38 the total amount of quinine products sold in Bengal was in the neighbourhood of 30,000 lb.

Ipecacuanha, which contains among other alkaloids the valuable emetine, largely used as a specific in cases of amoebic dysentery, is obtained from *Cephaelis* (*Psychotria*) *ipecacuanha* Rich. This plant has not had nearly so important a career as the noble *Cinchona*, as it only came to light through the travels of a Portugese priest, Manoel Tristaon, in Brazil during the sixteenth century. The properties of the drug became known in Europe during the 17th century. The plant was introduced into India in 1866.

In addition to *Rubia cordifolia*, mentioned above, another genus *Morinda*, is well known as a source of fast dyes. The colours obtained from it range from yellow to red, purple and chocolate.

Coffee is obtained from *Coffea arabica* Linn., an African plant, now largely cultivated in South India. The well-known tanning material, called Gambier in the trade, is furnished by the climber, *Uncaria gambir* Roxb.

The climbing genus *Uncaria* (from *uncus,* a hook) climbs by means of axillary hook-shaped tendrils. *Paederia foetida,* a common jungle plant in India, is a twiner.

Little is known of the mechanism of pollination of the species of this large family. In *Ixora* the anthers are pressed against the style in the bud. Before the flower bud opens the anthers dehisce and large quantities of pollen are left adhering to the tip of the 2-lobed style. The stigmatic surfaces of the stylar lobes are pressed together at this time and effectively prevent self-fertilisation. The unexpanded style is supposed to act as a perching rod for insects which carry away the pollen to older flowers. Subsequently the stigmatic lobes become recurved and expose the receptive surfaces to a visitor carrying pollen from another flower. This genus which has very highly coloured flowers would appear to favour cross fertilisation.

RUBIACEAE

Key to the Genera

Shrubs without thorns
  Flowers with one calyx lobe expanded into a white leaf-like
    structure; flowers orange or pale yellow .... *Mussaenda*
  Flowers with calyx lobes not so expanded
    Plant foetid when bruised; leaves small .... *Serissa*
    Plant not foetid when bruised
      Flowers white
        Flowers fragrant .... *Gardenia*
        Flowers not fragrant .... *Coffea*
      Flowers coloured
        Flowers red, yellow or orange
          Corolla salver-shaped
            Corolla lobes pointed .... *Ixora*
            Corolla lobes rounded .... *Rondeletia*
          Corolla tubular .... *Hamelia*
        Flowers mauve or white .... *Hamiltonia*
Thorny shrub with greenish flowers .... *Catesbaea*

Mussaenda Linn.

(*Mussaenda* is the Sinhalese name for the species of this genus.)

Shrubs or undershrubs, sometimes scandent, with opposite or ternately whorled leaves. Stipules free or joined together, persistent or deciduous. Flowers usually yellow, in terminal cymes. Calyx oblong, top-shaped or globose, with 5 lobes, one of which is expanded in a large white or coloured leaf. Corolla tubular below, funnel-shaped above; throat villous; lobes five, valvate in the bud, with everted margins, spreading in the open flower. Stamens in the tube or throat, filaments very short; anthers linear. Ovary 2-celled; ovules many on peltate fleshy placentae. Fruit a berry. Seeds many, pitted.

Key to the species of Mussaenda

  Flowers pale yellow .... *M. luteola*
  Flowers oranged-red .... *M. frondosa*

Mussaenda luteola Del.

(*Luteolus* means yellowish in Latin, and refers to the colour of the foliaceous sepal of this species.)

Description.—A small erect twiggy slender shrub. Branches green, somewhat compressed, covered with a hoary appressed pubescence. Leaves 1.5-2 in. long, petiolate, elliptic-lanceolate or oblong-lanceolate in shape, acuminate at the tip, narrowed at the base, membranous, olive-green above, pale below, sparsely hairy above, covered on the nerves below with short appressed pubescence; petiole short. Stipules subulate.

Inflorescence of few-flowered corymbs. Calyx tube 5-angled, 0.1

in. long, shortly hairy, 5-lobed; lobes subulate 0.2 in. long except one, which is often produced and expanded into an elliptic-acuminate or ovate foliar structure, 0.75-1 in. long, yellowish in colour, seated on a 'petiole' 0.4 in. long. Corolla tube 1 in. long, densely velvety pubescent at the mouth, greenish, pubescent, swollen about one-third the way down at the insertion of the stamens, 5-lobed; lobes pale yellow in colour, broadly ovate, long acuminate. Stamens 5, sessile; anthers linear,

Fig. 50.—*Mussaenda luteola* Del. × 1

included. Ovary 2-celled; ovules numerous. Style slender, glabrous, divided at the top into two stigmatic lobes.

Flowers.—Hot and rainy season. *Fruits.*—This shrub seldom fruits in Dehra Dun.

Distribution.—Native of tropical Africa, but now commonly cultivated in gardens throughout the plains of India.

Gardening.—A pretty bushy shrub with dense dull-green foliage and pale yellow foliaceous sepal. Propagated usually by layers, as cuttings are less successful. According to J. D. Hooker, it was first in-

PLATE 25

M. N. Bakshi

THE YELLOWISH DHOBIE'S TREE
Mussaenda luteola Del.

PLATE 26

DHOBIE'S TREE
Mussaenda frondosa Linn.

M. N. Bakshi

DHOBIE'S TREE
Mussaenda frondosa Linn

M. N. Bakshi

Plate 9

PAPERCHASE TREE OR DHOBY'S TREE
Mussaenda frondosa Linn.
(93.4 per cent nat. size)

troduced into Europe about 1860, by Capt. Grant from the headwaters of the Nile.

Mussaenda frondosa Linn.

Paperchase Tree or Dhoby's Tree

(*Frondosa* is a Latin word meaning leafy.)

Description.—An erect shrub. Branches green, angled when young, covered with a coarse brown pubescence. Leaves opposite, shortly and stoutly petioled, ovate or elliptic in shape, obtusely acute or acuminate at the tip, narrowed or rounded at the base, membranous, glabrous or sparsely hairy above, softly hairy on the lower surface, 6-9 in. long, by 3-4 in. wide; petiole stout, 0.5 in. long, coarsely hairy. Stipules large, broadly ovate-obtuse, often splitting at the apex into two lobes, covered with longish hairs.

Inflorescence in terminal cymes supported by two reduced leaves. Peduncles stout, very hairy; bracts and bracteoles present. Individual flowers subsessile. Calyx tube 0.2 in. long, adnate to the ovary, glabrous or sparsely hairy, oblong in shape, 5-lobed; lobes linear or subulate, 0.4 in. long, ciliate with long hairs. One lobe is occasionally produced into a hairy petiole, 0.75 in. long, to which is attached a leaf-like, white, elliptic-lanceolate appendage, 1 in. long or more. Corolla tube cylindrical, slender below, widened above the position of the stamens, covered with silky hairs, pale green in colour, yellowish hairy at the mouth, 5-lobed; lobes thick, fleshy, ovate-acute, yellowish green outside, inside of a beautiful orange-red colour, covered with a minute golden glandular excretion. Stamens 5, linear, sessile, included within the glandular hairy tube about half way down; the filaments are slightly adnate to the tube for the whole of their length, making the anthers sessile, but they can be easily separated. Ovary 2-celled; ovules numerous. Style short; stigmas 2.

Flowers.—Hot and rainy season. *Fruits* at the end of the rains.

Distribution.—Indigenous in Assam, Upper Burma, South India and Ceylon.

Gardening.—This is one of the commonest shrubs to be found in Indian gardens. The white foliaceous sepal in contrast to the deep green colour of the foliage makes it an ornamental and conspicuous shrub. It is advisable to prune it in the cold weather when it is deciduous. It is popularly known as the 'Paperchase' tree or Dhoby's tree on account of the fancied resemblance of the white foliaceous leaf to handkerchiefs or bits of paper. Easily raised by cuttings.

Serissa Comm.

(Bailey[1] remarks that the generic name comes from the Indian name

---
[1] The Standard Cyclopedia of Horticulture.

of this plant. On the other hand, Roxburgh says that this plant was introduced from China and hence there is no Indian name for it.)

A branchy shrub. Leaves opposite, subsessile, subcoriaceous, ovate-acute, foetid when bruised. Stipules cuspidate, interpetiolar, persistent. Inflorescence of solitary or fascicled flowers, axillary or terminal, sessile. Hypanthium obconic, ending in 4-6 lobes which are subulate-lanceolate in shape, persistent. Corolla infundibuliform, hairy outside and in the throat; lobes 4-6, short induplicate-valvate in the bud. Stamens 4-6, inserted at the base of the tube; filaments very slender; anthers linear-oblong, included. Ovary 2-celled; ovules 1 to each cell, basal. Style slender, ending in two subulate stigmatic lobes. Fruit fleshy, containing two seeds.

Serissa foetida Lamk.

(*Foetidus* means evil-smelling in Latin, and refers to the unpleasant odour of the leaves and flowers when bruised.)

Description.—A shrub with many glabrous branches. Young branches quadrangular, brown, covered with sparse hairs, becoming round, greyish in colour, and glabrous. Leaves small, 0.3-0.75 in. long by 0.2 in. wide, oblong or oblong-lanceolate in shape, acute at both ends, olive-green above, paler beneath, coriaceous in texture; nerves 3-4 main pairs, ascending obliquely, prominent below, hardly visible on the upper sur-

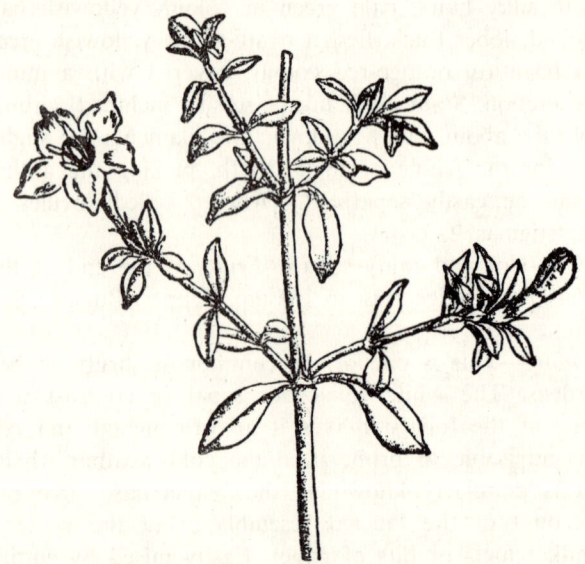

Fig. 51.—*Serissa foetida* Lamk. × 1

face; petiole .05 in. long. Stipules interpetiolar, sheathing at the base, divided above into several setaceous lobes.

Inflorescence of terminal fascicles of flowers, with an involucre of leaves. Flowers seated on very short pedicels; bracteoles broad and membranous at the base, connate into a sheath surrounding the flower, ending above in 2 long sharp setaceous ciliate lobes. Calyx tube very short, 4-5-(rarely 6-8-) lobed; lobes 0.1 in. long, lanceolate-acute, with ciliate margins. Corolla tube 0.2 in. long, cylindrical at the base, funnel-shaped above, hairy within; lobes 4-5, less than 0.1 in. long ovate-obtuse. Stamens 5, inserted in the lower third of the tube or above; filaments short; anthers linear-acute, about 0.1 in. long, included or slightly emergent. Ovary 2-celled; ovules 1 in each cell, basal; style slender, ending in 2 stigmatic lobes. Fruit a berry with 2 seeds.

Flowers.—Practically all the year round.

Distribution.—Indigenous to China and Japan. It has long been cultivated in gardens in India.

Gardening.—A small shrub about 2 ft. high. It is nearly always in bloom with its sparkling rather small white flowers which, like the leaves, when bruised emit a foetid stench. Single- and double-flowered forms are both quite common in gardens. According to Aiton's Hortus Kewensis it was introduced into England in 1787. Propagated by cuttings.

Medicinal uses.—The stem and leaves are said to be used by the Chinese for the treatment of carbuncles and cancer.

Gardenia Linn.

(A genus erected by Linnaeus in honour of Dr. Alexander Garden of Charleston S. C., one of his correspondents.)

Trees or shrubs, thorny or not. Leaves opposite, rarely ternate; stipules connate or not. Flowers solitary, fascicled or collected into cymes, usually large. Calyx lobes usually long, persistent in fruit. Corolla tube cylindrical or campanulate, 5-12-lobed; lobes twisted in the bud. Stamens as numerous as the lobes of the corolla and alternate with them; anthers linear, inserted on the tube, included, sessile or sub-sessile. Ovary 1-celled; ovules in two series on parietal placentae. Fruit fleshy with a leathery epicarp.

Gardenia florida Linn. (G. jasminoides Ellis)

Cape Jasmine

(*Florida* is a Latin word meaning many-flowered.)

Description.—A shrub reaching 6 ft. in height. Young branches brown, becoming covered with a greyish bark, striate. Leaves opposite or ternate, oblong-elliptic or obovate-lanceolate in shape, obtusely acuminate, wedge-shaped and decurrent at the base, shining, 3-6 in. long by 0.5-2 in. wide. Stipules about 0.5 in. long, acute, membranous.

Inflorescence of terminal solitary white very fragrant flowers, seated on 6-winged pedicels, 0.3 in. long. Calyx tube 0.3 in. long, bearing 6 longitudinal wings, 6-lobed; lobes up to 0.75 in. long, oblong, acute, persistent. Corolla tube 1.75 in. long, glabrous inside and out, 6-lobed; lobes up to 1 in. long, obovate, obtuse or rounded at the top, narrowed at the base, fleshy. Stamens 6, long exserted; filaments short; anthers 0.6 in. long, linear-obtuse. Ovary incompletely 2-celled; ovules very numerous on 2 placentas. Fruit 1-1.75 in. long, 0.5-0.75 in. wide, ovoid or elliptic, surmounted by the calyx-lobes, winged on 6 ridges.

Flowers.—March-April. *Fruits.*—Cold season.

Distribution.—Native of China, now widely cultivated in gardens throughout India.

Gardening.—A pretty shrub with handsome, dark green, glossy

Fig. 52.—*Gardenia florida* Linn. × 1

foliage. It produces at the beginning of the hot season numerous double creamy-white strongly and sweetly scented flowers, which resemble those of a double Camellia. It usually grows to about 6 ft. or so but may be kept to any convenient size by pruning. Easily propagated by cuttings during the rains. It is suitable for hedges and cut flowers.

Medicinal and economic uses.—The pulp of the fruit gives a yellow dye. A decoction of the leaves and roots with or without sugar, is used to alleviate fevers.

PLATE 10

SCARLET IXORA
Ixora coccinea R. Br.
(9/10 nat. size)

Ixora Linn.

(Van Rheede states in Hortus Malabaricus that the flowers of this genus are offered to the God, *Ixora,* a fact which has obviously given rise to the generic name.)

A genus of small shrubs or trees which are great favourites in Indian gardens, on account of their brilliantly coloured flowers and dark green handsome foliage. The leaves are opposite, glabrous, coriaceous, and reticulately nerved. The stipules are interpetiolar, simple, subpersistent. The inflorescence is terminal, paniculate or contracted, the branchlets being subtended by linear or subulate, free bracts and each pedicel provided with 2 bracteoles. Individual floweis bisexual, tetramerous. Calyx campanulate, adnate to the ovary, shortly 4-lobed. Corolla hypocrateriform (tube-like with spreading lobes); tube slender; lobes 4, twisted in the bud. Stamens as many as the lobes and alternate with them; filaments short, inserted in the throat. Ovary inferior with 2 cells; ovule 1 in each cell, attached to the centre of the septum. Disk annular. Style glabrous, shortly exserted and ending in two recurved stigmatic lobes. Fruit a globose drupe. Seed semiglobose, with a deep and wide excavation on the flat side.

Key to the Species

Flowers coloured
  Flowers yellow
    Corolla-tube 1.5 in. long; lobes broadly ovate, acute .... *I. lutea*
    Corolla-tube 1 in. long; lobes rotundate, obtuse .... *I. chinensis*
  Flowers red or reddish
    Flowers scarlet
      Corolla-tube up to 1 in. long
        Corolla-lobes rotundate, obtuse; corolla at first yellow then red .... *I. chinensis*
        Corolla-lobes ovate acute; corolla always scarlet .... *I. coccinea*
      Corolla-tube up to 1.75 in. long .... *I. fulgens*
    Flowers pink .... *I. rosea*
Flowers white
  Flowers less than 0.5 in. long
    Leaves undulate on the margin; calyx-lobes as long as the calyx-tube .... *I. undulata*
    Leaves not undulate; calyx-lobes much shorter than the calyx-tube .... *I. parviflora*
  Flowers over 0.5 in. long; throat of corolla woolly; corolla tube up to 1.5 in. long .... *I. barbata*

Ixora coccinea Linn.
Scarlet Ixora

(*Coccinea* is a Latin word meaning scarlet-coloured, and refers to the brilliant scarlet flowers of the commonly cultivated race of this species.)

Description.—A compact shrub or small tree, glabrous all over. Leaves sessile, opposite, elliptic, ovate or obovate in shape, apiculate, obtuse or mucronate at the tip, somewhat cordate at the base, thickly coriaceous in texture, 1.5-3.5 in. long, dark glossy green in colour, with 6-8 nerves on either side of the midrib. Stipules triangular, awned.

Inflorescence terminal, supported by two small, leaf-like bracts, strongly contracted, forming a compact corymb. Individual flowers with two bracteoles at the base. Calyx urceolate, 0.1 in. long, green, with 4 triangular acute lobes reddish or purplish at the tips. Corolla hypocrateriform, with a slender tube 1.25 in. long, surrounded by four spreading ovate-acute lobes 0.3 in. long, which eventually become reflexed, brilliant crimson in colour. In the bud the lobes are twisted to the left. Stamens 4, on short filaments attached to the throat, alternate with the corolla lobes. The stamens dehisce in the bud which, when just open, display the top of the style drenched with pollen. At this time the stigmas are immature and insects alighting on the style carry away pollen to an older flower.

Flowers.—Practically throughout the year, but is at its best during the rains. *Fruits.*—Cold season.

Distribution.—Native of the Western Peninsula; now widely cultivated throughout the tropics.

Gardening.—A shrub usually 3-4 ft. high, one of the commonest in cultivation, and certainly one of the most beautiful of the genus. It is practically in flower throughout the year, but develops to perfection during the rains, when it is a truly glorious object to behold, and few shrubs surpass the splendour of this Ixora when it is in full bloom. The bright scarlet flowers are arranged in large compact corymbs. During flowering time an occasional application of liquid manure will be found beneficial, and it is advisable to prune rather closely after flowering. Easily raised from seed or layers or cuttings during the rains. Like all other species of this genus it prefers full sunshine. Where frost is severe some damage may be expected.

Medicinal use.—This plant has been known in India since ancient times and the root has some repute in native medicine. It is said to act as a cholagogue and to give relief from pain to those suffering from dysentery.

Ixora rosea Wall.

Pink Ixora

(The specific name refers to the colour of the flowers.)

Description.—An untidy straggling shrub scarcely reaching 4 ft. in height at Dehra Dun. Older stems covered with a slate-grey brown glabrous bark; younger plants dark green, covered with a fine downy pubescence. Leaves subsessile, oblong, elliptic, elliptic-obovate in shape, obtuse or somewhat acute at the tip, rounded or slightly cordate at the base, coriaceous in texture, glossy green above, pale below, glabrous; stipules triangular, awned, pubescent.

Inflorescence peduncled; peduncles supported at the base by small bracts. Flowers shortly pedicelled, each with a pair of bracteoles below the calyx. Calyx about 0.1 in. long, urn-shaped, minutely pubescent; lobes 0.05 in. long, obtuse, tinted with red. Corolla hypocrateriform; tube 1.25, rose-coloured, minutely hairy, ending above in four elliptic-obtuse lobes, 0.3 in. long. Stamens 4; filaments 0.05 in. long, attached to the throat, alternate with the lobes; anthers 0.15 in. long, closely appressed to the style in the bud, reflexed in the fully opened flower. Ovary inferior, 2-celled; ovule 1 in each cell.

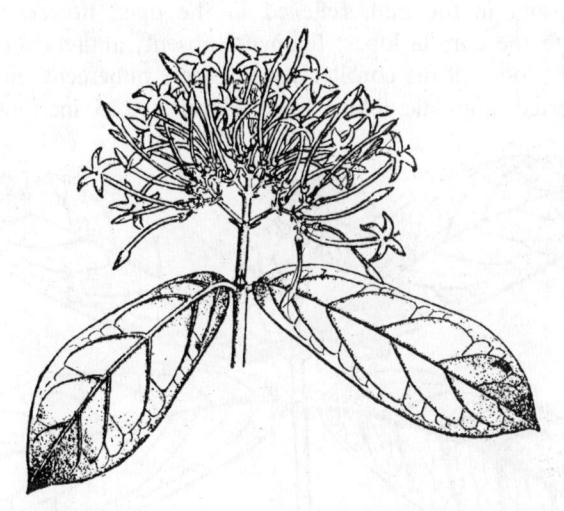

Fig. 53.—*Ixora rosea* Wall. ×½

Flowers.—August-September. The shrub seldom fruits in Dehra Dun.

Distribution.—According to Roxburgh it is a native of the Moluccas and China. Now widely cultivated in various parts of India.

Gardening.—An untidy straggling shrub with pink flowers in large round corymbs. It is hardly attractive when in flower, and far from being so at any other time of the year. Propagated by layers or cuttings during the rains.

Ixora parviflora Vahl (I. arborea Roxb. ex Sm.)
Small-flowered Ixora

(*Parviflora* means small-flowered in Latin.)

Description.—A small much-branched tree; bark thick, reddish brown, exfoliating in irregular patches; branchlets slightly compressed, glabrous. Leaves opposite, stipulate, ovate-oblong or slightly obvate, 3-4 in. long, by 1.5-2 in. wide, coriaceous, glabrous, dark green and shining above, pale when dry, rounded at the base, suddenly and shortly acuminate; venation pellucid, reticulate; petioles short and stout; stipules short, cuspidate, persistent.

Flowers white, sweet scented, in subglobose clusters, arranged in sessile, brachiate, pubescent or glabrous cymes with 3-5 pairs of opposite short branches; bracts and bracteoles subulate. Calyx minute, cup-shaped; lobes 4, very small, much shorter than the tube, subacute. Corolla tube short, glabrous, 0.3-0.4 in. long, 4-lobed; lobes linear-oblong, obtuse, ellipsoid in the bud, reflexed in the open flower. Stamens 4, alternate with the corolla lobes; filaments absent; anthers sessile, nearly as long as the lobes of the corolla. Style densely pubescent, ending above in two exserted stigmatic lobes. Fruit globose, 0.25 in. long.

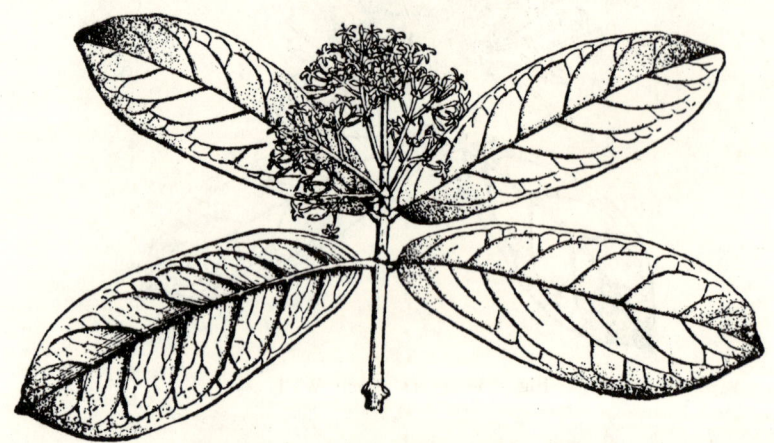

Fig. 54.—*Ixora parviflora* Vahl × 3/8

Flowers.—March-April. *Fruits.*—Cold season.

Distribution.—Native of the Western Peninsula, extending north to the Satpura range, Behar, Chota Nagpur, Orissa, Sunderbans, Chittagong, Burma and Nicobars.

Gardening.—A small much-branched tree which bears in March-April, dirty-white strongly sweet-scented flowers in profusion. It can hardly be said to be attractive at any period of the year. Easily raised from seed or layers or cuttings.

Economic uses.—Wood very heavy, hard, fine grained, yellow. It is used for turnery and furniture in Madras, but, as it never reaches any size, its use is restricted. The green branches are said to make excellent torches.

Ixora barbata Roxb.
Bearded Ixora

(*Barbata* means bearded in Latin and refers to the woolly mouth of the corolla.)

Description.—A large glabrous shrub. Trunk hardly any; branches numerous, opposite. Leaves opposite, short-petioled, elliptic or oblong in shape, somewhat acute, entire, shining on both surfaces, smooth, 6-9 in. long. The upper pairs of leaves are much smaller and are sessile and cordate. Stipules sheathing.

Inflorescence of terminal panicled corymbs, decompound, large, diffuse, often 1 ft across. Bracts and bracteoles decreasing in size upwards, ovate, acute. Calyx globose or ovoid, reddish green, adherent to

Fig. 55.—*Ixora barbata* Roxb. × 3/8

the ovary, 5-lobed; lobes small, erect acute. Corolla tube 1.5 in. long, slender, somewhat curved, greenish white, encircled at the mouth with a delicate fringe of hairs, 5-lobed; lobes obovate, pure white in colour, obtuse, reflexed when the flower is fully open. Stamens 5, alternate with the lobes; filaments short, recurved, so that the linear anthers hang down between the lobes. Style long, slender, exserted, glabrous. Stigma club-shaped, divided into two short stigmatic arms. Berry red, smooth, the size of a pea, 2-seeded.

Flowers.—April-May. *Fruits.*—Cold season.

Distribution.—Native of the Andamans and Nicobars. Commonly cultivated at Calcutta and elsewhere in India.

Gardening.—A large glabrous shrub with rich deep green leaves. It produces during the hot season large lax corymbs of long-tubed white fragrant flowers. Easily raised by layers or cuttings.

Ixora chinensis Lam.

Chinese Ixora

(*Chinensis* refers to the country of origin of this plant.)

Description.—A small shrub 3-4 ft tall, with straight branches covered with smooth dark-brown bark. Leaves opposite, stipulate, sub-sessile, obovate or obovate-oblong; entire, smooth on both surfaces, 5-6 in. long. Stipules interpetiolar, tapering, acute.

The inflorescence is terminal and consists of dense corymbs; branches of the inflorescence trichotomous, smooth, glabrous, reddish. Calyx tube adnate to the ovary, globose or ovoid, small, 4-lobed; lobes short, obtuse, reddish. Corolla tube cylindrical, slender, 0.75-1 in. long,

Fig. 56.—*Ixora chinensis* Lam. × ½

4-lobed; lobes almost orbicular. Stamens 4, alternate with the lobes; filaments short, inserted at the mouth of the corolla; anthers linear-acute, reflexed. The colour of the flowers is at first orange, becoming a deeper and deeper salmon-red, as the flowers become fully open. Style exserted; stigma 2-cleft; lobes reflexed. Berry smooth, succulent, red, 2-celled, with a single rugose seed in each cell.

Flowers.—July-September. Does not set fruit in this country.

Distribution.—Indigenous to the Malay Archipelago and China, now commonly grown in gardens in all tropical countries.

Gardening.—A shrub closely allied to *I. coccinea* and commonly cultivated in gardens for its showy rose-red flowers. Propagated by cuttings during the rains.

Economic uses.—According to Burkill a decoction of the roots of this plant is used after child birth by the Malays.

Ixora undulata Roxb.
Wavy-leafed Ixora

(*Undulata* means wavy in Latin and refers to the margins of the leaves.)

Description.—A large evergreen shrub. Leaves opposite, subsessile or distinctly petioled, stipulate, oblong or lanceolate in shape, 5-9 in. long by 1.5-3.3 in. wide, glabrous, usually tapering at both ends, acute or acuminate at the tip; margins undulate. Stipules interpetiolar, broad and rounded, but with a slender cusp.

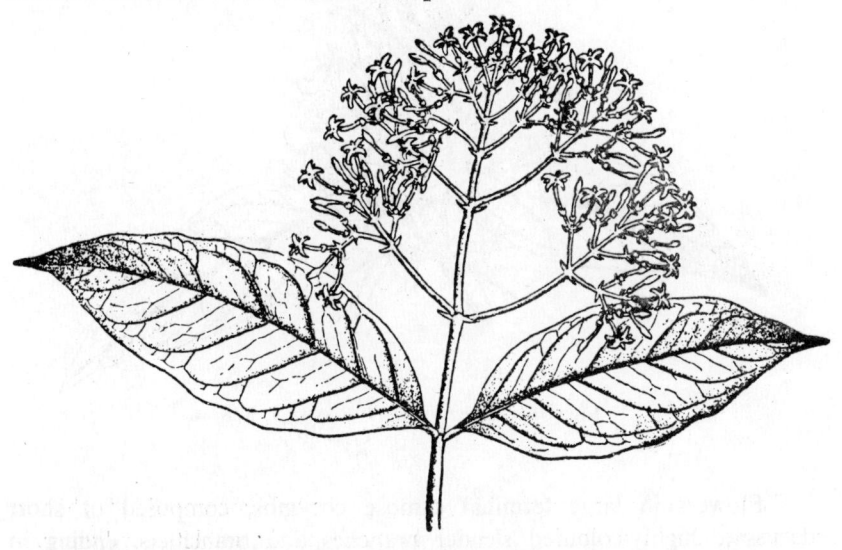

Fig. 57.—*Ixora undulata* Roxb. × ½

Flowers white, sweet-scented, in corymbs on the slender branches of long-penduncled brachiate panicles, which are up to 8 in. long. Calyx tube very short, 4-toothed; lobes acute as long as the tube. Corolla tube 0.28-0.32 in. long, white, 4-lobed; lobes 0.16-0.18 in. long, reflexed. Stamens 4; filaments short; anthers 2-cleft at the base. Style exserted from the tube, ending in two recurved stigmatic arms. Fruit succulent, 0.3 in. diameter, dull purple or slate-coloured with two plano-convex seeds.

Flowers.—March-April. *Fruits.*—Cold season.

Distribution.—Sikkim Terai and lower hills of Bhutan, Assam, Khasi hills and upper Burma.

Gardening.—A large evergreen shrub which produces in March-April, numerous small white flowers, having a powerful fragrance of jasmine. Easily propagated by seed or cuttings or layers.

Ixora fulgens Roxb.

(*Fulgens* is a Latin word meaning gleaming or shining and refers to the scarlet flowers.)

Description.—A short-trunked shrub dividing into many branches covered with dark brown bark. Leaves opposite, short-petioled, entire, smooth on both surfaces, linear-oblong to obovate-oblong in shape, acute at the tip, 6-8 in. long, 1-3 in. wide, with 20-30 pairs of nerves sunk in the upper surface. Stipules interpetiolar, awned.

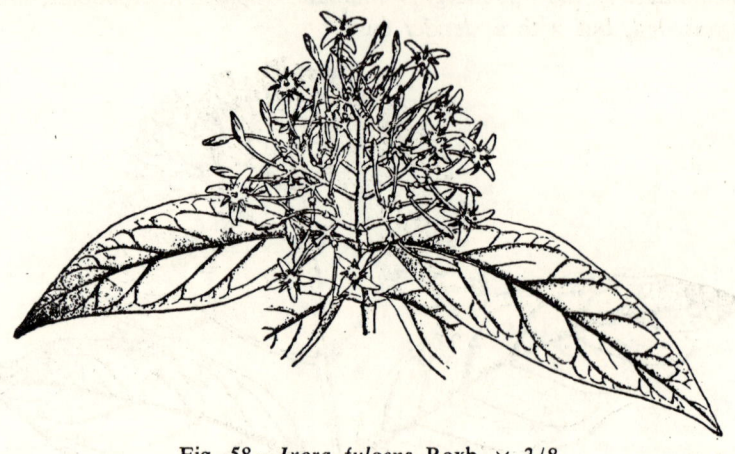

Fig. 58.—*Ixora fulgens* Roxb. × 3/8

Flowers in large terminal cymose corymbs, composed of short decussate highly-coloured slender branches and branchlets, ending in numerous short-pedicelled long-tubed orange-scarlet or scarlet flowers. Calyx tube adnate to the ovary, very short, 4-lobed; lobes obtuse. Corolla tube very slender, up to 1.75 in. long, cylindrical, glabrous, 4-lobed; lobes obovate, obtuse, 0.25 in. long, reflexed. Stamens 4, alternate with lobes; filaments short. Style slender exserted; stigmas 2. Berry 2-lobed, the size of a pea, deep purple when ripe, 2-celled.

Flowers.—Most of the year. Does not fruit in this country (?).

Distribution.—Native of Tenasserim. Widely cultivated throughout the tropics.

Gardening.—A highly ornamental and elegant shrub which bears corymbs of numerous long-tubed pretty large scarlet flowers. Easily multiplied by cuttings.

Ixora lutea Hutch.

Yellow Ixora

(*Lutea* means yellow in Latin.)

Description.—An erect shrub reaching 3 ft in height; branchlets covered with a fine pubescence, finally glabrous. Leaves opposite, stipulate, oblong-elliptic in shape, acute at the tip, unequally cordate at the base, 3.4 in. long by 1.5-2 in. wide, papyraceous in texture, with a slightly recurved margin, pale green in colour; petiole stout, 0.75 in. long, minutely pubescent. Stipules up to 0.5 in. long, awned,

Inflorescence a terminal lax corymbose cyme of sessile flowers; branches puberulous; bracts triangular-subulate, acute. Calyx-tube short,

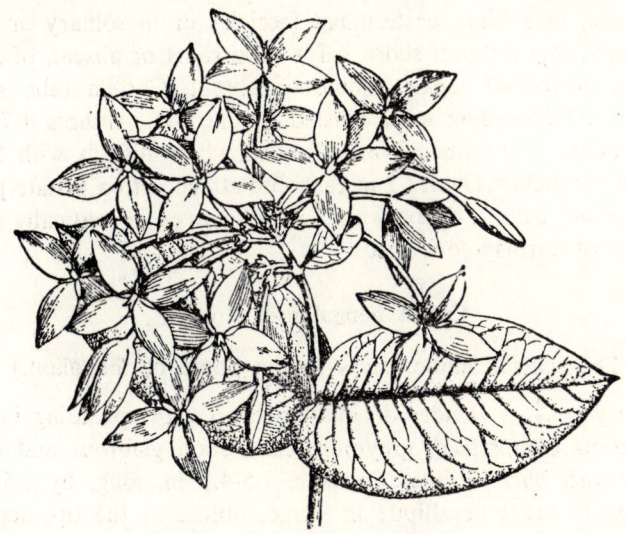

Fig. 59.—*Ixora lutea* Hutch. × ¾

4-lobed; lobes broadly ovate about 0.1 in. long, finely puberulous outside. Corolla ochre-coloured; tube cylindric, 1.5 in. long, very slender, glabrous outside, 4-lobed, lobes ovate-rhomboid, acute, up to 0.75 in. long, toothed, glabrous. Stamens 4, almost sessile, dark orange in colour; anthers acutely acuminate. Ovary 2-celled; style slender, glabrous, shortly exserted, arms about 0.1 in. long, slightly recurved and flattened on the inner surface.

Flowers.—Practically throughout the year. Does not set seed in this country.

Distribution.—Widely cultivated throughout the tropical and subtropical parts of the world. A plant of garden origin.

Gardening.—This beautiful Ixora, which in habit and foliage bears a close general resemblance to *I. coccinea,* can readily be distinguished from it not only by the colour of its flowers, but also by its laxer inflorescence and by the large ovate-rhomboid corolla lobes. It was introduced into the Royal Botanic Gardens, Kew, from Peradeniya. It is an exceedingly attractive plant when in flower, and will flourish under moist-tropical conditions. Raised by layers or cuttings.

Coffea Linn.

The generic name is derived from 'Kahwa', the Arabic name, for the beverage, itself supposed to be from Caffa, a district in southern Abyssinia.

Small bushy shrubs with opposite stipulate leaves. Flowers yellowish or white, in axillary or terminal fascicles or in solitary or axillary dense cymes. Hypanthium short, calyx tube short or absent, often glandular and persistent; sepals minute or absent. Corolla tube short or long, with 4-7 spreading lobes twisted in the bud. Anthers 4-7 sessile, often recurved or twisted. Ovary 2-celled, style filiform with 2 linear or subulate branches. Ovules 1 in each cell attached to a peltate placenta on the septum. Fruit a drupe with 2 plano-convex or vertically concave coriaceous or cartilaginous seeds.

Coffea bengalensis Roxb.

(The specific name refers to the home of the plant.)

Description.—A slender deciduous shrub with spreading branches. Young shoots compressed, grey-pubescent, soon glabrous and covered with a greyish bark. Leaves opposite, 1.5-4.5 in. long, by 0.5-2.5 in. wide, broadly ovate or elliptic in shape, obtuse at the tip, acuminate or subcaudate, rounded or acute at the base, dark green above, pale below, membranous; petiole up to 0.16 in. long.

Flowers appearing with the leaves, axillary, in groups of 1-3, white, sessile, fragrant, 1-1.5 in. across. Calyx tube short, many-toothed, pubescent; teeth rapidly deciduous. Corolla funnel-shaped, 0.5-0.7 in. long, 5-lobed; lobes ovate-oblong, slightly shorter than the tube, spreading or recurved, twisted in the bud. Stamens 5, alternate with the lobes, inserted in the mouth of the corolla tube, subsessile. Ovary 2-celled; ovules solitary in each cell. Style filiform, bifid. Fruit oblong, black when ripe, containing 2 plano-convex seeds.

The seed contains the alkaloids found in the true coffee plant, *C. arabica* L.

PLATE 28

SCARLET-IXORA
Ixora coccinea Linn.

M. N. Bakshi

PLATE 29

SMALL-FLOWERED IXORA
Ixora parviflora Vahl.

M. N. Bakshi

Flowers.—Feb.-April. *Fruits.*—Cold season.

Distribution.—Tropical Himalaya from Garhwal eastwards to Sikkim and Assam; also in Chittagong, central and south India, and Burma extending to Siam and Java.

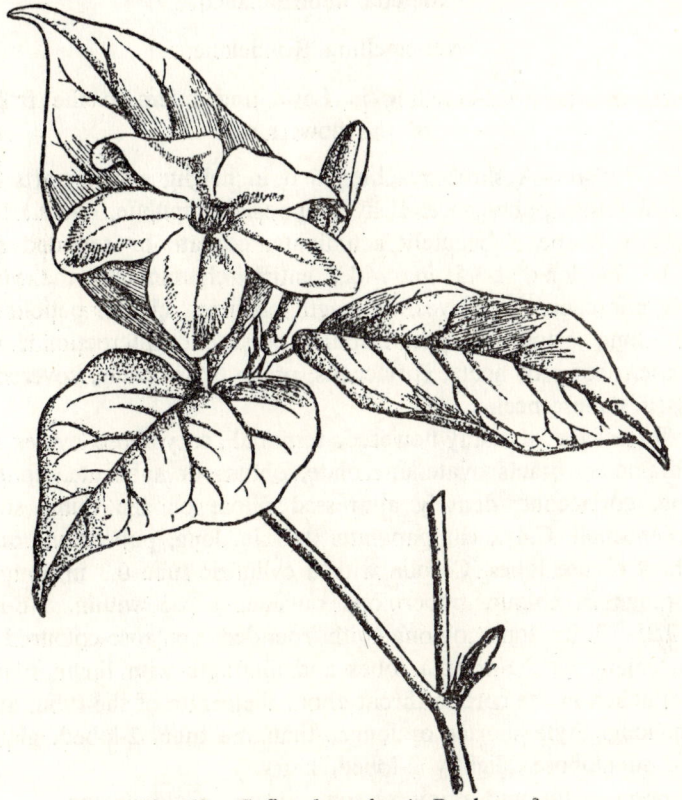

Fig. 60.—*Coffea bengalensis* Roxb. × 3/4

Gardening.—A small shrub, exceedingly beautiful when in full bloom during February, with its snow-white flowers produced in great profusion. Propagated by seed.

Rondeletia Linn.

(A genus erected by Linnaeus in honour of Guillaume Rondelet, a French naturalist, who died in 1566.)

This genus contains evergreen shrubs or trees. Leaves opposite, sessile or petiolate, stipulate sometimes ternately whorled, coriaceous, chartaceous or membranous. Inflorescence of terminal or anxillary corymbose cymes. Flowers coloured, white, red or yellow. Calyx adnate to ovary, shortly lobed. Corolla hypocrateriform; tube slender, usually short, glabrous or bearded in the throat; limb 5-lobed; lobes rounded. Style slender; stigmas exserted or included. Ovary 2-celled. Fruit a capsule.

Rondeletia odorata Jacq.

Sweet-smelling Rondeletia

(*Odorata* means sweet-smelling in Latin and refers to the fragrance of the flowers.)

Description.—A shrub reaching 6 ft in height; young parts covered with a setose pubescence. Leaves opposite, stipulate, ovate or elliptic-ovate in shape, subacutely acuminate, cordate or rounded at the base, 3-7 in. long, 1-3.5 in. wide, entire, chartaceous, the younger sparsely pilose on the nerves, at length glabrous, ciliate; petiole about 0.12 in. long setosely pilose. Stipules persistent, interpetiolar, ovate-lanceolate, obtuse or acute, coriaceous, up to 0.5 in. long, covered with appressed pubescence.

Inflorescence of many-flowered, terminal, corymbose cymes up to 5 in. diameter; bracts ovate-lanceolate, obtuse or subacute, up to 0.5 in. long, coriaceous, densely appressed pilose on the outer surface; bracteoles small. Calyx campanulate, 0.05 in. long, puberulous outside, with short obtuse lobes. Corolla with a cylindric tube 0.3 in. long, reddish orange in colour, puberulous outside, pilose within, 5-6-lobed; lobes 0.25-0.3 in. long oblong with rounded tip, rose-coloured. Stamens as many as the corolla lobes and alternate with them; filaments short, attached in the corolla throat about the centre of the tube; anthers 0.05 in. long. Style shorter or longer than the tube, 2-lobed, glabrous. Capsule subglobose, slightly 2-lobed, hairy.

Flowers.—Hot and rainy season. *Fruits.*—Cold season.

Distribution.—A native of the West Indies and Mexico, now commonly cultivated in the tropics of the whole world.

Gardening.—A handsome small hardwooded shrub about 3 ft high. It bears beautiful orange-scarlet flowers in constant succession, through the hot and dry seasons. The faded flowers remain on the plant for a long time giving it an unsightly appearance and should, therefore, be removed quickly. Propagated by layers during the rains, but it usually takes 3-4 months before they are ready for removal.

Hamelia Jacq.

(This genus was named in honour of H. L. du Hamel de Monceau, 1700-1782, a French botanist.)

Shrubs with slender glabrous or pubescent branches. Leaves opposite or in whorls of 3 or 4, petiolate, membranous. Stipules lanceo-

PLATE 11

SWEET SMELLING RONDELETIA
Rondeletia odorata Jacq.
(9/10 nat. size)

late, subulate, deciduous. Inflorescence in di- or tri-chotomous branched terminal cymes. Flowers shortly pedicellate. Calyx ovoid or top-shaped, 5-lobed. Corolla tube cylindrical 5-lobed; lobes imbricate in the bud. Stamens 5, inserted at the base of the tube; filaments very short; anthers fixed by the base, narrowly linear, appendaged at the apex. Ovary 5-celled; style filiform; stigma narrowly fusiform, grooved, somewhat twisted. Ovules numerous. Fruit a berry crowned by the remains of the disk

Hamelia patens Jacq.

(*Patens* means spreading in Latin and refers to the habit of the species.)

Description.—A shrub or small tree; branches reddish, covered with a short crisped pubescence, quadrangular when young, afterwards rounded. Leaves stipulate, petiolate, up to 5 in. long by 2.5 in. wide,

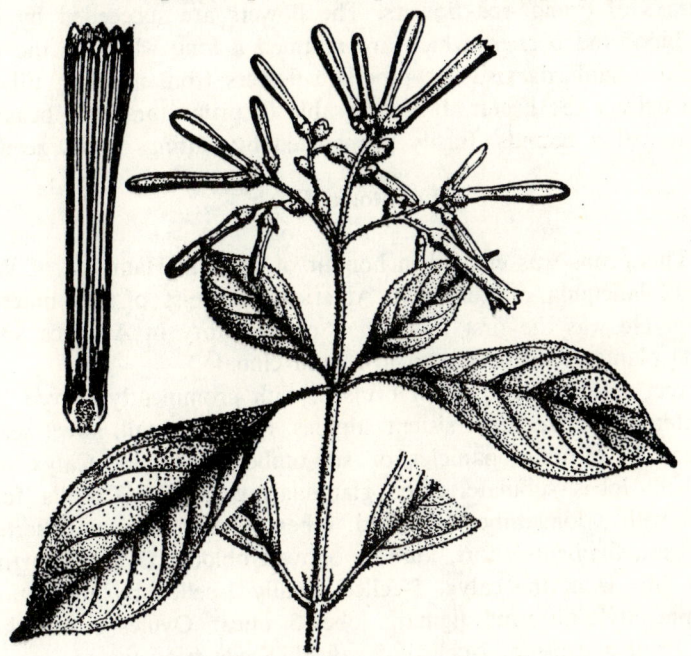

Fig. 61.—*Hamelia patens* Jacq. × ¾

elliptic or oblong in shape, narrowed at both ends, olive-green above, paler beneath, hairy on both surfaces, membranous, opposite or in threes; nerves 6-9 pairs, prominent and often tinged with red beneath; petiole 0.5 in. long, reddish, hairy; stipules broad at the base, produced into a stout linear awn.

Inflorescence terminal, in cymes up to 3 in. long; principal axis short; secondary axis up to 1 in. long bearing 3-5 flowers arranged in a bostryx or scorpioid cyme. Pedicels very short. Bracteoles minute. Calyx tube adnate to the ovary, campanulate, reddish in colour, produced above into 5 minute lobes, covered with very short hairs. Corolla tube cylindrical, 0.8 in. long, slightly attenuate just above the base, cylindrical, ridged (the ridges corresponding to the 5 lobes) shortly pubescent all over, 5-lobed; lobes very short, valvate in the bud. Stamens 5; filaments attached near the base of the tube, short; anthers about 0.5 in. long, linear acute at the tip, bifid at the base, included in the corolla tube. Ovary 5-celled, surmounted by a thick conical disk; style 0.8 in. long. Berry ellipsoid, 0.25 in. long, surmounted by the fleshy disk.

Flowers.—Hot and rainy seasons. *Fruits.*—Cold season.

Distribution.—Indigenous to tropical America; now commonly cultivated throughout the tropics.

Gardening.—A large evergreen shrub, prized for the profusion of its sprays of orange-red flowers. The flowers are succeeded by handsome blood-red berries, which are retained a long while on the plant. Numerous sunbirds visit the pipe-like flowers from morning till evening to extract the nectar. It is advisable to prune the plant heavily to keep it within bounds. Easily propagated by cuttings or by seed.

Hamiltonia Roxb.

(This genus was erected in honour of William Hamilton of Woodland, Philadelphia, an eminent American botanist of the nineteenth century. He was the first to build a conservatory in America so that tropical plants could be grown in a cold climate.)

Erect shrubs, foetid when bruised, with prominently nerved leaves and interpetiolar acute persistent stipules. Flowers small, sweet scented, arranged in terminal panicles or sub-umbellate cymes. Calyx ovoid, 4-5-lobed; lobes subulate, often glandular, persistent. Corolla funnel-shaped with a long tube, 4-5-lobed; lobes valvate. Stamens inserted in the throat, filaments short, anthers obovate-oblong. Ovary 5-furrowed, almost free from the calyx, 5-celled, finally 1-celled by absorption of the septa. Style filiform; stigmatic lobes 5, linear. Ovules 1 in each cell, basal. Fruit a capsule, 1-celled, 5-valved. Seeds triquetrous.

Hamiltonia suaveolens Roxb. (Spermadictyon suaveolens Roxb.)

(*Suaveolens* means sweet-smelling in Latin.)

Description.—Stem stout, shrubby, with ash-coloured bark covered with purple specks, reaching a height of 10 ft. Leaves opposite, 3-6 in. long, broad-lanceolate in shape, smooth, entire short-petioled. Stipules interpetiolar, ensiform.

PLATE 30

M. N. Bakshi

SWEET-SMELLING RONDELETIA
Rondeletia odorata Jacq.

PLATE 31

Sweet-smelling Rondeletia
Rondeletia odorata Jacq.

M. N. Bakshi

Flowers sessile in terminal corymbiform heads, on short trichotomous branches, pure white or mauve, delightfully scented, supported by linear glandular-villous bracts and bracteoles. Calyx tube ovoid; lobes 4-5, subulate, 0.1 in. long, covered with gland-tipped hairs. Corolla tube slender, 0.5-0.6 in. long, 4-5-lobed; lobes 4-5, oblong-obtuse, valvate in bud. Stamens 4-5; filaments very short, inserted in the throat of the corolla and alternate with the lobes. Ovary 5-furrowed, almost free from the calyx, 5-celled, but becoming 1-celled from absorption of the septa; ovules 1 in each cell, basal. Fruit a 1-celled capsule, 5-valved at the apex, 5-1-seeded.

Flowers.—Cold season. *Fruits.*—May-June.

Distribution.—Sub-Himalayan tract and outer hills from the Punjab to Bhutan, ascending to 6,500 ft, Bihar, and Western Peninsula.

Fig. 62.—*Hamiltonia suaveolens* Roxb. × $\frac{3}{4}$

Gardening.—A large stout shrub with lavender-blue or whitish sweet-scented flowers which are freely produced during the cold weather. It prefers a sheltered situation, and is greatly improved by being well cut in after flowering. Propagated by cuttings. The fragrant flowers are much frequented by Hummingbird Hawk-moths.

Medicinal uses.—A decoction of the root is said to be valuable in dysentery and cholera.

Catesbaea Linn.

(A genus erected by Linnaeus in honour of an English Botanist, Mark Catesby, 1679-1749, traveller and naturalist.)

Spinescent shrubs or small trees, with terete twigs and small, glabrous, often fascicled leaves. Stipules small, deciduous. Flowers white, solitary and short-pedicelled in the axils. Calyx subcampanulate, with 4 narrow subpersistent lobes. Corolla funnelform or campanulate, 4-lobed; lobes valvate. Stamens 4, attached near the base of the corolla. Ovary 2-celled; ovules few or many. Stigma 2-lobed. Fruit a berry.

A small genus of about 8 species.

Catesbaea spinosa Linn.

Prickly-apple; Spanish Guava

(*Spinosa* means spiny in Latin.)

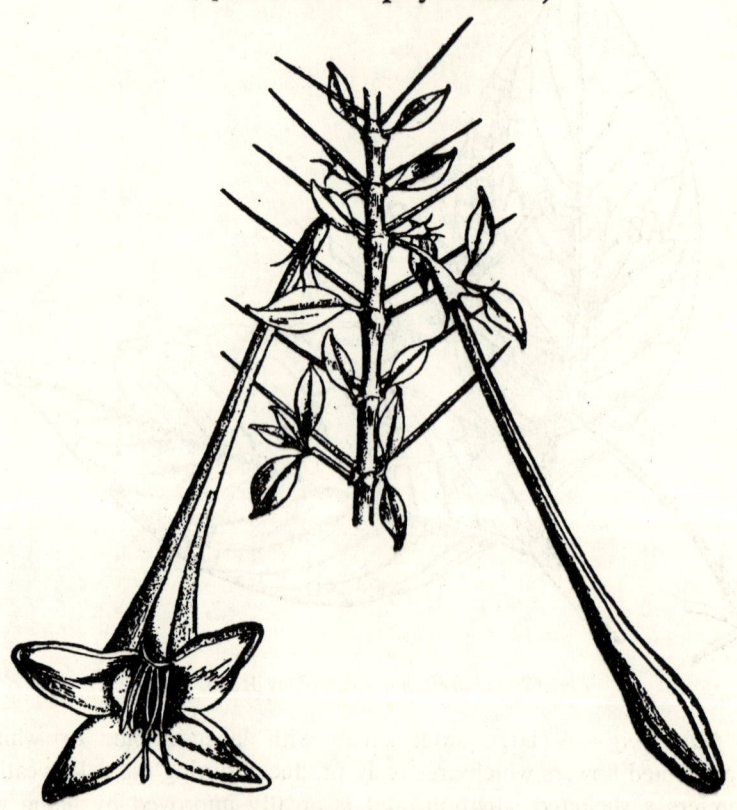

Fig. 63.—*Catesbaea spinosa* Linn. × ¾

Description.—A spinescent shrub reaching 4 ft in height at Dehra

THE SPREADING HAMELIA
Hamelia patens Jacq.

M. N. Bakshi

PLATE 33

M. N. Bakshi

THE SPREADING HAMELIA
Hamelia patens Jacq.

PLATE 34

M. N. Bakshi
THE SWEET-SCENTED HAMILTONIA
Hamiltonia suaveolens Roxb. (Cultivated form)

PLATE 35

THE SWEET-SCENTED HAMILTONIA
Hamiltonia suaveolens Roxb. (Cultivated form)

M. N. Bakshi

Dun. Old branches covered with corky bark, younger terete, green, minutely pubescent; spines axillary, opposite, stout, sharp, 0.5-1 in. long, pubescent, becoming glabrous. Leaves opposite, shortly petioled, stipulate, thin or somewhat fleshy, entirely glabrous and smooth, attenuate at the base into the short petiole, apiculate at the tip, ovate-elliptic or orbicular in shape, entire, glossy or dull green, 0.25-0.5 in. long; nerves inconspicuous; petiole very short; stipules interpetiolar, membranous, pubescent, rapidly deciduous.

Flowers solitary in the axils, pedicelled; pedicel 0.1 in. long, tinged with red. Calyx adnate to the ovary, green, sometimes reddish, oblong-campanulate, 4-lobed; lobes subulate. Corolla tube funnel-shaped, white turning yellow with age, 4-angled at the base, 3-4 in. long, pendulous, sparsely hairy outside, hairy and glandular inside, 4-lobed; lobes triangular, acute, valvate in the bud, 0.75 in. long. Stamens 4.5; filaments very long, attached at the very base of the corolla, rather stout, stiff, glabrous, attenuate at the tip; anthers linear, somewhat less than 0.5 in. long, divided at the base, apiculate at the tip. Ovary 2-celled. Style slender, long, 2-lobed. Fruit up to 2 in. long.

Flowers.—May-July. It seldom fruits in this country.

Distribution.—Indigenous to the West Indies, now commonly grown in gardens throughout the plains of India.

Gardening.—A rather slow-growing hardly armed shrub about 4-5 ft high. The creamy-white pendent flowers which are 3-4 in. long are outsize in proportion to the rest of the plant. Propagated by cuttings during the rains.

Chapter 6
ACANTHACEAE

The family gets its name from the Latin word for 'thorn', acanthus, which is also the name for one of the prickly genera of the family, a family which is mainly herbaceous but which does contain shrubs as well as a few small trees.

The characteristics of the family are as follows. Leaves opposite or rarely whorled, inserted on the swollen joints of the shoots, which are often quadrangular in section. The bisexual flowers are irregular with a 4-5-partite calyx. The corolla is obliquely 2-lipped or 5-lobed. The stamens number 4, or 2 with 2 staminodes. The 2-celled ovary is superior. The fruit is a woody capsule which often opens elastically with the two valves recurved. The orbicular seeds are seated on hard curved supports.

This is a large family of some 200 genera containing about 2,000 species. Many of the species are cultivated in gardens because of their beautifully coloured flowers. The leaves contain eystoliths or crystals of calcium oxalate which sometimes appear as translucent streaks in the fresh leaves, or raised lines in the dried leaves. It is, however, unusual to make these out in the living leaves.

Thunbergia Retzius

The genus *Thunbergia* was erected by Retzius in honour of Karl Peter Thunberg, professor of Botany at Upsala, who died in 1828.

The species of *Thunbergia* are mostly tall perennial climbers which are favourite garden plants on account of the beauty and profusion of their flowers. They are extensively cultivated in all parts of the world and in India, where several are indigenous. A good many species are hardy out of doors and are extremely decorative if grown to the best advantage.

CHARACTERS OF THE GENUS

The leaves are opposite, petiolate (sessile in *Thunbergia natalensis*) on the usually swollen joints of the stems, rarely subentire, more usually lobed or toothed in various ways; the nerves are, in most cases, palmate, i.e. they arise and spread from the top of the petiole. The flowers are large and showy and are arranged in terminal pendulous racemes or solitary or in pairs in the axils of ordinary leaves. Each flower is temporarily enclosed in a spathe consisting of two large bracteoles cohering along the margins. The calyx is very short, sometimes a mere rim, or, more often, crowned by 10 to 15 teeth. The corolla has a narrow or widely infundibuliform, often curved, tube ending in five

Plate 12

LARGE-FLOWERED THUNBERGIA
Thunbergia grandiflora Roxb.
(⅔ nat. size)

suborbicular lobes all spreading, or a companulate and strongly curved tube arising from the short cylindrical or conical base ending in five lobes of which the upper two are erect. The stamens are four in number, usually in pairs, with filaments of different lengths, inserted in the lower portion of the tube and included. The anthers are two-lobed and are glabrous or with a fringe of hairs along the margins, or the lobes are bearded at the base or furnished with a crest of hairs; sometimes one or both lobes spurred. The ovary is seated upon a large annular disk, which is often larger than the ovary. The ovary is 2-celled with 2 collateral ovules in each cell; style straight or curved, surmounted by a stigma which may be obconical in shape or, more often, distinctly

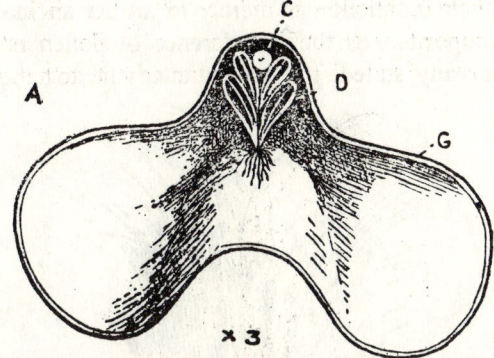

Fig. 64. A.—Cross section of corolla showing position of style and stamens; note the curvature of the corolla which ensures that the insect will touch the stamens.

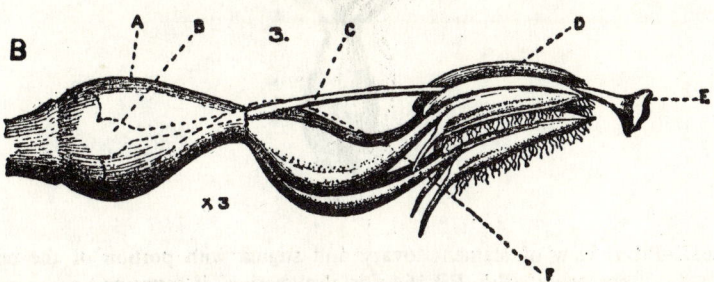

B.—Side view of stamens, ovary and style; corolla tube removed. A. base of corolla; B. ovary-seated on nectary disk; C. style; D. stamens; E. stigma; F. path to honey which must be followed by proboscis of insect; G. corolla.

2-lipped, with the upper lip erect and the lower spreading. The pollen grains are globose, without germ pores, but with one or more often spirally twisted grooves. The fruit is shaped rather like a bird's head and consists of a globose 2-celled capsule crowned by a stout beak. On dehiscence the beak splits violently from apex to base and the seeds are flung out. The seed is semiglobose in shape with a large pit on the inner side.

The brightly coloured flowers of the species of *Thunbergia* are an indication that the process of fertilisation is carried out through the agency of insects. The construction of the flowers, however, is such that only insects with certain characteristics are suitable for the purpose.

Taking the flower of *Thunberiga grandiflora* as an example, it will be observed that it can be divided into three portions; a lower conical portion which directly surrounds the ovary and nectary, a tubular-ventricose portion which carries the four stamens in a groove on its upper surface, and the spreading lobes. The spreading lobes are usually brightly coloured and their function is merely to attract an insect. The tubular portion is important as the transference of pollen must take place within it. As already stated, the four stamens lie together in a groove

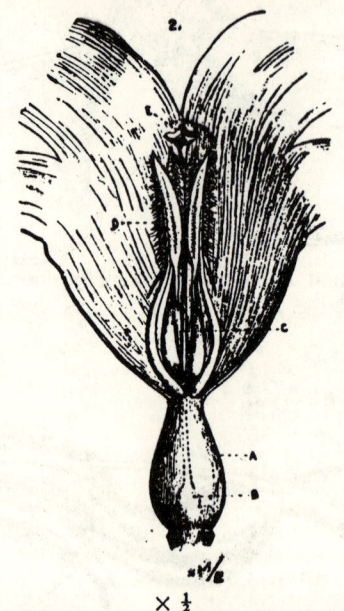

Fig. 65.—Front view of stamens, ovary and stigma with portion of the corolla cut away. (See Fig. 64 for explanation of lettering.)

on the upper surface of the corolla tube. On the lower surface of the tube inside is a bulge corresponding to a groove on the outside. This bulge ensures that the back of any visitor is pressed well up against the anthers on the roof. The anthers of the upper pair of stamens are pressed closely together and form a convenient groove in which lies the style. The anthers of the lower pair lie on either side of the upper pair. The style itself expands into the cup-shaped stigma a little in advance of the stamens. The filaments of the stamens though compressed are swollen at the base and the four of them are so placed that they practically block the entrance to the chamber in which the ovary is situated. A small foramen is left through which passes the style. The anthers are bearded with long thin hairs and bear one or two horny hooks on their posterior margins. The nectary which secretes the honey is a fleshy disk surrounding the ovary.

The first thing that happens when an insect visits the flower is that its dorsal surface touches the viscous lower lip of the stigma which is pressed down and any pollen which the insect is carrying is transferred. The insect pushes on still further and comes into contact with the hooks on the anther, which results in pollen being shaken down upon its back and thorax. Owing to the smallness of the hole at the base of the stamens and its distance from the honey only insects with a long proboscis can reach the booty. The proboscis must be passed upwards over the style and then downwards to reach the honey. As the insect retreats its pollen-dusted back presses up the lower non-viscous edge of the stigma and prevents self-fertilisation.

Burkill[1] observed the process of cross-fertilisation in this species in Calcutta and discovered that the agents were *Xylocopa latipes* and *X. aestuans*. The bodies of these bees just fit the antrum of the flower and the proboscis is sufficiently long to reach the honey. He observed that the dorsal surfaces of the insects were dusted with pollen on their emergence from the flower. A further refinement in the mechanism of pollination is that, when the stigma has two lips, the lower is non-receptive. This lip touches the back of the insect and drags down the upper receptive lobe which duly collects the pollen from the back of the visitor without any danger of self-fertilisation.

Key to the Species

Flowers axillary, solitary
  Petioles winged .... *T. alata*
  Petioles wingless
    Colour of flowers white .... *T. fragrans*
    Colour of flowers blue or purple
      Plant erect
        Leaves petiolate .... *T. erecta*
        Leaves sessile .... *T. natalensis*
      Plant climbing .... *T. grandiflora*
Flowers in terminal or axillary racemes
  Flowers blue
    Leaves broadly ovate, cordate at the base .... *T. grandiflora*
    Leaves ovate-oblong to lanceolate-oblong, rounded
      at the base .... *T. laurifolia*
  Flowers yellow .... *T. mysorensis*
  Flowers scarlet .... *T. coccinea*

---

[1] Burkill, I. H., *J.A.S.B.* (1906), pp. 511-14.

Thunbergia alata Boj.

(*Alata* is a Latin word meaning winged and refers to the wings on the petiole.)

Description.—A slender herbaceous twiner with very hairy shoots. Leaves opposite, seated on a narrowly winged petiole which is often as long as the blade, sagittate in shape, up to 2 in. long by 1.75 in. wide, acuminate, pubescent on both surfaces and hirsute on the nerves beneath; margin entire, undulate, with one or two broad teeth on either side. Flowers axillary, seated on peduncles which are shorter than the subtending leaves. Bracteoles spathe-like, ovate, apiculate, pubescent on both surfaces and hirsute on the margins, persistent, about 0.5 in. long. Calyx very short, bowl-shaped, surmounted by 10 subulate lobes, covered with glandular hairs. Corolla tube infundibuliform, 0.75 in. long and 0.2 in. wide in the throat, ending in five yellow orange or rarely white lobes, deep claret-coloured in the throat. Stamens four, one pair of filaments shorter than the other; anthers of the lower pair

Fig. 66.—*Thunbergia alata* Boj. × ¾

with two spurs, those of the upper with only one spur, densely fringed along the margin and at the base with long club-shaped hairs. Ovary seated in a cup-shaped disk, style 0.5 in. long; stigmas 2-lipped, the upper lip about twice as long as the lower but much narrower.

Flowers.—September-November. *Fruits.*—Cold season.

PLATE 36

M. N. Bakshi

THE WHITE THUNBERGIA
Thunbergia fragrans Roxb.

PLATE 37

M. N. Bakshi

THE WHITE THUNBERGIA
Thunbergia fragrans Roxb.

Distribution.—Native of south-east Africa. Cultivated or naturalised throughout India.

Gardening.—This twiner closely resembles *Thunbergia fragrans*, but is softly villous and has winged petioles and usually yellow flowers with a brown or purple eye or sometimes shades of buff or orange or even white. It is well suited for a small trellis-work. Propagated usually by seeds. There are several varieties but the following are more commonly seen in cultivation:

Thunbergia alata Boj. var. *alba* Paxt. has white flowers with a blackish centre.

Thunbergia alata Boj. var. *aurantiaca* Kuntze is a variety with bright orange flowers with a dark centre.

Medicinal use.—In Malaya the leaves are made into poultices which are applied to the head to relieve headaches.

Thunbergia fragrans Roxb.

(*Fragrans* is Latin for fragrant. Why this epithet should have been applied to this plant, which has odourless flowers, is not known.)

Fig. 67.—*Thunbergia fragrans* Roxb. × 1

Description.—A slender herbaceous twiner with scabrid, or more or less glabrescent, shoots. Leaves opposite on slender wingless petioles which are usually shorter than the blade, lanceolate, or triangular-ovate, cordate or subcordate at the base, almost hastate, up to 2.5 in. long by 1 in. wide, toothed on both sides at the base, more or less rough on both surfaces, palmately 5-nerved. Flowers axillary on peduncles which are usually longer than the subtending leaf. Bracteoles spathe-like, ovate or ovate-lanceolate in shape, less than 0.75 in. long. persistent. Calyx bowl-shaped, very short, surmounted by 15 subulate teeth, glandular-pubescent. Corolla tube infundibuliform, about 1 in. long, white and without fragrance; lobes spreading, about 0.75 in. in diameter. Stamens in pairs, the filaments of one pair as long as those of the other which are about 0.5 in. long; anthers entirely glabrous and without spurs. Ovary seated in a disk; style nearly 1 in. long; stigma 2-lipped, the upper lip slightly longer and narrower than the lower lip. Capsule scabrid. Seeds reticulate.

Flowers.—September-November. *Fruits.*—Cold season.

Distribution.—Throughout India and Malaya.

Gardening.—A herbaceous climber with beautiful snow-white flowers. Easily propagated by seed which it produces abundantly. It is suitable for growing over a screen. Contrary to what the specific name would seem to denote, the flowers are not fragrant.

Thunbergia erecta (Benth.) T. Anderson

(*Erecta* is Latin for upright and refers to the habit of the species.)

Description.—This species is a small shrub which in Indian gardens reaches a height of about 4 feet. The shoots are quadrangular in section

Fig. 68.—*Thunbergia erecta* (Benth.) T. Anderson × ½

and each angle bears a narrow wing, glabrous when old but the younger are covered with a mantle of short crisp hairs near the nodes; buds in the axils of the leaves densely covered with short golden or reddish hairs. Leaves opposite, exstipulate, but connected at the base by a raised ridge across the node, petiolate (petiole 0.1 in. long), ovate-elliptic, acute and apiculate, subacute at the base, glabrous, penninerved but the first pair often arise near the base; margin entire, undulate, or occasionally with a broad triangular tooth above the middle. Flowers axillary, solitary or paired, seated on peduncles up to 1.5 in. long, usually much less. At the top of the peduncle are a pair of greenish white bracteoles, spathe-like, caducous or deciduous. Calyx very short, bowl-shaped, surmounted by 15 subulate calyx-teeth, densely covered with gland-tipped hairs. Corolla tube 1.5-2.5 in. long, glandular outside, slightly conical at the base swelling above, infundibuliform, curved, ending in five subequal lobes; yellowish white at the base, central portion cream shading into the dense violet lobes and mouth, ochre-coloured in the throat. Stamens four, in pairs, the filaments of one pair 0.5 in. long, of the other 0.4 in. long, covered with short gland-tipped hairs. Anthers shortly awned or mucronate, with a thick fringe of hairs along the margins. Ovary seated in an annular disk. Style up to 1.5 in. long; upper lip of the stigma erect and folded, the lower spreading and very broad.

Flowers.—Chiefly during the cold weather.

Distribution.—Native of west tropical Africa, but is a very common cultivated plant in gardens throughout India.

Gardening.—A hardy shrub, very pretty when in full bloom during the cold season. It thrives best in bright sunshine and is well adapted to the topiarist's art. Easily propagated by cuttings during the rains. It was introduced into India from Kew in 1859.

Thunbergia erecta (Benth.) T. And. var. *alba* is a variety with white flowers, but it is by no means as beautiful as the type.

Thunbergia natalensis Hook.

(The specific name refers to the home of the plant.)

Description.—An erect shrub, 2 ft or more high, with four-angled stems, which are almost hirsute in specimens grown at Dehra Dun but are stated by Hooker, Bailey and others, to be glabrous except at the internodes. Leaves opposite, somewhat crowded, sessile or very shortly petiolate, ovate-acute or acuminate, sinuate-dentate, cordate at the base, palmately 3-nerved, slightly scabrous on both surfaces, glabrous above, hairy on the nerves below, 4 in. long by 2 in. broad. The axillary solitary flowers are seated on peduncles which are shorter than the subtending leaf-blade. Bracteoles at the top of the peduncles herbaceous, 3-nerved, hirsute on the nerves, ovate, spathaceous, nearly as long as the corolla tube, to which it is closely appressed. Calyx bowl-shaped with six blunt teeth. Corolla 2 in. long, shortly conical below, swelling above into an infundibuliform curved tube and ending in five obcordate nearly equal horizontally spreading lobes. Stamens four; filaments nearly equal, filiform, thickened towards the base; connective bluntly apiculate, anther cells shortly spurred and bearded below. Ovary seated in a fleshy disk. Style as long as the corolla tube, obconic, sup-

Fig. 69.—*Thunbergia natalensis* Hook. × ½

porting a concave stigma, glandular-hairy and bearded below the stigma. Capsule 1.25 in. long, densely hairy or glabrous.

Flowers.—April-May. Does not fruit in Dehra Dun.

Distribution.—Native of Natal; but now cultivated in gardens throughout India.

Gardening.—An undershrub up to 2 ft high with handsome pale blue flowers; the corolla tube yellow, 2 in. long. Propagated by cuttings.

Thunbergia grandiflora Roxb.

(*Grandiflora* means large-flowered in Latin.)

Description.—A large woody climber reaching a height of 20 feet or more. Shoots four-angled or -ribbed, shortly hairy, rough, becoming smoother and glabrous with age. Leaves opposite on the swollen nodes, petiolate (petioles up to half the length of the blade, twisted at the base and scabrid), cordate at the base, palmately 7-nerved, coarsely toothed or lobed, 4 in. long by 3-4 in. broad, ovate-acute or ovate-acuminate in shape, very rough on both surfaces. Flowers sometimes solitary in the axils of ordinary leaves, but more often, appearing as

PLATE 38

THE UPRIGHT THUNBERGIA
Thunbergia erecta (Benth.) T. Anders.

M. N. Bakshi

PLATE 39

M. B. Raizada

THE UPRIGHT THUNBERGIA
Thunbergia erecta (Benth.) T. Anders.

terminal pendulous racemes, pedunculate (peduncles up to 2 in. long). At the top of the peduncle will be found two velvety obliquely obovate or oblong bracteoles which are persistent. The calyx is reduced to a velvety rim. Corolla tube shortly conical at the base, then very widely and obliquely campanulate, up to 2.5 in. long, ending in five lobes, the upper two of which are erect and the others spreading. The upper surface of the corolla tube is grooved to take the stamens and style. Corolla whitish in the lower half shading into the blue lobes, yellow inside with blue stripes in the throat. Stamens subequal; filaments about 0.4 in. long, flattened and corrugated at the base where they almost occlude the entrance to the lower conical portion of the tube. Anthers of the lower stamens 0.3 in. long, both cells spurred; those of the upper 0.4 in. long with only one cell spurred; margins densely fringed with long thin hairs. Style 1 in. long; upper lip of the stigma folded and erect, the lower spreading. Capsule shaped like a bird's head, 1-2 in. long.

Flowers.—March-November. *Fruits.*—Cold season.

Distribution.—Native of Eastern Bengal. Commonly grown in gardens in the plains throughout the country.

Gardening.—A very extensive climber with pendent branches, and dark green heart-shaped leaves. It bears large bluish flowers from March right through the rains. It is of very luxuriant growth, and if allowed to climb a lofty tree will cover it with a dense green curtain of foliage. It can, however, be made to flower when small, by judicious close pruning. Easily propagated by cuttings or layers during the rains.

Medicinal use.—A decoction of the leaves is said to be used in Malaya for stomach complaints. The leaves are used as a poultice.

Thunbergia laurifolia Lindl.

(*Laurifolius* means having leaves like a laurel.)

Description.—A shrubby climber with terete smooth and glabrous stems. Leaves opposite, petiolate, oblong-lanceolate, acuminate, rounded at the base, glabrous and smooth or slightly rough on both surfaces, palmately 3-nerved at the base; reticulation prominent below, up to 7 in. long 2 in. broad; margins entire or distantly toothed; petioles up to 3 in. long, thickened at the apex and base. Inflorescence in pendulous axillary or terminal racemes. Individual flowers pedicellate. Bracteoles spathaceous, cohering along the upper margin, herbaceous. Calyx very small, cup-shaped; margin crenulate. Corolla tube cylindrical or broadly conical at the base, swelling above, obliquely funnel-shaped, very wide at the mouth. Lobes five, rotundate, emarginate, spreading. Stamens four, inserted near the base of the tube; filaments broad subulate curved. Anthers oblong, apiculate, with two subulate spurs at the base; margins fringed. Ovary globose, sunk in the crenatelymargined fleshy disk. Style long, included; stigma two-lobed; lobes channelled.

Flowers and Fruits.—Cold season.

Distribution.—Upper and Lower Burma, Andamans and the Malay Peninsula, now common in gardens throughout the country.

Gardening.—A large climber which bears during the cold season large lavender-blue flowers in profusion. It is suitable for growing over

Fig. 70.—*Thunbergia laurifolia* Lindl. × ¾

walls or strong trellis work. Yields seeds abundantly. Propagated by seed or layers.

Medicinal use.—In Malaya the juice of the leaves is said to be efficacious in cases of menorrhagia. It is also applied to the ears for deafness.

Thunbergia mysorensis T. Anderson

(*Mysorensis* refers to the home of the plant.)

Description. A glabrous twining shrub with slender glabrous often twisted stems. Leaves opposite, ovate-lanceolate or narrowly elliptic-acuminate in shape, 4-6 in. long by 1.25-2 in. broad, broadly cuneate at the base, sinuate, entire or toothed on the margins; basal nerves

3, prominent with conspicuous venation between; petioles 0.5-1.5 in. long. Flowers large, in long pendulous interrupted racemes. Bracteoles spathaceous, enclosing the corolla tube, ovate-oblong. 1 in. long, parallel-nerved, margins cohering in the bud. Calyx very shallow, salver-shaped; margins obscurely lobed or crenulate. Corolla tube purple, 2 in. long, shortly conical below, swelling above and ending in four lobes, the upper lobe erect with reflexed side lobes, the lower lip of 3 subequal spreading lobes. Lobes bright yellow or maroon, spotted with yellow or brown. Stamens four; filaments hairy at the base; anthers bearded with a short spur at the base of each cell. Ovary immersed in the thick fleshy disk; style long; stigma cup-shaped. Capsule 1.25 in. long, glabrous; seeds rugose.

Fig. 71.—*Thunbergia mysorensis* T. Anderson × ½

Flowers.—Cold season. Seldom fruits in Dehra Dun.

Distribution.—South India in the Western Ghats, South Canara and Mysore to Travancore and Tinnevelly up to 3000 ft. Cultivated in the plains throughout India.

Gardening.—An extensive glabrous climber with long slender branches; flowers in long pendent racemes bright yellow or orange, or maroon with a purplish tube. Usually shy of seeding and has to be propagated by layering.

Thunbergia coccinea Wall.

(*Coccineus* is a Latin word meaning scarlet-coloured, and refers to the bright scarlet flowers of this species.)

Description.—A slender climbing and widely spreading twiner. Stems and shoots 4-angled; the angles shortly winged, smooth and glabrous. Leaves opposite, shortly petioled (petiole twisted), cordate at the base, ovate-lanceolate or ovate, acuminate, the lower variously toothed, the upper entire on the margin, undulate, palmately 5-nerved, smooth and glabrous, green above, somewhat glaucous beneath, up to 5 in. long, 3 in. broad; nerves prominent on the lower surface.

Flowers in terminal or axillary pendulous racemes, up to 3 feet long. Individual flowers shortly or long-pedicelled. Bracteoles 0.5-0.75

Fig. 72.—*Thunbergia coccinea* Wall. × ¾

in. long, large, spathaceous, cohering along the margins, concave, ovate-acuminate in shape, including the whole flower except the lobes which are reflexed back over them, brown in colour. Calyx bowl-shaped sur-

PLATE 40

THE LARGE-FLOWERED THUNBERGIA
Thunbergia grandiflora Roxb.

M. N. Bakshi

mounted by 10 or more blunt triangular teeth. Corolla tube infundibuliform, orange in the throat, enclosed in the bracteoles, 5-lobed, emarginate, scarlet, lobes being reflexed over the exterior margins of the spathe. Stamens four, in pairs, the filaments dilated at the base; anther cells spurred, the lower with two spurs, the upper with one only. Ovary seated in a disk. Capsule globular; beak short, blunt.

Flowers.—December-March. *Fruits.*—March-April.

Distribution.—Outer Himalayas from Kumaon eastwards, Khasia hills, Tenasserim, commonly cultivated in gardens throughout India.

Gardening.—A widely spreading climber with long pendent branches. It bears during the cold weather scarlet or orange flowers in lax pendulous racemes about a foot or more in length. Usually multiplied by layers during the rains.

Chapter 7
SOLANACEAE

The Deadly Nightshade Family

Solanaceae is derived from *Solanum,* a name used by Pliny to designate the well-known Bitter-sweet (*Solanum dulcamara* Linn.), a poisonous plant found in waste places in Europe.

The *Solanaceae* comprise herbs, erect or climbing shrubs, or even small trees. The leaves are usually alternate, entire, or lobed or cleft in various ways, exstipulate. The inflorescence is either lateral or terminal and the flowers are cymose or panicled or even solitary, regular, pentamerous or tetramerous. The corolla is gamopetalous, short and rotate, or long and funnel-shaped, rarely campanulate, more or less deeply lobed, sometimes entire; lobes 4-5-10. Stamens 4-5, inserted on the corolla tube; the filaments are very short and the anthers are either attached by the base or by the dorsal surface. Dehiscence is by pores or longitudinal slits. The ovary is in most cases 2-locular, rarely with 3-4-5 compartments; ovules many, on prominent peltate placentas; style linear; stigma capitate or shortly lobed. Fruit a berry or capsule, many-seeded. Seeds compressed, discoid or kidney-shaped.

The family contains many plants cultivated for food, e.g. potato, tomato, Cape gooseberry, capsicum and tree tomato. The tobacco plant, *Nicotiana tabacum* Linn., is widely cultivated in this country. The family also contains many well-known ornamental plants among which are *Browallia, Petunia, Solanum, Brunfelsia, Datura, Cestrum, Schizanthus* and *Salpiglossis.*

Apart from the value of the family to horticulturists, many species found wild in nature are prized as drug plants. The Deadly Nightshade, *Atropa belladonna* Linn., derives its specific name from the fact that in bygone days ladies in Italy used the leaves to impart lustre to their eyes and so enhance their beauty. The plant actually contains a drug (atropine) which is used for ophthalmic complaints, for neuralgia and as a valuable antidote in opium poisoning. One of its effects is to enlarge the pupil of the eye. The berries are sweet and poisonous and are often eaten by children with fatal results. The name *Atropa* is from the Greek, *Atropos,* one of the Fates who cut the thread of life, and has reference to its deadly poisonous nature. *Belladonna* is found wild in India on the Himalayan ranges of the Punjab, Kashmir and Lahul.

The Henbane (*Hyoscyamus niger* Linn.) is a common plant of rather high altitudes in the western Himalayas. It is supposed to be fatal to fowls hence the trivial name. In the middle ages in Britain it was used for toothache. The alkaloids it contains are narcotic and antispasmodic.

Two species of *Withania* Pauq., *W. coagulans* Dunal and *W. somnifera* Dunal, are found in India. The former is found in the Punjab, Sind and Baluchistan. The seeds have the interesting property of being able to coagulate milk, and hence can be used as a substitute for animal rennet. The fresh fruit can be used as an emetic. *W. somnifera* Dunal, found in Bombay, possesses an alkaloid which is used in rheumatism, as an aphrodisiac and as a cure for scorpion sting.

Datura is another genus with poisonous properties. An account of these properties will be found under the genus.

No account of the *Solanaceae* would be complete without some mention of the Mandrake, a plant famous in legend and history. This genus of plants, *Mandragora*, has been for many centuries, and is even

Fig. 73.—Harvesting the Mandrake
(*From a drawing in the Nürnberg Museum*)

today, regarded with superstitious awe. All this superstition has arisen because the thick fleshy root bears a remote resemblance to a human figure. To that resemblance can be directly traced the belief that the plant shrieks when pulled from the earth.

Apart from this, however, it was known in ancient times as an aphrodisiac and also as an anaesthetic. According to F. W. Jones[1], 1935, it is a well-attested fact that Hua T'o, the great Chinese surgeon, who was born in the first century after Christ, performed major operations (such as the removal of the spleen) upon patients to whom he had administered a drug that produced insensibility. Part of the

---

[1] Jones, F. W., Syme Oration, 'The Master Surgeon', 1935

fame of that fabled plant, the mandrake, rests upon the fact that it was the first anaesthetic discovered by man. Its properties have been known for over three thousand years, but only comparatively recently has it been shown that it contains hyoscine, hyosyamine and scopolamine as well as other soporific alkaloids. In ancient medicine this plant was used to produce deep sleep and complete insensibility.

The roots of this plant, and the tuberous roots of certain other plants, are even at the present day carved to resemble more nearly a human figure, and are hoarded by the superstitious as powerful magical charms.

As can be expected, a plant which had these valuable medical properties and which also possessed certain human characteristics, acquired in olden times the most extravagant reputation for magical powers. Theophrastus, for instance, in his history of plants, says that it is dangerous to gather it unless a sword is swung in a circle three times, and the face turned towards the west before the plant is lifted from the ground.

But listen to what Josephus wrote in the first century after Christ in his history of the Jewish war: 'In the valley in which the town of Macharus is situated, is found a wonderful root called Baara. In colour it is a fiery-red and in the evening it emits rays of light. It is difficult to pluck it because it avoids the hand of anyone who goes near, and in any case to touch it means certain death. There is a way, however, in which it can be collected quite harmlessly. You must dig away the soil from around the plant so that it is only attached to the earth by a single rootlet. A dog is taken and tied to the root. Its owner then calls it and it will run to him, at the same time severing the rootlet. It immediately falls dead and then the plant can be handled without danger. Men take such great trouble because of its wonderful properties. It can even drive out evil spirits if it is brought near a sick person.' Baara was an early name of the mandrake. The accompanying sketch shows the conception of a mediaeval artist of the way in which the plant should be collected. It follows very closely the procedure recommended by Josephus.

Key to the Genera

Flower trumpet-shaped or if salver-shaped with a short tube
    Fruit a drupe
        Corolla tube longer than the lobes .... *Cestrum*
        Corolla tube shorter than the lobes .... *Solanum*
    Fruit a capsule .... *Datura*
Flower salver-shaped with a long tube .... *Brunfelsia*

PLATE 41

THE GOLDEN CESTRUM
Cestrum aurantiacum Lindl.

M. N. Bakshi

PLATE 42

THE GOLDEN CESTRUM

M. N. Bakshi

Cestrum Linn.

The name *Cestrum* is derived from the Greek word *kestron*, which was used by Dioscorides in his manual on medicinal plants to designate a Labiate, *Stachys officinalis* or *S. alopecuroides*. The word itself means pointed iron, or the style of the ovary, and may refer to the pointed anthers of these plants. Why Linnaeus applied the name *Cestrum* to this genus is not known.

This genus comprises shrubs or small trees. Leaves alternate, entire, petioled or subsessile, exstipulate, often with an unpleasant odour when bruised. Inflorescence in axillary or terminal cymes or panicles. The calyx is usually short, campanulate or cylindric, 5-lobed or 5-partite. Corolla usually fragrant, green, white, yellow, orange or even reddish; tube elongate, cylindrical or narrowly infundibuliform, sometimes bell-shaped at the mouth and then contracted, 5- or more-lobed or 5-partite; lobes spreading, much shorter than the tube, with the margins often incurved, glabrous or hairy. Stamens 5, included. Filaments adnate to the corolla for a large portion of their length, often dilated or hairy at the point of insertion and provided with an appendage of variable shape; anthers globular, dehiscing longitudinally. Ovary 2-celled; ovules few; style filiform, more or less glandular at the summit; stigma capitate. Fruit an oblong or ovoid berry, black, purple or rarely white.

The Cestrums are natives of South and Central America, but are now cultivated in the open in many parts of the warmer regions of the world or under glass in temperate latitudes.

When flowering the flowers are usually to be found pendulous, a position which ensures that the pollen is not wetted and rendered ineffective by rain.

Owing to the fact that the flowers of *Cestrum* are fragrant, it is to be inferred that this is a device to secure cross-fertilisation. In the case of some species the flowers are only fragrant at night which leads one to believe that cross-pollination is effected by a night-flying insect.

If cross-fertilisation is not accomplished the flowers still possess a mechanism which ensures self-pollination. When the stigma has been fertilised the style decays. On the other hand when the anther cells have opened the corolla becomes detached from the receptacle. Should the style have decayed as a result of fertilisation the corolla can drop off. If, however, the style is still intact the corolla is prevented from dropping off by the appendages at the base of the filaments. These appendages vary in shape but are often tongue-like processes which curve inwards round the style. The corolla on becoming detached is prevented from dropping off immediately by the tips of the appendages becoming locked under the capitate stigma. There is, therefore, every likelihood of self-fertilisation taking place even though cross-fertilisation has not been accomplished through the agency of insects.

Key to the Species

```
Flowers reddish .................................................. C. elegans
Flowers white, green or varying shades of yellow
    Lobes of corolla blunt, becoming distinctly reflected
        Corolla orange-yellow .......................... C. aurantiacum
        Corolla white ...................................... C. diurnum
    Lobes acute, erect or spreading
        Leaves oblong-ovate, shortly acuminate, not foetid
            when bruised ................................. C. nocturnum
        Leaves lanceolate, very long tapering, very foetid
            when bruised ..................................... C. parqui
```

Cestrum elegans Schlecht

Description.—A large shrub with pendulous leafy branches; branches cylindrical, covered with a soft velvety pubescence. Leaves alternate, exstipulate, shortly petioled, lanceolate or oblong-lanceolate in shape,

Fig. 74.—*Cestrum elegans* Schl. × ¾

quite entire, tapering to a point, 3-4 in. long, up to 1 in. wide, covered on both surfaces with short, crisped hairs, particularly so on the nerves beneath, deep green in colour, membranous in texture, rounded, acute

PLATE 13

THE GOLDEN CESTRUM
Cestrum aurantiacum Lindl.
(nat. size)

or obscurely cordate at the base; petiole 0.1 in. long, pubescent.

Inflorescence in axillary or terminal dense pendulous thyrsoid compound racemes. Individual flowers subtended by hairy bracteoles, sessile or subsessile. Calyx 0.2 in. long, top-shaped, green, hairy, 5-lobed; lobes broadly triangular-ovate, acuminate, erect. Corolla purplish red, tubular, gradually inflated then contracted below the mouth, glabrous, 0.75 in. long. Lobes 5 or more, triangular, acute, very shortly ciliate on the margins; mouth and inner surface glabrous. Stamens as many as the corolla lobes and alternate with them; filaments inserted two-thirds the way down the tube, filiform; anthers included, small, yellow. Ovary globose, shortly stipitate, seated on a disk, 2-celled, ovules few on the axile placentas; style slender; stigma capitate. Fruit a berry, 0.5-0.75 in. in diameter, fleshy, deep red-purple in colour, 2-celled, many-seeded.

Flowers.—September-December.

Distribution.—Native of Mexico, largely grown in gardens in temperate countries.

Gardening.—A shrub eminently suitable for growing in hill stations. The red-purple flowers which are borne in dense, terminal drooping cymes are produced almost continuously throughout the year and make this shrub a very attractive object. It prefers partial shade and a rich well-drained soil and is propagated by cuttings.

Cestrum aurantiacum Lindl.

(*Aurantiacum* is a Latin word meaning orange-red, and refers to the colour of the flowers.)

Description.—An erect or scrambling shrub, glabrous or young parts puberulous. Leaves alternate, stipulate, petioled, ovate-acute or ovate-acuminate in shape, membranous in texture, entire, glabrous, dark green in colour, somewhat undulate on the margins, 3-4 in. long, up to 2.5 in. broad; petiole 0.75 in. long.

Inflorescence axillary or terminal in pedunculate racemose or panicled clusters; peduncles often pubescent. Lower flowers pedicelled, the upper sessile or subsessile, bracteolate; bracteoles leaf-like, narrow, lanceolate, puberulous. Calyx gamosepalous, 0.25-0.3 in. long, cylindrical-campanulate, 5-ribbed, each rib continued above into a linear awl-like lobe. Corolla tube constricted at the base, obconical in shape, orange-red in colour, 0.9 in. long, 0.3 in. wide at the top, ribbed; ribs double the number of the lobes. Lobes ovate-obtuse, completely reflexed at full anthesis, 5-7 in number. Stamens corresponding in number to the lobes and alternate with them; filaments inserted in the tube, adnate to the lower half of the tube, free in the upper half, included, appendaged; appendage bracket-like, glandular. Ovary seated on an obscure fleshy disk; style slender; stigma capitate.

Flowers.—October-December. Does not fruit in this country.

Distribution.—Native of Guatemala, now commonly cultivated in the warmer and temperate regions of the whole world.

Gardening.—A pretty bushy shrub with scentless orange flowers borne in panicles. It will grow up to 10 ft or so and can withstand considerable moisture. Six weeks after flowering the plant should be well cut back in order to obtain a shapely bush and induce profuse flowering. It prefers rich soil and is propagated by cuttings. This plant forms a very pretty green-house shrub and as its flowers do not drop off easily it has become a great favourite in all European gardens.

Cestrum diurnum Linn.

The Day Jasmine

Description.—An erect shrub with numerous leafy branches. Branches green (fawn in age) with well-marked white lenticels; young parts covered with a very sparse glandular scruf. Leaves alternate, ex-

Fig. 75.—*Cestrum diurnum* Linn. × 1

stipulate, petiolate, dark green above, pale below, glabrous, entire, obtuse at the apex, obtusely wedge-shaped below, ovate-lanceolate in shape, up to 5 in. long by 1.5 in. wide; petioles up to 0.5 in. long.

PLATE 43

M. N. Bakshi

THE DAY JASMINE
Cestrum diurnum Linn.

PLATE 44

THE DAY JASMINE
Cestrum diurnum Linn.

M. N. Baksh

Inflorescence consisting of a long axillary peduncle bearing short clusters of white sweet-smelling flowers, each cluster supported by a leaf-like bract. Individual flowers sessile, with or without bracteoles. Calyx gamosepalous, about 0.15 in. long, somewhat puberulent, obtusely 5-ribbed, 5-lobed; lobes obtuse, ciliate. Corolla tube narrowly infundibuliform, white, sweet-scented, about 0.5 in. long, 5-lobed; lobes very obtuse, completely recurved when the flower is fully open. Stamens oblong, 5 in number, alternate with the corolla lobes, brown in colour, included; filaments adnate to the tube, free for a very short distance. Ovary seated on a nectary-secreting disk; style filiform, glabrous; stigmas truncate-capitate. Berry nearly globular, black.

Flowers.—Rainy season. *Fruits.*—Cold season.

Distribution.—Native of the West Indies, widely cultivated in gardens throughout the country.

Gardening.—A quick-growing evergreen shrub with dark green foliage and white flowers that are very sweet scented during the day. It is well suited for screens and borders. Easily propagated by seed which it produces abundantly. It is advisable to prune it after flowering so as to prevent it from becoming ragged.

Cestrum nocturnum Linn.

Lady of the Night, Night Jessamine, Rat-ki-rani (Hind.)

(The specific name refers to the fact that this plant opens its flowers at night.)

Description.—A glabrous shrub reaching a height of 9 ft. Branches slender, smooth, often yellowish with numerous lenticels. Leaves alternate, exstipulate, ovate-oblong, acute at the apex, cuneiform or rounded at the base, membranous in texture, entire, glabrous, up to 4 in. long by 1.5 in. wide; petioles up to 0.5 in. long.

Inflorescence in many-flowered pedunculate axillary or terminal panicles, exceeding by much the length of the leaves; common peduncles up to 1 in. long, erect or spreading. Individual flowers supported by a lanceolate bract, pedicelled or subsessile. Calyx campanulate, sparsely glandular, 0.15 in. long, glabrous, with 5 small triangular teeth, which are acute, scarious and short ciliate at the tips. Corolla greenish, sweet-smelling at night; tube cylindrical or narrowly infundibuliform, about 0.75 in. long, 0.1 in. broad at the top; lobes 5, ovate-acute, margins incurved; stamens 5, alternate with the lobes of the corolla; filament adnate to the corolla tube for a quarter of its length from the top, provided at the junction with a linear-acute appendage. Ovary globular, seated on a fleshy disk which probably secrets nectar; style filiform, glabrous; stigma capitate, above the stamens. Fruit a blue or blackish berry, ovoid in shape, about 16 in. long. Seeds numerous compressed.

Flowers.—Practically throughout the year, but most profusely during the rains. *Fruit.*—Cold season.

Distribution.—Indigenous to the West Indies. Largely cultivated throughout the plains of this country.

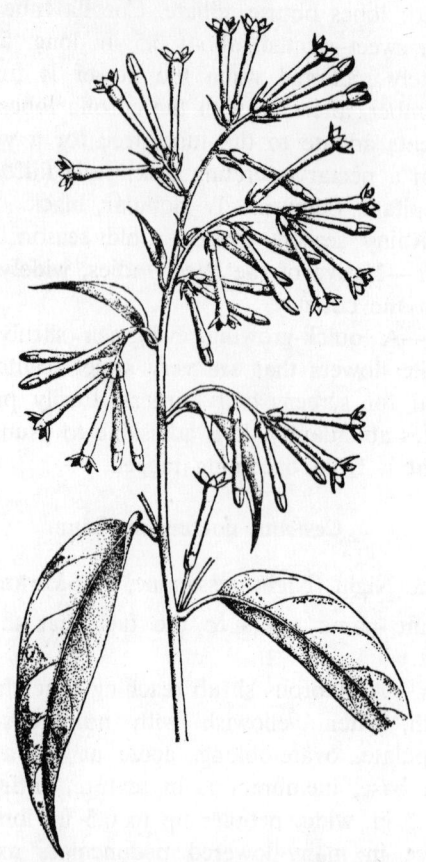

Fig. 76.—*Cestrum nocturnum* L. × ¾

Gardening.—A hardy subscandent, quick-growing shrub about 10 ft high. The small, greenish and rather inconspicuous flowers, which are produced in great profusion open at night, and are strongly sweet-scented. It is a great favourite with Indians and is extensively cultivated in their gardens. This shrub is well adapted for tall borders and screens and can easily be trained on a trellis. It is hardy and drought resistant. Very easily propagated by cuttings.

Cestrum parqui L'Héritier
Willow-leaved Jasmine

(*Parqui* is the Chilean name of this plant and was once proposed as the generic name.)

PLATE 45

LADY OF THE NIGHT
Cestrum nocturnum Linn.

M. N. Bakshi

PLATE 46

LADY OF THE NIGHT
Cestrum nocturnum Linn.

M. N. Baks

PLATE 47

M. N. Bakshi

WILLOW-LEAVED JASMINE
Cestrum parqui L'Herit.

PLATE 48

M. N. Bakshi

WILLOW-LEAVED JASMINE
Cestrum parqui L'Herit.

SOLANACEAE 133

Description.—A shrub reaching 4-5 ft in height. Branches cylindrical, glabrous, greenish, covered with small greenish white lenticels. Leaves petioled, alternate, exstipulate, lanceolate, long tapering at the apex, up to 6 in. long, 0.5-0.75 in. wide, very foetid when bruised, entire, glabrous, margins undulate; petioles short, scarcely 0.1 in. long.

Inflorescence in pedunculate umbelliform cymes or panicles, many-flowered, not exceeding the leaves in length. Individual flowers sessile, supported by a linear bract. Calyx gamosepalous, cylindrical, sparsely glandular without, about 0.1 in. long, with 5 acute, triangular, ciliate teeth. Corolla gamopetalous, fragrant at night, glabrous, pale yellow in colour, about 0.75 in. long, narrowly infundibuliform in shape, 0.15

Fig. 77.—*Cestrum parqui* L'Hérit. × ½

in. wide at the top. Lobes ovate-acute, ciliate on the margins which become revolute with age, slightly tinged with purple on the outside. Stamens 5, alternate with the lobes of the corolla; filaments inserted on the corolla about one-third its length from the top; stamens globose; appendage absent. Ovary globose, seated on an inconspicuous disk; style elongate, glabrous; stigmas truncate-capitate.

Flowers.—More or less throughout the year. *Fruits.*—Cold season.

Distribution.—Native in the mountains of Chile, South America, now largely cultivated in the tropical and sub-tropical parts of the globe.

Gardening.—A shrub about 4-5 ft high, with leaves that have a foetid smell when bruised. The greenish yellow flowers are very fragrant at night. It was introduced into England in the year 1787. Easily propagated by seed.

Solanum Linn.

(This plant name is mentioned by Celsius in his *De Medicina*, published in the first century. In the Middle Ages such names as *Solatrum, Solaticum*, etc. were applied to *Solanum nigrum*. Other authors derive it from *Solamen*, a Latin word meaning solace or quieting, which may have a reference to the narcotic properties of some species of the *Solanaceae*.)

This genus contains a vast group (over 2000) of temperate and tropical herbs, erect or climbing shrubs and even trees, glabrous or woolly, often spiny. Leaves usually alternate, entire or lobed, rarely pinnate. Inflorescence of solitary flowers or of pedunculate or subsessile fascicles of terminal or axillary cymes. Calyx funnel-shaped, angled or not, rarely entire, usually lobed, slightly or not enlarging in the fruit. Corolla rotate with a very short tube, 5-10-(4-6-) lobed; lobes glabrous or more often woolly, flat or shallowly channelled. Stamens the same number as the corolla lobes; filaments short, inserted in the throat; anthers erect, contiguous, forming a tube around the style, dehiscing by pores or by very short slits at the top of the anthers. Fruit a two-celled round berry; seeds numerous.

The genus *Solanum* contains a number of food plants of which *S. tuberosum*, the potato, is perhaps the best known. The egg-plant, *S. melongena*, and the tomato, *S. lycopersicum*, also belong to the genus. It may be, perhaps, well to mention here that brinjal and tomato are the actual fruits of the plant, while the potato is a tuber developed on the roots.

Several Indian species of *Solanum* are poisonous on account of the presence of an alkaloid, solanine, in their tissues. The drug is said to be present in the potato close to the so-called 'eyes'. It is found, however, that cooking renders these poisonous plants innocuous. The leaves of several species of *Solanum* are smoked in a *chilam* like tobacco and are considered to give relief from toothache, the belief being that the smoke kills the germs which are supposed to cause the pain.

A good many species, both indigenous and introduced, are cultivated in gardens for their handsome flowers and often spiny foliage.

KEY TO THE SPECIES

Shrubs, young parts more or less pubescent
    Flowers small, less than 0.5 in. in diameter, white ...... *S. pseudo-capsicum*
    Flowers 0.75 in. or more in diameter, violet with a central
        yellow spot .... *S. rantonnetii*
Climbers
    Flowers 1.5-2 in. in diameter .... *S. wendlandii*
    Flowers under 1.5 in. in diameter
        Lower leaves pinnatisect .... *S. seaforthianum*
        Lower leaves lobed, but not pinnatisect .... *S. jasminoides*

SOLANACEAE

Solanum pseudo-capsicum Linn.
Jerusalem Cherry

(*Pseudo-capsicum* means false *Capsicum* and bears witness to this plant's similarity to that genus.)

Description.—A small leafy shrub reaching a height of 3-4 feet. Branches slender, cylindrical, glabrous, green. Leaves alternate, bright

Fig. 78.—*Solanum pseudo-capsicum* Linn. × 1

green and shining, shortly petiolate, lanceolate, oblong-lanceolate, oblanceolate or narrowly elliptic, narrowed gradually at the base into the short petiole, obtuse or obtusely acute at the apex, membranous in texture, glabrous or minutely pubescent and smooth on both surfaces, sinuate on the margins; nervation prominent below.

Flowers solitary or in few-flowered fascicles, seated on pedicels up to 0.5 in. long. Calyx 0.25 in. long, turbinate, glabrous, 5-lobed; lobes linear-acute, 0.1 in. long or more. Corolla white, up to 0.5 in. in diameter, star-shaped, rotate, 5-lobed; lobes apiculate. Stamens 5, alternate with the corolla lobes; filaments short; anthers erect, oblong, opening by apical pores, at any rate at first. Style slender, glabrous. Fruit a yellow or scarlet globose berry, 0.5-0.75 in. in diameter, which persists for a long time.

Flowers.—August. *Fruit* ripens in May.

Distribution.—Native country uncertain. Indigenous or naturalised in Australia, South Africa, China, Madeira and Brazil. Comparatively recently it has also been naturalised in various parts of the Dehra Dun.

Gardening.—A low erect undershrub with white or purple flowers borne extra-axillary, solitary or in few-flowered, umbellate cymes. The flowers are followed by scarlet berries which persist on the plant for a long time. It is an old-fashioned plant and is grown usually in pots for its scarlet fruit. Cultivation is by seed.

Solanum rantonnetii Carr.

Fig. 79.—*Solanum rantonnetii* Carr. × 1

(This plant was named in honour of Rantonnet, a French horticulturist, who published several articles on species of *Solanum* about the middle of the nineteenth century.)

Description.—An erect shrub reaching 3-4 feet in height. Stem covered with a dense short greyish or yellowish tomentum, almost quadrangular from the raised lines upon it which appear to be decurrent from the base of each side of the leaf-petiole. Leaves alternate, petiolate, lanceolate to ovate, up to 2.5 in. long, 1.5 in. wide, acute at the tip, attenuate at the base, covered with a short stiff pubescence on the upper surface or sub-glabrous; pubescence marked on the prominent nervation of the lower surface; petiole pubescent, 0.25-0.3 in. long.

Flowers solitary or fascicled in the axils of the leaves, 1 in. or more in diameter, voilet in colour with a central yellow spot. Calyx 0.1 in. long, turbinate, pubescent, truncate at the top and furnished with 5 linear processes. Corolla rotate, 5-lobed; lobes apiculate, covered with a short pubescence on the lower surface near the margins. Stamens 5; filaments slender, alternate with the corolla lobes; anthers thick, oblong, bright yellow, erect. Style slender, glabrous. Fruit about 1 in. in diameter, drooping, scarlet.

Flowers.—May-August. Does not fruit in Dehra Dun.

Distribution.—Indigenous to the Argentine, widely cultivated in the plains throughout the country.

Gardening.—An attractive erect shrub growing up to 6 ft. It has a pretty deep green foliage and violet coloured flowers with a yellow centre. The flowers are followed by red fruits an inch in diameter which make the plant very ornamental. The plant is not particular as to soil conditions. Frequent pruning will improve its appearance as the old branches become straggly and leafless. Easily grown from cuttings.

Solanum wendlandii Hook.

Giant Potato Vine

(This magnificent climber was named in honour of Dr. Wendland, Director of the Botanical Gardens, Herrenhausen, Hanover, who first sent it to Kew in 1882. Its home, however, is in Costa Rica.)

Description.—A climbing shrub. Branches rather fleshy, cylindric, green, glabrous. Prickles on the stems, branches and petioles few in number, stout, recurved. Leaves alternate, very variable, the uppermost simple, oblong or elliptic, acuminate or apiculate at the tip, rounded, cuneate or cordate at the base, entire on the margin, membranous, glabrous, 1-4 in. long by 2 in. wide; the lower up to 10 in. long by 4 in. wide or more, divided or cut in various ways, pinnate with the terminal leaflet very large or deeply lobed, membranous, glabrous; midrib with a few recurved prickles below; petiole glabrous, 2-4 in. long, with 1 or 2 prickles.

Inflorescence of terminal pendulous cymes, 6 in. wide or more. Calyx campanulate, 0.3 in. long, glabrous outside, 5-lobed; lobes 0.1 in. long, ovate or oblong-apiculate, spreading in the opened flower, minutely hairy on the margins. Corolla lilac-blue or mauve in colour, up to 2 in. across, rotate, shallowly 5-lobed; lobes apiculate, hairy, induplicate in the bud. Stamens 5, alternate with the corolla lobes; filaments short, flat, inserted in the throat; anthers bright yellow, opening by slits, loosely connivent round the style. Ovary hairy, 2-celled. Fruit globose.

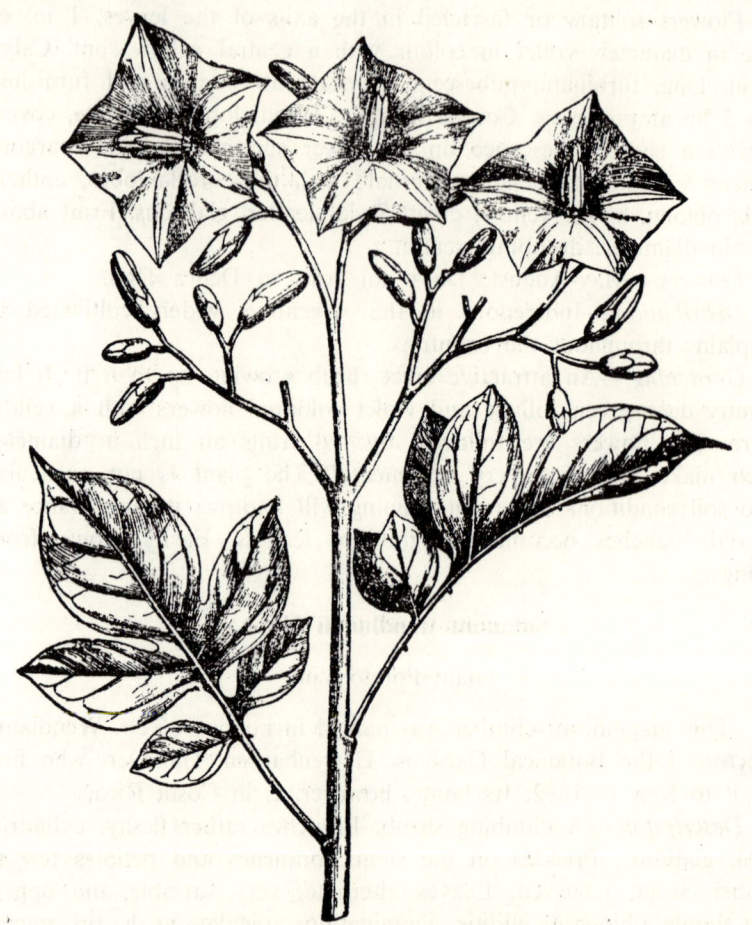

Fig. 80.—*Solanum wendlandii* Hook. f. × 1

Flowers.—Rainy season. Does not fruit in Dehra Dun.

Distribution.—Native of Costa Rica, now cultivated in all tropical

Plate 14

THE POTATO CREEPER
Solanum seaforthianum Andr.
(nat. size)

and temperate parts of the world.

Gardening.—A magnificent climber with very large pale lilac blue flowers in pendulous cymes about 6 in. across. It grows luxuriantly, but if neglected becomes ragged. The flowers which are mostly produced during the rainy season are very effective. This is perhaps the most showy of the cultivated Solanums. It prefers partial shade and is a splendid object for a green house. It was introduced into the Royal Gardens, Kew, in 1882. Propagated easily by cuttings.

Solanum seaforthianum Andrews
Potato Creeper

(The plant was named in honour of Francis Lord Seaforth, a patron of Botany.)

Description.—A glabrous climber. Stems and branchlets slender, cylindrical or angled, glabrous. Leaves alternate, the uppermost simple, elliptic-lanceolate, acute, seated on a petiole, 0.5 in. long, entire or undulate on the margins, membranous, glabrous on both surfaces; the lower trifoliate or pinnatisect, the leaf being cut to the midrib into a number of opposite segments which simulate leaflets; upper leaflets elliptic, narrowed at both ends; the next pair broad at the base and the rest almost petioled, all glabrous, acute or accuminate, entire or undulate on the margins; petiole up to 2 in. long.

Flowers arranged in axillary, rarely terminal, cymose panicles. Pedicels slender, up to 0.5 in. long. Calyx about 0.1 in. long, tube very short, obconical, glabrous, 5-lobed; lobes triangular-acute. Corolla rotate, 0.5 in. or more in diameter, blue or rose-coloured, shallowly 5-lobed; lobes hairy on the margins, induplicate-valvate in the bud. Stamens 5, alternate with the lobes; filaments short; anthers bright yellow in colour, oblong, connivent above the hairy style; ovary hairy, 2-celled. Fruit a globose berry, glabrous, scarlet.

Flowers and Fruits.—July-September.

Distribution.—A native of Brazil now extensively cultivated in all tropical and sub-tropical parts of the globe.

Gardening.—A pretty somewhat woody climber with bluish purple flowers borne in axillary compound cymes. This 'vine' does well in the sub-Himalayan tract but not in the plains. It prefers cool situations and begins to flower even when very young. Multiplied by seed or cuttings. It was introduced into England in the year 1804.

Solanum jasminoides Paxt.
Potato Vine

(*Jasminoides* means jessamine-like, and refers to the leaves of this twiner.)

Description.—A slender twining shrub. Branches cylindrical, glabrous, smooth. Leaves alternate, petiolate, 1.5-2 in. long, lanceolate or ovate-lanceolate in shape, round at the base, acute or obtuse at the tip, glabrous on both surfaces, somewhat coriaceous in texture, entire on the margins; nervation obscure; petiole about 5.5 in. long, terete, glabrous. Lower leaves sometimes lobed or pinnatisect.

Flowers white with a tinge of blue, star-shaped, arranged in terminal cymose panicles, about 3 in. across. Calyx 0.1 in. long, tube very short, 5-lobed; lobes nerved, ovate-acute or lanceolate-acute, glabrous. Corolla rotate, glabrous, 5-lobed; lobes spreading, apiculate, shortly hairy towards the margins on the under surface. Stamens 5, alternate with the corolla lobes; filaments short; anthers oblong erect about the hairy style. Ovary hairy, 2-celled.

Fig. 81.—*Solanum jasminoides* Paxt. × 1

Flowers.—March-October. Does not fruit in Dehra Dun.

Distribution.—Native of South America, commonly grown in gardens in the plains and hills throughout the country.

Gardening.—A pretty climber with ovate-lanceolate leaves. Flowers star-shaped, white, about 0.8 in. across, in many-flowered cymes. It prefers well-drained soil and partial shade and is a useful climber for the cool house. Propagated by cuttings.

PLATE 49

GIANT POTATO VINE
Solanum wendlandii Hook.

N. L. Bor

Datura Linn.

(*Datura* is the Arabic name for the plant *Datura stramonium* Linn.)

The genus comprises herbs, rarely shrubs or small trees, branchy, glabrous or slightly hairy, with a foetid odour. Leaves alternate, entire, or shallowly incised. The flowers are large, axillary and solitary. Calyx tubular, herbaceous, 5-lobed, rarely 3-lobed, splitting horizontally into 2 portions, the lower of which is persistent and surrounds the capsule, the upper deciduous. Corolla large, elongate, cylindrical at the base, widened at the top, 5-6-10-lobed or -toothed. Stamens 5; filaments inserted towards the base of the corolla tube and attached to it for half their length; anthers included, linear, dehiscing by longitudinal slits. Ovary 2-celled or imperfectly 4-celled; style long, filiform; stigma small, lobed. Fruit an ovoid or globular capsule, spinescent, rarely smooth, opening by 4 valves or irregularly. Seeds numerous.

This genus contains from 20 to 25 species which are generally confined to the tropics, especially of South America, and of which several are cultivated for their large flowers which become scented towards the evening. The large flowers close during rainy weather so that the stamens and nectaries are protected from wetting. Several of the species are said to be cross-fertilised through the agency of birds. Other agents of cross-pollination are beetles and moths.

The seeds, leaves and other parts of several species of this genus are very poisonous and also medicinal, and have been known as such from very early times. The alkaloids to be found in the plants are usually hyoscyamine and scopolamine. An interesting fact about *Datura* (quoted by Burkill) is that when the plants are deflowered it is found that the amount of alkaloid is greatly increased.

Key to the Species

Flowers erect or inclined, often purplish outside; fruit globose,
  armed with conical prickles .... *D. metel*
Flowers drooping, greenish white outside; fruit unarmed ...... *D. suaveolens*

Datura metel Linn. (D. fastuosa Linn.)

(*Fastuosus* is a Latin word meaning proud; why it has been applied to this plant we are unable to state unless it refers to the erect flowers which in this genus are usually pendulous.)

Description.—An annual herb, glabrous, reaching a height of 4-5 feet. Branches purplish and spotted with white patches. Leaves alternate, long petiolate, obliquely and broadly ovate-acute, toothed or entire, 2.5-6 in. long by 2.5 in. wide; petiole 2-3 in. long. Flowers axillary, solitary, shortly pedunculate; peduncle 0.5 in. long, more or less. Calyx cylindrical, 2-3.5 in. long, 5-lobed; lobes about 0.5 in. long, lanceolate-acute. Corolla large, 6-10 in. long, trumpet or funnel-shaped, white, violet or tinged with purple, with 5-6 folds and 5-6 lobes which are more or less cuspidate. Stamens 5, filaments attached to the corolla tube for half their length; anthers 0.3 in. long. Ovary glabrous; style as long as the corolla; stigma capitate, bifid. Fruit a globose capsule, spiny, 1 in. in diameter, glabrous, indehiscent, opening irregularly at the top. Seeds flattened, furrowed, tubercled.

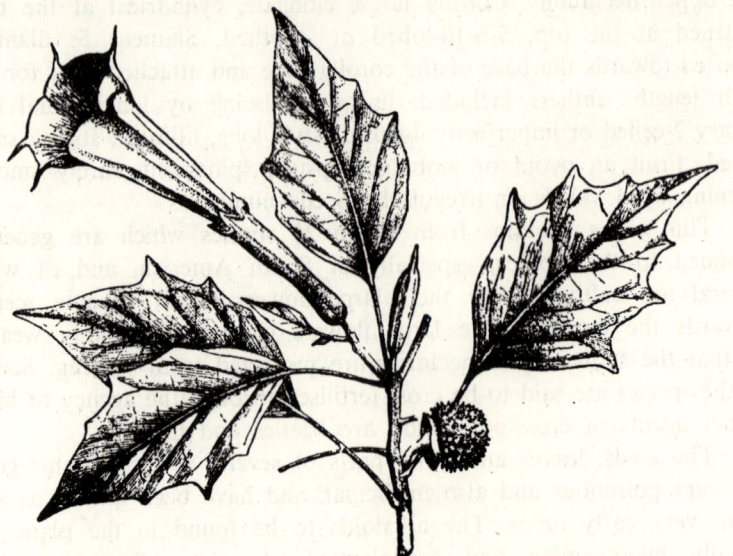

Fig. 82.—*Datura metel* Linn. (*D. fastuosa* Linn.) × ½

Flowers and Fruits.—Chiefly during the rains.

Distribution.—Native of India, wild or naturalized throughout the tropics of both hemispheres.

Gardening.—An erect herb 4-5 ft. high The double and triple forms of this species which are common in cultivation, are extremely handsome when in bloom during the rains. Easily grown from seed.

Medicinal and Economic Uses.—The seeds of this plant are highly poisonous and have been in use for centuries for criminal purposes. The victim is drugged by means of food or drink to which the ground seeds have been added, as a preliminary to the theft of his property. The action of the poison is, however, so powerful that sometimes the victim does not recover. The leaves are of use for making cigarettes and fumigating powders for the relief of asthma. Ganja eaters often adulterate the ganja with datura leaves. The juice of this plant is believed to be a cure for hydrophobia. The leaves are boiled and used as a poultice to relieve pain.

PLATE 50

THE WHITE DATURA
Datura fastuosa Linn.

N. L. Bor

PLATE 51

ANGEL'S TRUMPET
Datura suaveolens Humb. & Bompl.

SOLANACEAE

Datura suaveolens Humb. et Bonpl. [**Brugmansia suaveolens** (Humb. & Bonpl. *ex* Willd.) Bercht. & Presl.]

Angel's Trumpet

(*Suaveolens* means sweet-smelling in Latin.)

Description.—This shrub reaches 10-15 ft in height. Ultimate shoots green, terete, smooth and eventually glabrous, when young

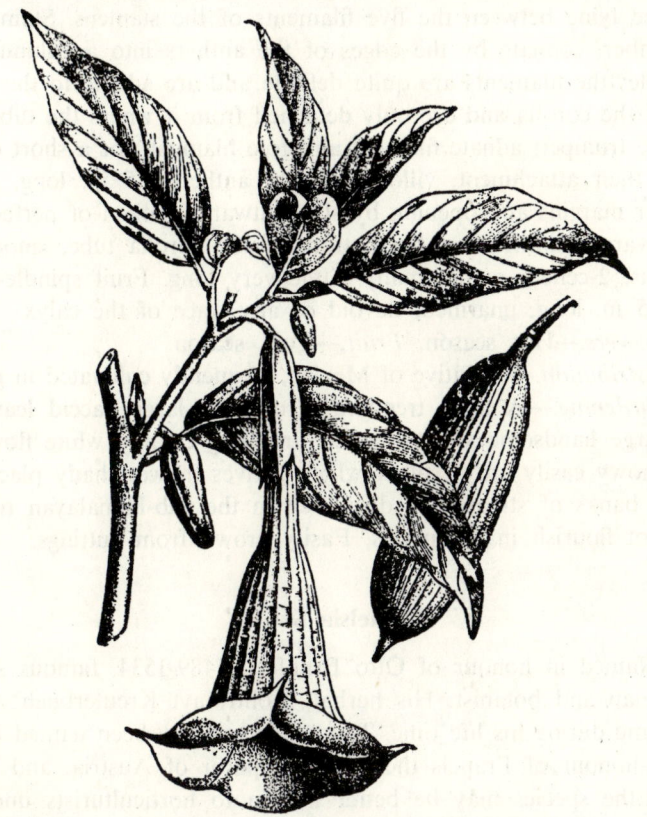

Fig. 83.—*Datura suaveolens* Humb. & Bonpl. [*Brugmansia suaveolens* (Humb. & Bonpl. *ex* Willd.) Bercht. & Presl.] × ½

covered with a minute evanescent tomentum. Leaves opposite, 6-12 in. long, 2-4 in. broad, one of the pairs about one-third shorter than the other, petiolate, ovate-oblong, quite entire on the margins, oblique at the base, glabrous, smooth on both surfaces, darker above, pale below, foetid when crushed; nervation prominent below; petiole up to 2 in. long, terete, finally glabrous, nodding.

Flowers axillary, solitary, seated on peduncles up to 1.5 in. long;

peduncles smooth and glabrous, terete. Calyx 4-5 in. long, yellowish green in colour, tubular, 5-veined, ending above in five oblong-acute lobes which are valvate in the bud, and through each of which passes a vein, glabrous within and without, smooth. Corolla greenish white or creamy white, up to 12 in. long, trumpet-shaped, with a spreading plicate limb, shortly broadly and acuminately lobed, 5-angled and narrow below, swelling above into the trumpet, which has three nerves to each lobe, minutely hairy outside, becoming glabrescent. Inside the corolla the three nerves of each lobe are deeply impressed, each group of three lying between the five filaments of the stamens. Stamens five in number, connate by the edges of the anthers into a column round the style; the filaments are quite definite and are adnate to the tubular part of the corolla and are only detached from it where the tube swells into the trumpet; adnate filament and free filament, for a short distance above their attachment, villously hairy; anthers 1.5 in. long, connate by their margins and opening by slits outwards; pollen of perfect spheres. Ovary cylindrical at the bottom of the corolla tube, smooth and glabrous, 2-celled; ovules many. Style very long. Fruit spindle-shaped, up to 5 in. long, unarmed, devoid of any trace of the calyx.

Flowers.—Hot season. *Fruit.*—Cold season.

Distribution.—A native of Mexico, frequently cultivated in gardens.

Gardening.—A large tree-like shrub with large flaccid leaves and very large handsome sweet scented drooping creamy white flowers. A very showy easily grown plant which thrives in wet shady places such as the banks of streams. It does well in the sub-Himalayan tract but does not flourish in the plains. Easily grown from cuttings.

Brunfelsia Sw.

(Named in honour of Otto Brunfels, 1489-1534, famous surgeon, theologian and botanist. His herbal 'Kontrafayt Kreuterbuch' brought him fame during his life time. The genus has also been named *Franciscea* in honour of Francis the First, Emperor of Austria, and at least one of the species may be better known to horticulturists under this name.)

This genus contains mostly shrubs, but a few are small trees. The leaves are alternate, petiolate, entire, elliptic-obovate or narrowly elliptic, exstipulate. The flowers are usually large and showy and have the strange property of changing colour as they grow older; some change from mauve or blue to white, others from yellow to white, or white to yellow, arranged in terminal cymes or clusters, sometimes solitary, sometimes sweetly fragrant. Calyx gamosepalous, 5-lobed. Corolla salver-shaped; tube narrow, long; lobes five, rounded. Stamens four, didynamous; filaments attached to the throat of the corolla.

SOLANACEAE 145

Ovary superior; style included or slightly exserted; stigma 2-lobed. Fruit a berry.

KEY TO THE SPECIES

Flowers mauve .... *B. latifolia*
Flowers yellow or white
    Small tree .... *B. undulata*
    Shrub .... *B. americana*

Brunfelsia latifolia Benth.

(*Latifolia* in Latin means broad-leaved.)

Description.—A small shrub with spreading branches, and green smooth glabrous twigs. Leaves elliptic or oblanceolate in shape, 2-4 in. long, acute or obtuse, glabrous and dark green above, slightly pubescent and lighter green below, quite entire; nervation obscure. Flowers

Fig. 84.—*Brunfelsia latifolia* Benth. × 1

arranged in terminal clusters, or few-flowered axillary cymes. Calyx erect, campanulate, 0.5 in. long, glabrous and smooth, ending above in five short blunt triangular lobes. Corolla hypocrateriform, pale violet or mauve in colour with a white centre, fading a few days after opening to a pure white; tube 1.5 in. long, slender, slightly curved; lobes 5, orbicular, about 0.75 in. long, somewhat undulate. Stamens four, two long, two short, each pair connate by the kidney-shaped anthers. Filaments expanded and flattened below the anthers and attached to the corolla tube. Ovary superior, 2-celled. Ovules many, on axile placentas; style just exserted; stigma two-lobed.

Flowers.—Profusely in the hot weather, but is in bloom practically throughout the year. Does not fruit in Dehra Dun.

Distribution.—Native of tropical America; commonly grown in all tropical and subtropical countries of the world.

Gardening.—A small low bush with slender branches and soft green leaves, which it sheds in the cold weather but which are replaced by the end of February, producing at the same time numerous sweetly scented flowers which are at first pale violet with a white centre but change in a day or so to white. For this reason it is often called by the popular name of 'Yesterday, To-day and Tomorrow'. It is rather a slow growing plant and usually cultivated in large pots, though it flourishes equally well in the ground. It is slightly affected, especially when young, by frost when it is severe. It flowers very freely in the early hot weather, the pale blue flowers against the dark green foliage producing a marvellous effect. It is a great favourite with Indians who invariably call it by the name of *Franciscea*. It prefers full sunshine or partial shade. Easily propagated by layers.

Brunfelsia undulata Sw.

(*Undulata* is a Latin word meaning wavy and refers to the margins of the leaves and petals.)

Description.—An evergreen shrub or small tree which reaches 20 ft in its own home. Leaves variable in shape, ovate-lanceolate, narrowly elliptic, oblanceolate or elliptic-oblanceolate, narrowly cuneate at the base, obtuse or acute at the tip, entire on the margins, exstipulate, quite glabrous, 2-7 in. long, up to 1.5 in. wide, nervation rather obscure, finely reticulate, petiole 0.25 in. long.

Flowers solitary and terminal in wild plants, in cultivated plants forming clusters in the upper axils. Calyx about 0.25 in. long, campanulate, somewhat irregularly 5-lobed, shortly glandular-pubescent externally. Corolla hypocrateriform, pure white fading to a creamy white with age; tube cylindric, slightly curved, pubescent externally, up to 3.5 in. long, 5-lobed; lobes 5, 1 in. long, rounded, with undulate margins. Stamens 4, as long as the tube. Ovary oblong; style as long as the tube, slightly exserted ending in a 2-lobed stigma. Fruit berry-like.

Flowers.—Rainy season. *Fruits.*—Cold season.

Distribution.—Native of Jamaica, now frequently cultivated in all tropical and subtropical parts of the world.

Gardening.—An upright shrub very similar to *B. americana* Sw.

It prefers a loamy soil and a sunny site. Propagation is by seeds or layers. It was introduced into England about the year 1800 from

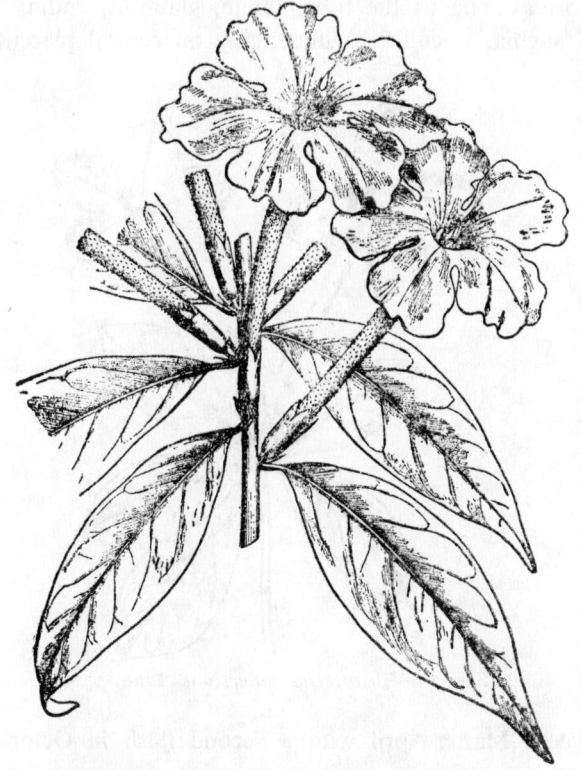

Fig. 85.—*Brunfelsia undulata* Sw. × 1

Jamaica. It is a magnificent free-flowering species and is a great acquisition to any garden.

Brunfelsia americana Linn.

Description.—A dwarf shrub with smooth glabrous or minutely pubescent branches and branchlets. Leaves 2-4 in. long, alternate, exstipulate, elliptic, elliptic-oblanceolate, or elliptic-obovate in shape, cuneate at the base, obtuse or acute at the tips, dark green and glabrous above, paler and pubescent below, petiolate; petiole 0.1-0.25 in. long; nervation obscure.

Flowers arranged in few-flowered terminal clusters or solitary in the axils of the leaves. Salyx campanulate, smooth and glabrous divided above into 5 triangular-acute lobes, the whole including the lobes about 0.25 in. long. Corolla hypocrateriform, minutely pubescent outside, pure white fading to yellow, sweetly scented, especially at night, about 2-2.5 in. long and 1.5-2 in. wide at the mouth; tube very slender; lobes 5, broadly elliptic or orbicular in shape. Stamens 4, attached to the tube, didynamous; anthers reniform. Ovary superior, style almost as long as the tube, terete, glabrous, ending above in a two-lobed stigma. Loculi 2, ovules many on central placentae. Fruit a berry.

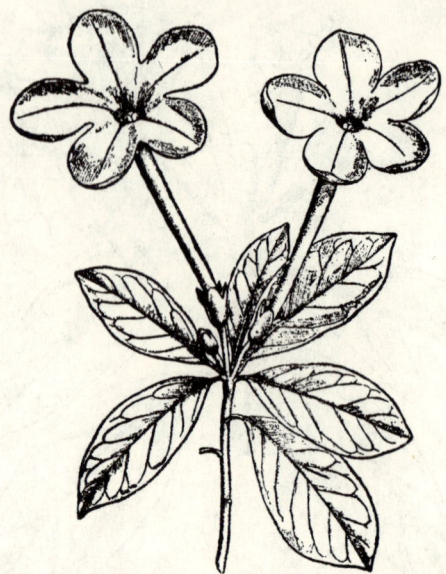

Fig. 86.—*Brunfelsia americana* Linn. × ½

Flowers.—March-April with a second flush in October. *Fruits*—cold season.

Distribution.—A native of tropical America now commonly cultivated throughout the tropical and subtropical parts of the globe.

Gardening.—An erect dwarf shrub about 3 ft high with slender branches. The flowers are at first pure white, fading with age to pure yellow and sweetly scented specially at night. It is not so ornamental as *B. latifolia* Benth. except when in full bloom during March and October. It prefers full sun and is multiplied by layers as well as by seed.

Chapter 8

VERBENACEAE

This family takes its name from one of its genera, the well-known garden plant, *Verbena*. About 70 genera comprising some 800 species are included in the family, which is for the most part confined to the tropics and subtropics of south-east Asia and the Malayan archipelago. Few genera penetrate to the cooler regions of the earth and the *Verbenaceae* are poorly represented in Africa.

The family contains all types of plants from tiny herbs up to giant trees. The ultimate shoots are often quadrangular. The leaves are usually opposite, compound or simple, rarely whorled or alternate, exstipulate. The flowers are arranged in panicles or cymes. The calyx is inferior, bell-shaped, cylindrical or platter-like, 4-5- rarely 8-cleft, -lobed, or -toothed, occasionally entire. Corolla gamopetalous, cylindrical, expanded towards the top, straight or curved, 4-5-lobed, often 2-lipped; lobes imbricate. Stamens 4, rarely 2, inserted on the corolla; anthers 2-celled; cells often divergent, opening lengthwise. Ovary superior, seated on an obscure disk, 2-8-celled, often 4-celled; ovules solitary or paired, erect or rarely pendulous. Fruit a drupe or berry.

Many genera of this family, *Verbena, Clerodendrum, Stachytarpheta, Congea, Petrea,* and *Holmskioldia* are favourite garden plants. *Duranta plumieri* is grown everywhere as a hedge. The best known species of the family, however, is *Tectona grandis,* the teak, whose timber is famous all over the world. Another tree is *Gmelina arborea,* which furnishes a timber of some value as a furniture wood.

Many species of *Verbenaceae* contain bitter and astringent principles and often a volatile oil as well. A decoction of the root and leaves of *Verbena officinalis* is a powerful astringent and is applied to inflamed and bleeding piles. A decoction of the leaves of several species of *Lantana* is used as an expectorant. *Vitex agnus-castus,* a European species, produces berries which were supposed to be an aphrodisiac, these when roasted were supplied to the monasteries in mediaeval times under the delightful name of 'Monk's pepper'. *Vitex peduncularis,* a tree common in Assam, contains a light yellow crystalline substance which is identical with vitexin, the active principle of *Saponaria officinalis* and *Vitex littoralis*. Dr. S. Krishna (Biochemist, Forest Research Institute) informed us that the mature leaves of *Vitex peduncularis* lose their vitexin. In Assam a decoction of the leaves is always used in cases of black-water fever and many cases are reported to have been cured by its use. Its reputation is so great that a large quantity of seed and even seedlings have been sent from Assam to other provinces.

The corolla of the species is usually tubular and is often brightly coloured and fragrant. Such an arrangement points to cross-pollination by insects or birds. In some species of *Clerodendrum* the corolla tube is up to 6 in. long. In such cases the agents of cross-fertilisation are honey-birds and long-tongued *Lepidoptera*.

According to Koorders a species of *Clerodendrum*, endemic in the Celebes, possesses a calyx which is always full of water, produced from numerous hydathodes seated upon the inner surface of the calyx (hydathodes are specialized epidermal cells which secrete water). The ostensible object of this production of water is to prevent robber insects from boring through the base of the corolla to the honey without cross-pollinating the flowers. This object is also achieved by certain species which develop extra floral nectaries on the calyx. The nectaries are occupied by ants which vigorously attack insects which attempts to get honey by illegitimate methods. Some *Clerodendrum*s go so far as to allow ants to take up quarters in their stems in order to attain this object.

An interesting feature in certain genera is the colour change which takes place as a flower matures. In one species of *Lantana* the change is from orange-yellow to dark crimson. This change has doubtless some connection between the ripening of the pollen and the receptivity of the stigma and may act as a guide to insects.

Key to the Genera

Flowers in a narrow spike .... *Stachytarpheta*
Flowers panicled or cymose
  Flowers clustered in a pink or mauve involucre .... *Congea*
  Flowers not clustered in an involucre
    Calyx blue, lobed .... *Petrea*
    Calyx green, white or red
      Calyx red .... *Holmskioldia*
      Calyx white or green .... *Clerodendrum*

Stachytarpheta Vahl

(The generic name comes from two Greek words which mean 'spike' and 'thick' and together give a word picture of the inflorescence.)

Hairy or glabrous shrubs and herbs. Leaves opposite or alternate, toothed, exstipulate. Inflorescence a long terminal spike. Flowers solitary in the axils of bracts, sessile or half immersed in the rhachis; bracts lanceolate; bracteoles absent. Corolla tube slender, cylindric; limb oblique with 5, equal or unequal, flat, spreading lobes. Perfect stamens 2 (the lower pair) included within the corolla tube; staminodes 2,

PLATE 52

M. N. Bakshi

THE VARIABLE STACHYTARPHETA
Stachytarpheta mutabilis Vahl.

PLATE 53

THE VARIABLE STACHYTARPHETA

M. N. Bakshi

minute or absent; filaments short; anther cells vertical, divaricate. Ovary 2-celled; ovules solitary in each cell; style long filiform. Fruit enclosed in the calyx, linear-oblong, separating into 2, hard, 1-seeded bodies.

Key to the Species

Corolla blue	*S. indica*
Corolla crimson or rose	*S. mutabilis*

Stachytarpheta indica (Linn.) Vahl

Description.—A branchy annual herb, 1-3 ft tall. Branches almost quadrangular, sparsely pubescent. Leaves opposite, petiolate, ovate or obovate, obtuse or acute at the tip, cuneate and long-decurrent at the base, dentate in the upper two-thirds, chartaceous or membranous in texture, glabrous on both surfaces or with a few hairs below, 1-4 in.

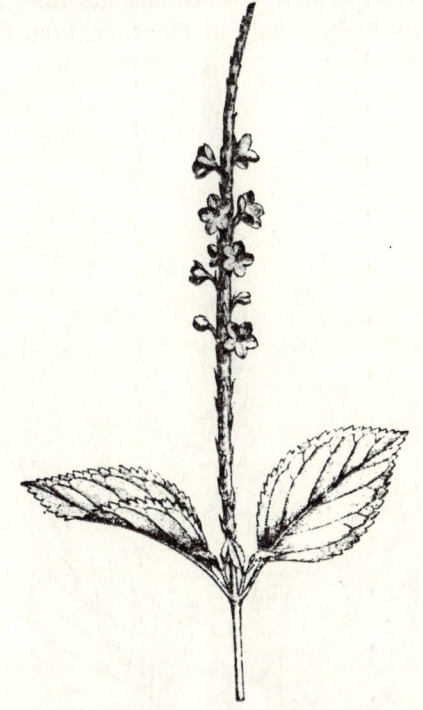

Fig. 87.—*Stachytarpheta indica* (Linn.) Vahl. [*S. jamaicensis* (L.)] Vahl × ½

long by 1-3 in. wide; petiole slender, winged, about 0.3 in. long.

Inflorescence of terminal, curved, slender long spikes, 4-13 in. long, rhachis about 0.12 in. broad, excavated at the insertion of the flowers and the cavities covered by the bracts. Bracts lanceolate-acuminate, ciliate, scarious, 0.2-0.4 in. long; bracteoles absent. Flowers sessile, blue, 0.6-0.8 in. long. Calyx cylindrical, narrow, glabrous, membranous,

5-nerved, 0.25 in. long, split towards the rhachis, 4-5-toothed; glabrous, curved, hairy above and within. 0.3-0.5 in. long, 5-lobed; lobes oblique, spreading, rounded, about 0.4 in. long. Stamens 2, inserted above the middle of the tube. Ovary glabrous. Fruit a pear-shaped capsule, included in the calyx, up to 0.2 in. long.

Flowers and Fruits.—Rainy season.

Distribution.—Native of tropical America, cultivated and naturalized in various parts of Asia and Africa.

Gardening.—A herb with deep blue flowers in long spikes. Propagated by seed.

Medicinal uses.—Said to be used in Malay as an abortifacient.

Stachytarpheta mutabilis (Jacq.) Vahl

(*Mutabilis* is a Latin word meaning changeable or variable, and refers to the colour of the flower which changes from crimson to rose.)

Description.—A herb, woody at the base, branches quadrangular;

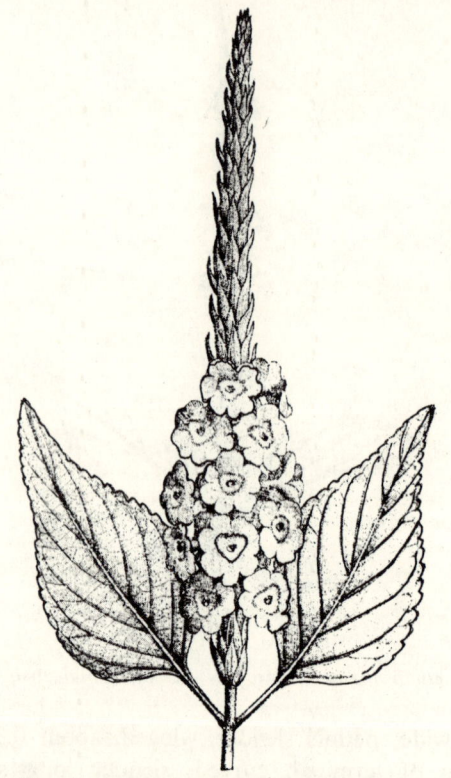

Fig. 88.—*Stachytarpheta mutabilis* (Jacq.) Vahl × ½

covered with a crisped pubescence. Leaves opposite or alternate, when opposite joined by a stipular line, petiolate, ovate, ovate-oblong or elliptic, acute or acuminate at the tip, cuneate and long decurrent at the base, serrate-dentate on the margins, rather rough on the upper surface, softly pubescent or scabrid on the lower surface; petiole up to 0.5 in. long, slender, grooved above, pubescent.

Inflorescence a stout terminal spike up to 1 ft long. Rhachis terete, pubescent, excavated at the insertions of the sessile flowers; excavations covered by the oblong, acuminately awned bracts, bracts 0.3 in. long without the awn, pubescent; bracteoles absent. Calyx narrow-tubular, 4-ribbed, 4-toothed, 0.3-0.5 in. long; teeth short, acute. Corolla tube cylindrical, curved, 0.4-0.5 in. long; limb 5-lobed; lobes orbicular, irregular in size, 0.2-0.4 in. wide. Flowers crimson, fading to rose. Stamens 2, included in the corolla tube. Ovary glabrous; style long slender; stigma capitate, exserted.

Flowers and Fruits.—Rainy season.

Distribution.—Native of tropical America, now cultivated in the plains throughout the country.

Gardening.—A shrub 3-5 ft high with 4-angled tomentose villous branches. The flowers are scarlet, fading to rose and are borne in long spikes. To keep it in good shape it is advisable to trim it after the rainy season. Easily propagated by seed.

Medicinal uses.—The leaves are powdered and mixed with lime and applied to wounds and sores. It is also said to be used as an abortifacient.

Congea Roxb.

Climbing shrubs with opposite leaves. Inflorescence of small involucral cymes arranged in terminal panicles. Involucre 3-phyllous, spreading, supporting a sessile 6-9-flowered cyme. Calyx tubular, widened at the mouth, 5-cleft, persistent. Corolla 2-lipped, the tube as long as the calyx; limb very unequal, the upper lip elongate, erect and 2-cleft, the lower spreading and shortly 2-lobed. Stamens 4, didynamous, inserted in the throat and long exserted; filaments filiform; anthers dorsifixed, 2-celled. Ovary 2-celled; ovules 2 in each cell, pendulous; style capillary with a 2-cleft stigma. Fruit a coriaceous capsule, indehiscent, 1-seeded by abortion.

Congea tomentosa Roxb.

(*Tomentosa* is derived from the Latin word *tomentum*, the stuffing of a pillow, and is usually used to describe a shaggy pubescence.)

Description.—A large deciduous scandent shrub. Branches and young parts covered with a soft or rough short shaggy pubescence. Leaves opposite, exstipulate, ovate to ovate-oblong in shape, petiolate, 4-5 in. long, 2-4 in. wide, obtuse or rounded at the base, acute or acuminate, entire; petiole about 0.2-0.3 in. long, pubescent; upper surface very rough to the touch; under surface softly hairy to almost hirsute on the mid-rib; nerves prominent below.

Inflorescence simulating a large compound terminal panicle; main branches pubescent, opposite. On the branches are opposite, pedunculate cymes of flowers. Peduncles supported at the base by a velvety bract, 0.25 in. long. Peduncle up to 1 in. long, hairy. At the top of the peduncle is an involucre of 3 bracteoles connate at the base, forming a very small cup; bracteoles elliptic-obtuse, oblanceolate, obovate in shape, velvety-tomentose on the inner surface, tomentose and veined outside, pale purple, pink or whitish in colour, up to 1 in. long. Flowers in cymes sessile in the involucre. Calyx gamosepalous, covered with dense hairs, 0.25-0.3 in. long, 5-lobed; lobes acute. Corolla gamopetalous, as

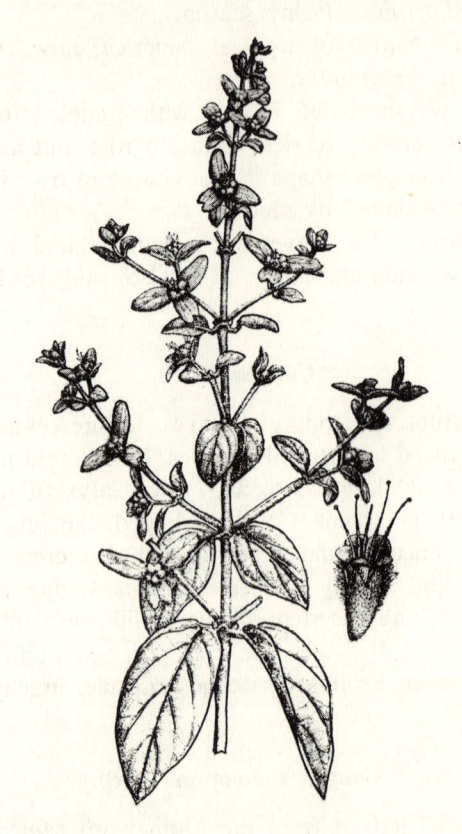

Fig. 89.—*Congea tomentosa* Roxb. × ½

long as the calyx tube, tubular, constricted above the ovary, 2-lipped, glabrous, whitish. Stamens 4, didynamous, filaments as long as the corolla. Ovary glabrous; style filiform. Fruit an indehiscent capsule.

Flowers.—December-March.

PLATE 54

M. N. Bakshi

PURPLE WREATH
Petrea volubilis Linn.

PLATE 55

M. N. Bakshi

PURPLE WREATH
Petrea volubilis Linn.

PLATE 15

THE PURPLE WREATH
Petrea volubilis Linn.
(½ nat. size)

Distribution.—Native of Manipur, Chittagong, Burma and Siam. Frequently cultivated in gardens in India.

Gardening.—A strong deciduous climber allied to *Petrea volubilis*, and conspicuous for the persistent pinkish mauve bracts which cover this vine during spring and remain on the plant for a long time. It is usually propagated by seed, as cuttings take root with difficulty. The plant commonly cultivated in gardens is var. *azurea* Wall.

Petrea (Petraea) Houst. ex Linn.

(This genus was named in honour of Lord Petre (1713-1743), Thorndon, Essex, who in his short life had managed to put together the finest collection of exotic plants then existing in Europe.)

Shrubs, trees, or woody vines. Leaves opposite or whorled, deciduous, exstipulate, rough. Inflorescence an axillary or terminal many-flowered raceme. Calyx inferior, gamosepalous; tube campanulate, usually ribbed, 5-lobed, lobes longer than the tube, coloured. The rim of the calyx often bears a 5-lobed calicinal crest. Corolla gamopetalous, inferior, darker blue or purple than the calyx or white, slightly irregular, 5-lobed; lobes collected into 2 lips; one of 2, the other of 3 lobes. Stamens 4, didynamous, included in the corolla tube and inserted on it close together near the middle, filaments short and slender, anthers oblong. Ovary subglobose, 2-celled; ovules 1 in each cell. Fruit a drupe enclosed by the the mature calyx.

Petrea volubilis Linn.

Pulple Wreath

(*Volubilis* means twining in Latin and refers to the habit of the plant.)

Description.—A woody vine or undershrub with greyish bark on the old wood. Branches and branchlets slender, prominently lenticellate, shortly hairy. Leaves opposite, exstipulate, petiolate, firmly membranous, rather dull green above, brighter green beneath, very rough to the touch, 2-8 in. long by 1-4 in. wide, acute or shortly acuminate at the apex, cuneate at the base; petiole up to 0.5 in. long densely or sparsely pubescent.

Inflorescence a drooping many-flowered raceme, solitary in the axils, up to 12 in. long; rhachis puberulent. Individual flowers pedicellate; pedicels 0.3 in. long, obscurely hairy. Calyx gamosepalous; tube cylindric, about 0.12 in. long, not ribbed, densely pubescent outside, 5-lobed; lobes membranous, oblong, 0.75 in. long by 0.25 in. wide, lilac or bluish in colour, rounded at the apex; calicinal crest composed of five membranous, triangular-ovate lobes. Corolla hypocrateriform,

0.3 in. long, deep blue or purple, with an infundibuliform tube, lower three quarters glabrous, upper quarter densely pubescent, 5-lobed; lobes broadly elliptic. Inner surface of the corolla tube puberulent, densely pilose at the base of the stamens. Stamens 4, didynamous, inserted on the tube; filaments filiform, glabrous; anthers oblong; staminodes obsolete. Ovary oblong-obovoid, glabrous, seated on a conspicuous disk; style filiform; stigma obliquely capitate. Fruiting calyx up to 0.17 in. long, densely pubescent.

Flowers.—March-April. Sometimes with a second flush in October. It does not fruit in this country.

Distribution.—Native of tropical America. Now commonly cultivated throughout the tropics.

Gardening.—The 'purple wreath' is one of the most distinct and beautiful of cultivated climbers. It bears long spikes of cloudy blue star-like flowers which are extremely effective when massed. It can be used not only as a climber but can be trained as a standard. The flowers begin to open at the base of the spike-like raceme and the showy 5-pointed star is the calyx, whose sepals are coloured like petals. The calyx spreads open while the corolla is still like a bud in the middle and remains long after the corolla has fallen, so that the vine, at first glance, appears to bear two kinds of flowers. Propagated by layers and gooties or from the suckers which it frequently sends up.

Holmskioldia Retz.

(A genus erected in honour of Theodor Holmskiold, 1732-1794, a Danish scientist.)

Scandent shrubs with opposite, exstipulate, simple leaves. Flowers arranged in terminal short panicles or spurious axillary racemes. Calyx membranous, usually red, the tube very short, the limb large, rotate and spreading, entire or nearly so. Corolla gamopetalous, tubular, the tube elongate, slightly curved, somewhat dilated at the throat, 2-lipped; upper lip 2-cleft with the lobes erect spreading, the lower 3-cleft with short reflected lobes. Stamens 4, didynamous, exserted. Ovary 4-celled, with a solitary ovule in each cell; style filiform. Fruit a 4-3-2-lobed capsule surrounded at the base by the large flat calyx.

Holmskioldia sanguinea Retz.

Cup-and-Saucer Plant

(*Sanguineus* means blood-coloured and refers to the normal colour of the flowers.)

Description.—A large, scrambling shrub; young parts pubescent. Leaves opposite, 2-4 in. long, ovate or elliptic-ovate in shape, rounded

PLATE 16

CUP AND SAUCER PLANT
Holmskioldia sanginea Retz.
(nat. size)

or truncate at the base but acute at the insertion of the petiole, shallowly crenate-serrate on the margin, acuminate at the tip, membranous in texture, dark green above, pale beneath, glabrous on the upper surface, minutely pubescent below when young; petiole slender, up to 1 in. long.

Inflorescence a terminal puberulous little-branched panicle or reduced to short racemes at the tips of axillary short shoots. Calyx gamosepalous, almost orbicular in shape from a very short basal bell-shaped portion, 0.6-0.75 in. in diameter, net-veined, reddish or orange in colour, finally turning brown in fruit. Corolla about 1 in. long, gamopetalous, tubular; tube curved, 2-lipped; upper 2-lobed, lower 3-lobed. Stamens 4, didynamous, exserted; anther cells parallel. Ovary 4-celled; ovule 1 in each cell; style slender; stigma bifid. Fruit a drupe, obovoid in shape, deeply 4-lobed at the apex, separating into 4 seeds.

Flowers.—October-December.

Distribution.—Native of the sub-tropical Himalayas. Often cultivated in gardens throughout India.

Gardening.—A straggly shrub, remarkable for its large subrotately campanulate red calyx reaching an inch in diameter in fruit. The plant will grow even in poor soil and does best in full sunshine. It is advisable to prune it closely after flowering so as to keep it compact and within bounds. This plant is very popular for cut flowers for even when the corolla has fallen the calyx is pretty and very effective. Usually propagated by layers, as cuttings are sometimes hard to start.

A variety with orange flowers has recently been introduced from Assam where it is wild. It is prettier than the type.

Clerodendrum Linn.

The name of the genus comes from two Greek words, *kleros,* lot, fate, and *dendrum,* tree. The combination may mean 'tree of fate' or 'tree of chance'. The reason for the application of the name to the genus is obscure, but it may well come from the fact that some of the species are considered to have healing properties while others act in exactly the reverse way. According to Burkill[1] 'Clerodendrons are, par excellence, of Malay magic'. He then proceeds to give examples and mentions that *C. siphonanthus* is the chief magical species of northern India. The alleged magical properties of the plant may have also been a reason why the genus was called 'tree of fate'.

A large genus, distributed more or less throughout the tropics, it possesses the following characteristics:

Trees, shrubs, undershrubs or rarely herbs. Leaves simple, opposite,

---

[1] A Dictionary of the Economic Products of the Malay Peninsula.

rarely verticillate (see *C. siphonanthus*). Inflorescence usually a terminal panicle, less often of axillary cymes, usually lax, elongate or umbelled or capitate; bracts often foliaceous; flowers mostly large, white, blue, yellow or red. Calyx cup-shaped, campanulate or funnel-shaped, herbaceous or coloured, truncate or with 3-5-lobes, often enlarged in the fruit. Corolla with a slender tube, often very long, cylindrical, recurved, 4-5-lobed; lobes spreading, almost equal. Stamens 4, exserted, inserted in the throat or in the upper portion of the tube, coiled in the bud; anthers with parallel cells, opening by longitudinal slits. Ovary incompletely 4-celled; ovules 4, pendulous or laterally attached. Style slender, often as long as the filaments, stigma shortly bifid. Fruit a 4-grooved or 4-lobed drupe.

The Clerodendrums are great favourites in Indian gardens on account of their showy flowers and often handsome foliage. They never show to such an advantage as they do in their natural homes in the gloom of the evergreen forest where they develop their beautiful panicles to perfection.

Key to the Species

Tube of the corolla 4 in. long .... *C. siphonanthus*
Tube of the corolla 2 in. long or less
  Climbing plants .... *C. thomsonae*
  Shrubs or small trees
    Calyx truncate or very shortly toothed .... *C. inerme*
    Caiyx distinctly lobed or dentate
      Flowers red or scarlet .... *C. squamatum*
      Flowers white or rose
        Calyx with peltate glands
          Calyx lobes triangular-acute .... *C. fragrans*
          Calyx lobes broadly ovate, foliaceous .... *C. infortunatum*
        Calyx without peltate glands
          Flowers in a pendulous panicle .... *C. nutans*
          Flowers in erect inflorescences
            Flowers in trichotomous panicles; leaves large,
              6 in. long or more .... *C. trichotomum*
            Flowers in dichotomous panicles, leaves small,
              2 in. long or less .... *C. phlomidis*

Clerodendrum siphonanthus R. Br. [**C. indicum** (Linn.), O. Ktze.]

Turk's turban; Tube-Flower

(The specific name is derived from the Greek word, *siphon*, tube, and refers to the hollow stems of this plant.)

Description.—An annual shrub reaching 9 ft in height. Branches channelled; bark smooth and shining. Leaves verticillate in threes or fives, or opposite, sessile, narrowly lanceolate or oblong-lanceolate, attenuate at both ends, glabrous on both surfaces, entire with recurved margins, membranous, 4-8 in. long by 0.5-1 in. wide.

Flowers arranged in a leafy terminal panicle, 8-10 in. long, made up of verticelled few-flowered cymes; bracts linear-lanceolate, up to 0.5 in. long; bracteoles subulate; pedicels slightly hairy, up to 0.75 in. long. Flowers white, turning a cream colour, 4-5 in. long. Calyx campanulate, hairy, 0.6 in. long; tube narrow 0.2 in. long, 5-lobed; lobes ovate-lanceolate 0.4 in. long. Corolla glabrous; tube cylindrical, recurved, dilated towards the top, up to 4 in. long, 5-lobed; lobes spreading, obovate or oblong, obtuse, 0.5 in. long or less. Stamens 4; filaments, long-exserted, slender, glabrous; anthers oblong. Ovary glabrous; style very long, slender; stigma shortly bifid. Fruit a bluish black drupe enclosed in the enlarged calyx.

Flowers.—Rainy season. *Fruits.*—Cold season.

Fig. 90.—*Clerodendrum siphonanthus* R. Br. × ½

Distribution.—Common in many parts of India and Burma extending to Malaya, Indo-China and Java. Often cultivated in gardens for its flowers.

Gardening.—A shrub which grows 4-8 ft tall in Dehra Dun, with a slender upright form which makes it attractive when grown against a wall. The long white tubular flowers hanging bell-like from an upright stalk make this a very striking plant during the rains. The flowers are followed by conspicuous dark blue fruits supported by the persistent spreading red calyx. The plant prefers partial shade and is propagated by seed.

Medicinal uses.—The root is considered useful in the treatment of asthma, cough and scrophulous affections. Pieces of the wood are tied round the neck as a charm against various ailments. The juice expressed from the herbaceous portions mixed with ghee is used as remedy for skin diseases.

Clerodendrum thomsonae Balfour

(The specific epithet commemorates the name of the wife of the Rev. W. C. Thomson, a missionary on the West Coast of Tropical Africa, from where the plant was introduced into England in 1861.)

Description.—A climbing shrub. Branches 4-angled, minutely

Fig. 91.—*Clerodendrum thomsonae* Balfour × ½

pubescent. Leaves **opposite**, broadly ovate, acute at the tip, subcordate, truncate or abruptly attenuate at the base, minutely and sparsely pubescent below, entire on the margins, chartaceous, 2-3 in. long by 0.5-2 in. wide; petiole up to 0.6 in. long.

Flower arranged in axillary cymes; peduncles 1.5-2.5 in. long; bracts foliaceous, pubescent, lanceolate; pedicels bracteolate, up to 0.6 in. long; flowers scarlet. Calyx inflated, membranous, white or grey, 0.75-1.25 in. long by 0.4-1 in. wide; tube very short; lobes 5, broadly ovate-lanceolate, acute, glabrous or minutely pubescent near the margins. Corolla scarlet or red; tube widened at the throat and hairy, 5-lobed; lobes ovate, 0.2 in. long. Stamens 4; filaments long-exserted, slender; anthers oblong. Ovary glabrous; style not as long as the filaments.

Flowers.—Rainy season.

Distribution.—Native of west tropical Africa; commonly cultivated in gardens in the tropical and subtropical parts of the world.

Gardening.—A most ornamental climbing shrub, very showy during the rains with its scarlet flowers which are in sharp and striking contrast to the white persistent calyx. Propagated by seed or by cuttings in sand during the rains.

Clerodendrum inerme Gaertn.

(*Inermis* means unarmed in Latin, but to what this refers we are unable to say, as none of the species of this genus are armed.)

Description.—A shrub reaching a height of 6-9 ft. Branches slender, greyish brown, lenticellate, glabrous or finely pubescent. Leaves opposite, elliptic-ovate or ovate-lanceolate, rounded or shortly and obtusely acuminate at the tip, cuneate or acute at the base, entire, coriaceous, rarely membranous, 1.5-5 in. long by 0.75-3 in. wide, when young sparsely pubescent; petiole slender, 0.3-0.75 in. long.

Inflorescence of umbelled, axillary cymes, each of 3 pedicelled flowers, seated on a peduncle, 0.3-1.5 in. long. Flowers on pedicels 1.3-4 in. long, white, supported by small linear bracts. Calyx cupular, pubescent or glabrous, often glandular without, 5-toothed; teeth very small. Corolla glabrous or sparsely pubescent, glandular; tube slender, 0.75-1.5 in. long, villous inside, 5-lobed; lobes ovate, obtuse; 0.2-0.3 in. long. Stamens 4; filaments slender, red, long-exserted, inserted in the villous part of the tube; anthers oblong. Ovary glabrous; style long, slender, equalling the filaments in length; stigma shortly bifid. Fruit a glabrous, 4-lobed drupe, up to 0.5 in. long, enveloped in the striate, enlarged calyx.

Flowers.—Throughout the year but chiefly from July to November.

Distribution.—Indigenous to the sea coast of India extending to Polynesia.

Gardening.—A straggling, almost scandent, evergreen shrub with dark green foliage and white flowers. It is a very hardy and quick growing shrub and is suitable for covering banks, walls, etc. If kept nicely trained it will form a pretty fence and is consequently very suitable for hedges. Owing to its rapid growth it might perhaps be useful for afforestation work.

Fig. 92.—*Clerodendrum inerme* Gaertn. × ¾

Medicinal uses.—The wood, root and leaves are bitter and are said to be of use in treating fevers. It has a reputation as an alterative and tonic.

Clerodendrum squamatum Vahl [C. japonicum (Thunb.) Sweet]

(*Squamatum* means scaly in Latin and refers to the peltate glands on the under surface of the leaf.)

Description.—A shrub attaining 6 ft in height. Branches 4-angled, channelled, glabrous or finely pubescent, furnished at the nodes with a line of interpetiolar woolly hairs. Leaves opposite, ovate-rounded, shortly acuminate at the tip, deeply cordate at the base, covered with stiff, sparsely arranged hairs, dark green above, pale and covered with numerous peltate glands below, chartaceous in texture, toothed on the margin; petiole glabrous, 0.2-9.4 in. long.

Flowers pedicelled, scarlet, sometimes white or rose-coloured, arranged in pyramidal terminal broad glabrous or pubescent panicles seated on long peduncles; bracts and bracteoles small; pedicels short. Calyx glabrous, red, 0.2-0.3 in. long, 5-lobed; lobes much longer than the very short tube, broadly lanceolate-acute. Corolla almost glabrous; tube cylindrical, 0.5-0.6 in. long, 5-lobed; lobes spathulate, 0.15-0.2 in. long; stamens 4; filaments slender, long-exserted: anthers oblong; ovary glabrous; style very long, slender, stigma shortly bifid. Fruit a blue-black drupe, up to 0.5 in. broad, within the accrescent calyx.

Fig. 93.—*Clerodendrum squamatum* Vahl × ½

Flowers.—March-April.

Distribution.—Native of China, extending to the Himalayas. Japan to Sumatra, cultivated throughout the tropical and subtropical parts of the globe.

Gardening.—This is one of the most showy of shrubs, having great clusters of scarlet flowers which appear during March-April. It should be cut back after flowering, otherwise it becomes bare and scraggy. The plant prefers partial shade and is often attacked by insects, especially mealy bugs and scales. It has been in cultivation in European gardens since 1790. Propagation is by seed.

Medicinal uses.—An infusion of the leaves in vinegar is said to be used for gonorrhoea, and the chewing of the bract for passing blood.

Clerodendrum fragrans (Vent.) Willd.

Description.—A shrub. Branches terete, somewhat quadrangular towards the tips, pubescent, almost tomentose. Leaves opposite, broadly ovate, nearly round, shortly acuminate at the tips, almost cordate, but

Fig. 94.—*Clerodendrum fragrans* (Vent.) Willd. × ½

sometimes truncate, at the base, covered with rather sparse, stiff hairs above, pubescent, especially on the nerves below, glandular near the petiole, membranous, regularly toothed on the margins; 4-6 in. long and as much wide; petiole pubescent-tomentose, up to 2 in. long.

Inflorescence of numerous flowers arranged in terminal compact almost sessile panicles; bracts foliaceous, persistent, lanceolate-oblong, attenuate at both ends, pubescent and bearing on the back a nectariferous gland. Individual flowers seated on very short pedicels, white or rose in colour, very fragrant, 0.75-1.25 in. long. Calyx conical, narrow, finely puberulent, with nectariferous glands, up to 0.5 in. long, 5-toothed; teeth as long as the tube, lanceolate-subulate, reflexed. Corolla double, glabrous; tube 0.5-0.8 in. long, 5-lobed; lobes obovate, reflexed, 0.25-0.4 in. long. Stamens 4; filaments long exserted, slender; anthers oblong. Ovary glabrous, truncate; style slender, very long; stigma shortly bifid. Fruit a drupe, enveloped at the base by the accrescent calyx.

Flowers.—Hot and rainy season.

Distribution.—Indigenous to China, cultivated or wild throughout India and most parts of the tropics.

Gardening.—A vigorous low growing plant; flowers double, appearing like little roses, white tinged with pink, fragrant. It spreads readily, throwing out suckers here and there and trespassing beyond its allotted space. It prefers a moist shady place but is quite hardy and drought resistant. Easily propagated by suckers or cuttings. It was brought into cultivation in European gardens about 150 years ago.

Medicinal uses.—According to Burkill it is used medicinally by the Malays externally, either as a fomentation for rheumatism and ague, or with other substances in the treatment of skin diseases.

Clerodendrum infortunatum Linn.

(*Infortunatus* means unhappy in Latin; but we do not know to what the epithet refers. Trimen says that this name is due to Hermann (Mus. 25, 45) who translates the Sinhalese name 'pinna kole' as 'infortunatus'.)

Description.—A shrub. The quadrangular branches are channelled and covered with a yellowish pubescence. Leaves opposite, ovate, acuminate at the tip, rounded or cordate at the base, more or less covered with rough hairs, with small round glands on the lower surface, toothed on the margins, rarely entire, 4-10 in. long by 3-8 in. wide; petiole densely pubescent, 1-4 in. long.

Flowers arranged in a broad and long terminal panicle; panicle-branches trichotomous, each ending in three flowers, covered with a yellowish pubescence; bracts foliaceous, ovate-lanceolate, falling early; bracteoles lanceolate; pedicels slender, up to 0.5 in. long. Flowers white, tinted with red, pleasantly scented. Calyx silky pubescent, covered with peltate glands, up to 0.4 in. long; calyx tube very short, with acute, erect, lanceolate lobes. Corolla densely hairy without; tube cylindrical, slender, up to 0.75 in. long, 5-lobed; lobes spreading, as long as the tube, the 2 upper spathulate, the 3 lower ovate-oblong, rounded at the top. Stamens 4; filaments long-exserted, slender, glabrous; **anthers** oblong. Ovary glabrous; style slender, stigma shortly bifid. Fruit globular, shining, black, seated upon the enlarged fleshy red calyx.

Flowers.—January-April. *Fruits.*—Rainy season.

Distribution.—Throughout India; often as an undergrowth in sal forest.

Gardening.—A deciduous shrub reaching 12 ft in height in favourable situations. It is usually considered to be a 'weed' but attractive

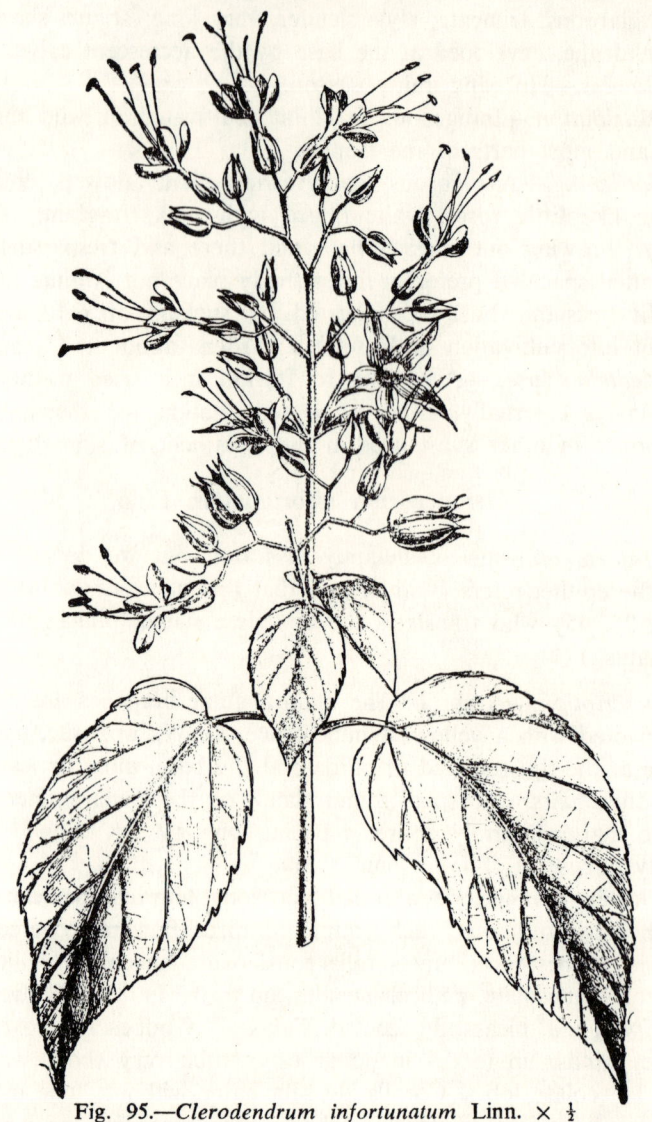

Fig. 95.—*Clerodendrum infortunatum* Linn. × ½

during the cold season with its large head of pinkish white, sweet-scented flowers. Propagated by seed.

Medicinal uses.—The juice of the leaves is stated to be anthelmintic, and is used as a bitter tonic in malarial fevers. The leaves warmed with ghee are applied to the head for catarrhal colds.

Clerodendrum nutans Wall. ex Don. (C. wallichii Merr.)

(*Nutans* means nodding in Latin and refers to the terminal racemose panicles of this plant.)

Description.—A shrub reaching 6 ft in height. Branches quadrangular, glabrous. Leaves opposite or ternate, oblong-obovate or oblanceolate, shortly acuminate at the tip, long attenuate at the base, entirely glabrous, entire or distantly toothed on the margins, membran-

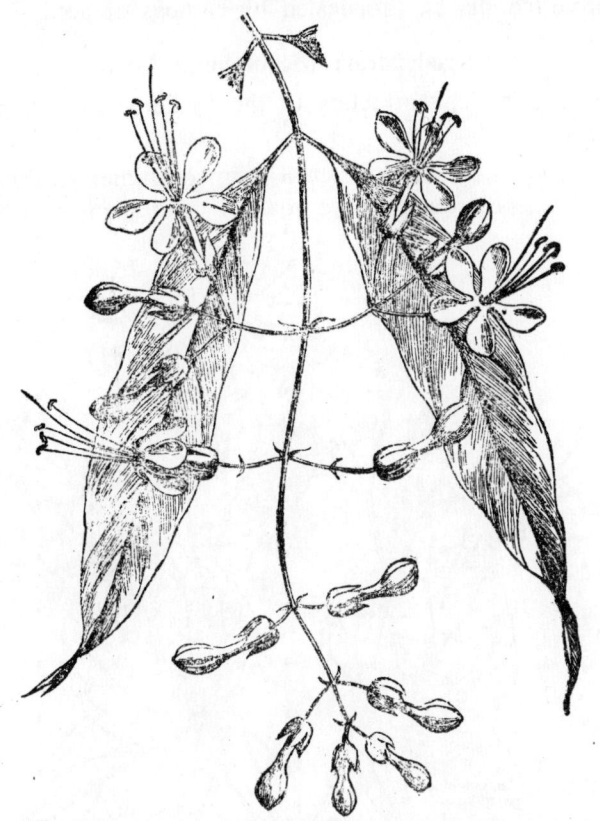

Fig. 96.—*Clerodendrum nutans* Wall. ex Don. × ½

ous, 5-8 in. long, by 1.5-2 in, wide; petiole channelled, 0.5-1 in. long.

Flowers arranged in pendulous slender terminal panicles. Individual flowers pedicelled, white; pedicels 0.75 in. long, with 2 bracteoles near the middle. Calyx ovoid, inflated, purple, glabrous, up to 0.5 in. long; tube very short, 5-lobed; lobes ovate-lanceolate, acute, reddish purple. Corolla glabrous or sparsely and finely pubescent; tube narrowed at the throat, up to 0.5 in. long, 5-lobed; lobes ovate, rounded. Stamens 4, didynamous; filaments slender, long-exserted, inserted in the throat; anthers small, ovate. Ovary small, glabrous; style shorter than the filaments. Fruit a globose drupe, smooth, shining, seated on the enlarged calyx.

Flowers.—September-November.

Distribution.—Sikkim, Assam, Chittagong and Burma, extending to the Malay Peninsula.

Gardening.—A small, hardy shrub with opposite, dark-green, elliptic leaves and pure white flowers produced on long drooping racemes. It is suitable for culture in the ground or in pots, especially in moist, protected places. Propagated by cuttings or seed.

Clerodendrum trichotomum Thunb.

(The specific epithet refers to the trichotomous arrangement of the flowers.)

Description.—A shrub or small tree, sometimes reaching a height of 10 ft. Branches round or the younger quadrangular, soft, glabrous,

Fig. 97.—*Clerodendrum trichotomum* Thunb. × ½

pubescent or covered with a fulvous, crisped tomentum; bark longitudinally fissured. Leaves opposite, petioled, lower very large and 3-lobed, the upper broadly ovate or orbicular-ovate, 6 in. long by 3 in. broad, rounded or truncate at the base, shortly cuneate at the petiole, long acuminate at the tip, soft and flaccid, sparsely hairy on both surfaces, especially on the nervation, entire on the margin, petiole terete, up to 3 in. long, slender, tomentose.

Calyx 0.5 in. long, reddish brown in colour, sparsely hairy; tube obconical, 5-lobed; lobes triangular or ovate, acute, 0.25 in. long or longer. Corolla white, tube more or less exserted, very slender, up to 1 in. long, slightly curved, 5-lobed; lobes nearly equal, 0.5 in. long, elliptic, obtuse or subacute, horizontally spreading. Stamens 4; filaments long-exserted, slender; anthers oblong. Ovary glabrous; style slender; stigma shortly bifid. Fruit 4-lobed, included in the scarcely enlarged calyx.

Flowers.—Rainy season.

Distribution.—A native of Japan, now commonly cultivated in the subtropical parts of the whole world.

Gardening.—A hardy shrub, quite handsome during the rains when it is in flower. It was introduced into England about 1880 and is propagated by seed.

Clerodendrum phlomidis Linn. f. (C. phlomoides Willd.)

Description.—A scandent bush or small tree. Branches pubescent, whitish grey. Leaves opposite, readily disarticulating, 1-2 in. long, deltoid-ovate, often broader than long, truncate or subcordate at the base,

Fig. 98.—*Clerodendrum phlomidis* Linn. f. × ½

obtuse or acute at the tip, coarsely crenate-dentate or subentire on the margin, glabrous above, more or less puberulous beneath; petiole 0.5-1 in. long.

Flowers arranged in small dichotomous axillary cymes on a rhachis so as to form a rounded, terminal panicle; bracts small, leafy, lanceolate, acute. Calyx 0.4 in. long or more, divided about half way down, glabrous, not enlarged in the fruit, 5-lobed; lobes ovate, acutely acuminate, veined. Corolla white or pinkish; tube 0.7-1 in. long, slightly pubescent outside, glabrous inside; lobes nearly equal, 5 in number, 0.3 in. long, elliptic, obtuse, veined. Stamens 4; filaments much exserted, slender, slightly pubescent. Ovary glabrous; style slender; stigma bifid. Fruit a drupe, 0.25 in. long, broadly obovoid in shape, depressed, seated within the persistent calyx lobes.

Flowers.—November-December and also at other times of the year.

Distribution.—Throughout the drier parts of India from the Punjab and Bengal to South India and Ceylon.

Gardening.—A shrub with white or pinkish fragrant flowers. It is a common jungle plant and hardly deserves to be grown in a garden. Propagated by seed.

Medicinal uses.—The leaves of this plant are given to cattle for diarrhoea and worms.

THE TRUMPET HONEY-SUCKLE
Lonicera sempervirens Linn.
(nat. size)

Chapter 9

CAPRIFOLIACEAE

The name of this family comes from the name *Caprifolium*, first used in the seventeenth century by Dodoens, the Flemish herbalist, to designate a climbing plant. The word itself seems to mean 'plant which climbs like a goat'.

This family is not of much importance in Indian gardens for, apart from the genus *Lonicera*, few if any of the other genera are found. In European countries, however, many species of *Viburnum* are grown, as well as other beautiful shrubs belonging to the genera *Diervilla, Leycesteria, Symphoricarpus, Dipelta* and others.

The leaves of this family are opposite, mostly exstipulate. The flowers are bisexual, regular or zygomorphic and generally of five combined parts. The stamens are inserted on the corolla tube. The ovary is inferior and the fruit is a berry or drupe. The resemblance to the *Rubiaceae* will be noticed at once—in fact, the only difference on paper seems to be that the leaves in *Rubiaceae* are stipulate.

The species occur mainly in the temperate regions of the northern hemisphere but they occur in the tropics, at the higher elevations.

One genus is found in many Indian gardnes.

Lonicera

(A genus named in honour of Adam Lonitzer, Professor of Medicine in Mainz, 1528-86.)

Lonicera L. comprises about 160 species which are distributed in the subtropical and temperate regions of the northern hemisphere, from the arctic circle southwards to the Malayan Archipelago, Southern Asia, North Africa, Madeira and Mexico.

Characters of the genus.—Upright to twining, rarely creeping shrubs. Stems glabrous or hairy, often glandular, with fistulose (hollow) or solid branches. The leaves are opposite, petioled or sessile and often with opposite pairs connate at the base. Stipules usually opposite, but occasionally interpetiolar stipules are present. Flowers white or yellow, purple or scarlet, usually arranged in simple cymes which are 2-flowered by the suppression of the middle flower, or three-flowered, and sessile or whorled and collected in terminal spikes or panicles. Individual flowers subtended by bracts and bracteoles. Calyx tube short, adnate to the ovary, ovoid or sub-globose with a short 5-toothed limb. The corolla is tubular to funnel-form or sometimes campanulate with a regular five-lobed limb or 2-lipped, in which case the upper lip is 4-lobed; lobes long or short, imbricate in the bud. Stamens five, inserted in the tube, usually near the mouth, usually exserted. Ovary inferior, 2- to 3-celled; each cell with 3-8 ovules pendulous from the inner angles. Style elongated, with a capitate stigma, usually exserted from the tube. Fruit a red or yellow berry, few-seeded. The receptacle at the base of the corolla tube secretes a nectar. Cross-fertilisation is accomplished by insects which come to take the nectar.

Before we show how this cross-fertilisation takes place we shall describe the mechanism of a typical specimen of the genus, *Lonicera periclymenum* L., a European species.

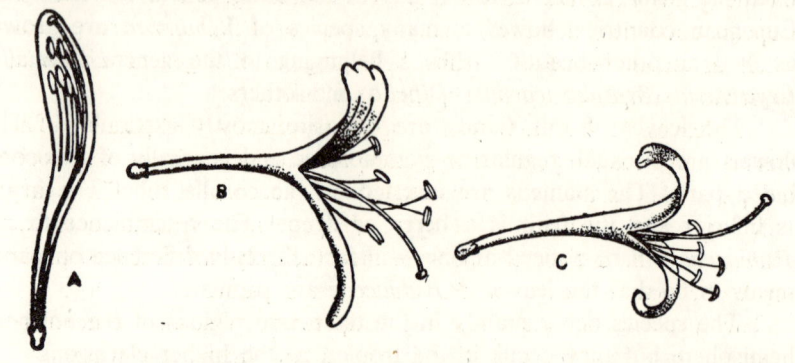

Fig. 99.—Pollination in *Lonicera periclymenum* Linn. after Kunth. × ¾

Just before the erect flower buds open the stigma is receptive and the anthers have already dehisced. The pollen is sticky and as the stigma stands some distance above the anthers, self-fertilisation is avoided (A). When the bud opens the flower sinks through an angle of 90° so that it comes to rest in a horizontal position. The positions of the stamens and style are shown in Fig. B. The style is well out of the way below the stamens whose dehisced anthers are turned upwards. Any insect visitor which now comes to get nectar from the flower is bound to touch the dehisced stamens and carry away pollen on its under surface. After pollen has been removed the filaments wither and sink down. The style at the same time curves upwards and comes to occupy the position indicated in C. It is obvious that in this state of the flower it is the stigma and not the stamens which will touch the undersurface of an insect-visitor. In stage B the flowers are large and brightly coloured; in stage C they have become smaller and duller.

The process of cross-fertilisation is as follows:—On any plant of *L. periclymenum* there will be found flowers in all stages of development from the bud to the fading fertilised bloom. When the bud opens an attractive scent is emitted and the flower itself is fresh and

PLATE 56

M. N. Bakshi

THE TRUMPET HONEYSUCKLE
Lonicera sempervirens Linn.

PLATE 57

M. N. Bakshi

JAPANESE HONEYSUCKLE
Lonicera japonica Thunb.

brightly coloured. Hence insects (hawk-moths) visit these flowers, which are in stage A, first. From the relative positions of stamens and style it is certain that the visitor will carry away quantities of pollen. In stage B, the flower has faded and the scent is not so strong, therefore these flowers are visited later, and as may be seen from Fig. C it is the stigma this time which must touch the insect. In this way cross-fertilisation is ensured.

The Loniceras are deservedly popular on account of the fragrant pretty flowers and handsome foliage.

Key to the Species

Leaves connate .... *L. sempervirens*
Leaves distinct
  Bracts subulate; ovary hairy;
    leaves and flowers larger .... *L. confusa*
  Bracts ovate, leafy; ovary glabrous;
    leaves and flowers smaller .... *L. japonica*

Lonicera sempervirens Linn.

The Trumpet Honeysuckle

(*Sempervirens* means evergreen in Latin.)

Description.—An extensively climbing glabrous woody twiner. Leaves oblong or ovate, rounded at the tips, wedge-shaped or rounded below, opposite, exstipulate, glossy, green on the upper surface, covered with a white bloom below, 2-3 in. long. The upper one or two pairs of leaves are connate by their bases.

Inflorescence a terminal interrupted spike, the individual flowers being whorled in groups of 4-6, supported by bracts and bracteoles. Calyx adnate to the ovary ending above in five short blunt teeth. Corolla tube, 1-2 in. long, seated on top of the ovary, slightly ventricose, glabrous outside, sparsely pubescent with white hairs within, scarlet, orange or sometimes yellow on the outer surface, yellow within, five-lobed; lobes reflexed. On the inner surface near the base, is an oval-shaped area which is slightly thicker than the rest of the corolla and is covered with stalk-like glands.

Stamens five inserted on the tube below the mouth, alternate with the lobes. Ovary inferior, 2-3-celled. Fruit a red berry. The flowers of this species are not fragrant.

Flowers.—February-March. Does not fruit in this country.

Distribution.—Indigenous to Southern United States. Cultivated in gardens in the plains.

Gardening.—A beautiful twiner when in full bloom with its scarlet, though scentless flowers. It is best adapted for a trellis or pergola. It is a common twiner in English gardens and is readily propagated by layers.

Lonicera confusa DC.

(*Confusa* means confused or uncertain in Latin, and refers to the fact that this plant has often been mistaken for a closely allied species, *L. japonica* Thunb.)

Description.—An extensively climbing hairy species. Stems cylindrical, fistulose, covered with short, dense, brownish hairs. Leaves opposite, ovate-lanceolate in shape, rounded at the base, petioled, 2-3 in. long, rather rough, glabrous or sparsely hairy on the upper surface, ciliate on the margins, softly and shortly hairy on the prominent nerves and reticulation of the under surface.

Flowers in axillary pedunculate pairs, or collected into terminal panicles or spikes of whorled pairs. Pairs of flowers supported by subulate hairy bracteoles. Calyx adnate to the ovary, very hairy, ending in five subulate hairy teeth. Corolla 1 in. long, two-lipped, softly and

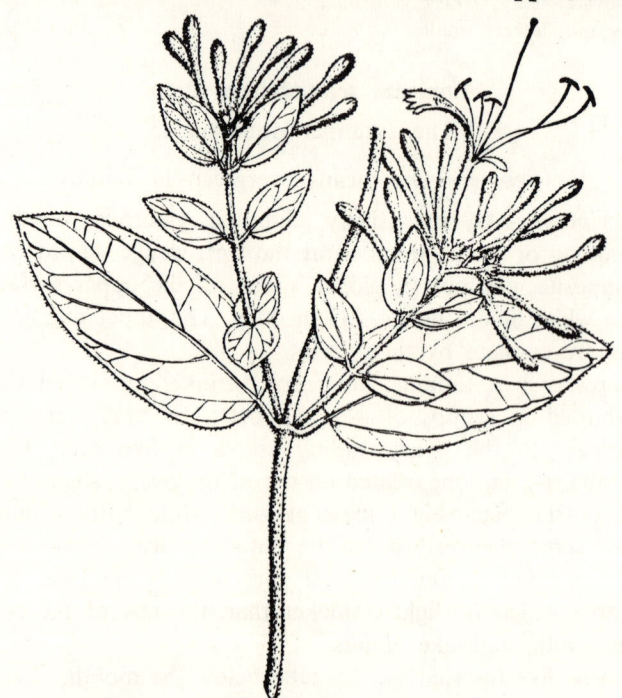

Fig. 100.—*Lonicera confusa* DC. × ¾

shortly hairy within and without, white fading to yellow; tube as long as the lips; upper lip 4-, the lower 1-lobed. Stamens five, alternate with the lobes of the corolla, long-exserted. Style long, filiform. Ovary inferior.

Distribution.—Found wild in China, Java and Borneo. Cultivated in the plains and hills.

Flowers.—March-April. Does not fruit in this country.

Gardening.—An evergreen twining shrub similar to *L. japonica*. The flowers, at first white, afterwards fading to yellow, are produced in March and are quite effective. It is suitable for growing over an arch or trellis. Propagation is by cuttings or layers during the rains.

Lonicera japonica Thunb.

Japanese Honeysuckle

(The specific epithet indicates the origin of Thunberg's specimen.)

Description.—A widely climbing twiner. Stems glandular-hairy, fistulose, cylindrical. Leaves opposite, petioled; ovate to oblong-ovate, 1.75-2.75 in. long, rounded at the base, acute or obtuse at the tip, glabrous or sparsely hairy on the upper surface, glabrous or softly hairy on the undersurface, ciliate on the margins; petiole 0.25 in. long, covered with soft yellow hairs.

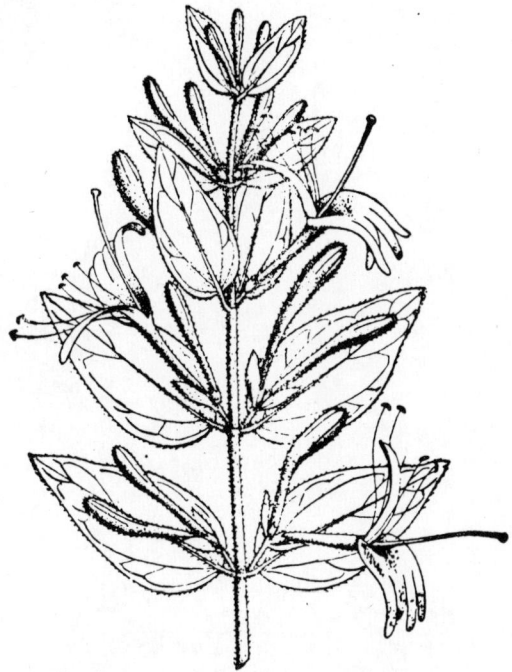

Fig. 101.—*Lonicera japonica* Thunb. × ¾

Inflorescence similar to that of *L. confusa*. The bracts are, however, foliaceous, instead of subulate, in shape. The corolla is somewhat larger and is hairy and glandular without. The calyx is glabrous.

Flowers.—Cold season. Does not set seed in this country.

Distribution.—Indigenous to China, Japan and Formosa, now commonly grown in gardens in the plains and in hill stations.

Gardening.—An evergreen twiner with white, red or purple-tinged flowers which are very fragrant. It is suitable for trellis work and grows readily from cuttings and layers. The form commonly cultivated in gardens is var. *chinensis*.

PLATE 18

THE CAPE PLUMBAGO
Plumbago capensis Thunb.
(⅔ nat. size)

Chapter 10

PLUMBAGINACEAE

The family *Plumbaginaceae* contains besides the genus *Plumbago*, several genera which are valued in temperate countries for their pretty flowers. *Limonium* and *Statice* may be mentioned in this connection, both of which are represented by many species in the gardens of America and Europe.

The family contains for the most part perennial shrubs or half-shrubs and also rosette plants. The leaves are entire, rarely lobed, and characterised by the possession of many-celled glands which secrete water, slime or carbonate. The inflorescence may be simple or compound with the flowers arranged in heads or racemes. Flowers gamopetalous; corolla tubular with a spreading 5-lobed limb. The calyx is gamosepalous, shortly 5-lobed and furrowed between the lobes. Stamens 5. The ovary consists of 5 carpels of which one only has a pendulous ovule on a basal funicle.

Plumbago Linn.

The name *Plumbago* seems to be derived from the Latin *plumbum*, meaning lead. Some authors assert that the name was given to the genus on account of the lead-coloured flowers of some species. There is, however, another explanation. Pliny and Dioscorides, who wrote on medicinal plants in ancient times, prescribed decoctions of the plant *Plumbago europaea* as treatment for an affection of the eyes known in Latin as *plumbum*. The sap of the roots is also said to leave lead-grey coloured flecks if applied to the skin.

The genus *Plumbago*, of which several species are commonly cultivated in the gardens of India, comprises shrubs and undershrubs sometimes scandent. The leaves are alternate, entire, with the petiole sometimes auricled at the base. The flowers are arranged in terminal racemes. The corolla tube is long and slender and is divided at the top into 5 spreading lobes. The colour of the corolla may be white, rose-coloured or blue. The stamens are five in number, opposite to the petals, and are free with oblong, dorsifixed anthers. Ovary superior, 1-celled, 5-angled, and contains one ovule. The fruit is a capsule.

The flowers are protogynous, that is, stigmas are thrust forth from the corolla tube well above the maturing anthers. This is a device to ensure cross-fertilisation which in the case of *Plumbago* seems to be accomplished by flying insects. Unwelcome visitors such as ants, are prevented from reaching the corolla by the rows of viscid glands found on the calyx tube.

A bitter substance known as plumbagin is contained in the tissues of the species of this genus. It is said to be identical with ophioxylin, which occurs in *Rauwolfia serpentina*, a species of Apocynaceae, which is well known in the Hindu pharmacopoeia. Plumbagin has vesicant properties, and is said to be widely used in tropical countries as an abortifacient. As the juice causes large and painful blisters, its use for this purpose seems to be not unaccompanied by danger. More legitimate uses for the drug are in its application to stimulate stagnant ulcers, as a counter-irritant for toothache, as a purgative and as a palliative for rheumatism, glandular swellings and the like.

This genus affords a good example of dispersal through the viscid hairs on the calyx. The calyx is persistent and contains the 1-seeded capsule. Any animal which brushes against a *Plumbago* bush is certain to carry away ripe fruits, transporting the seed to great distances.

Key to the Species

Flowers white or blue
   Calyx wholly covered with stalked glands .... *P. zeylanica*
   Calyx glandular in the upper part only .... *P. capensis*
Flowers red .... *P. rosea*

Plumbago zeylanica Linn.

The Ceylon Leadwort

(*Zeylanicus* is a Latin adjective, meaning of Ceylon, and refers to Southern India or Ceylon as the country of origin of the species.)

Description.—A shrub or semi-scandent shrub with diffuse branches and ribbed stems, covered with a scurfy, glandular exudation. Leaves alternate, petiolate; petiole up to 1 in. long, slightly auriculate at the base; blade up to 3 in. long by 2 in. broad, ovate, acute or obtuse, suddenly narrowed into a cuneate base, dark green above, rather pale below, glabrous; margin entire.

Flowers arranged in terminal simple or branched racemes up to 12 in. long; rhachis very glandular and covered with a viscid exudation. Flowers seated on a very short pedicel supported by a bract and two bracteoles. Calyx 0.5-0.6 in. long, conspicuously 5-ribbed, each rib ending in an acute calyx-lobe. The ribs are covered all over with numerous stalked glands with spherical tips, very viscid. Corolla gamopetalous, white; tube slender, 0.75 in. long, ending in 5 ovate lobes 0.3 in. long. Stamens 5; filaments green, as long as the tube, slightly dilated at the base. Ovary shortly stipitate; style long, divided at the tip into 5 stigmas. Fruit a capsule contained in the viscid glandular persistent calyx.

Flowers and Fruits.—Cold season.

Distribution.—Tropics of Asia and Africa, and east to Australia and Hawaii. Wild and extensively cultivated throughout India.

Fig. 102.—*Plumbago zeylanica* Linn. × ½

Gardening.—A rambling sub-scandent untidy shrub with star-like white flowers. It seeds abundantly but is usually propagated by suckers. It is capable of much improvement by judicious pruning and training.

Medicinal uses.—The medicinal properties of the root of this plant are well known to Indian practitioners. It is said to excite digestion and promote appetite. It is also used in piles, diarrhoea, and skin diseases (Dutt).

Plumbago capensis Thunb.

The Cape Leadwort

(This species, which is indigenous in South Africa, was first collected by Dr. Carl Thunberg.)

Description.—A woody, sub-scandent shrub with striate stems. The leaves are entire, oblong or oblong-spathulate in shape, obtuse or acute at the apex, glabrous, tapering downwards into a very short petiole, up to 2 in. long by 1 in. wide at the broadest part. The undersurface is covered with a white scurfy glandular excretion.

The inflorescence is a terminal raceme; rhachis puberulous. The flowers are seated on very short pedicels supported by bracts and bracteoles. The calyx is gamosepalous, about 0.4 in. long, with 5 acute triangular lobes at the top, 5-ribbed, each rib rather fleshy and ending in the tip of a lobe. The calyx between the lobes is somewhat membranous. In the upper half the calyx bears numbers of stalked glands with swollen purple viscid tips. The corolla is pale blue in colour; tube about 1.5 in. long, 5-lobed, each lobe about 0.5 in. long, abovate-obtuse in shape with a well-marked central vein of a darker blue. Stamens 5, blue, on long filaments the length of the corolla tube. Filaments dilated at the base into what appears to be five nectar-secreting glands. The ovary is shortly stipitate, 1-celled, with 1 pendulous ovule. Style simple, as long as the tube, ending in 5 feathery stigmas, from a basal funicle. Capsule oblong-clavate, rounded above, tapering and pentagonal below.

Flowers.—Practically throughout the year, but in profusion during July-October and December-March. Does not set seed in this country.

Distribution.—Native of the Cape of Good Hope; now widely cultivated in gardens throughout the plains of India.

Gardening.—An evergreen sub-scandnt or straggly upright bush with beautiful azure-blue flowers. It is one of the commonest, as also one of the most ornamental plants, of our Indian gardens. It is advisable to remove all dead wood and old stems, as it is on the new shoots that the best flowers are produced. In South California and elsewhere if planted against a tree it will climb up to 15 ft or so. It is said to make a thick and decorative hedge in America and South Africa. It is best suited for borders or as a bedding plant or even for pot planting. Usually propagated by suckers or division of roots, as cuttings are less successful. It is apt to be cut back by frost where it is severe. There is a white variety similar in every way to the blue except for the colour of the flower.

Medicinal uses.—A dust of the powdered roots is smeared on warts to make them disappear.

Plumbago rosea Linn.

Rose-coloured Leadwort

(*Roseus* refers to the colour of the flowers of this species.)

Description.—A perennial shrub with obscurely striate scandent branches. Leaves alternate, shortly petiolate, broadly ovate or elliptic in shape, up to 8 in. long by 3 in. broad, glabrous, entire, dark green above, pale below; petiole short, dilated and half-amplexicaul.

PLATE 58

M. N. Bakshi

THE CAPE LEADWORT
Plumbago capensis Thunb.

THE CAPE LEADWORT
Plumbago capensis Thunb.

Inflorescence a terminal raceme up to 2 ft long. Individual flowers spaced, often secund, seated on short pedicels, with bracts and bracteoles; rhachis glabrous. Calyx gamosepalous, 0.3 in. long, 5-ribbed and 5-lobed, the lobes short, triangular-acute, covered with very numerous gland-tipped stalked hairs. The corolla tube is 1.25 in. long, slender, with 5 rounded lobes about 0.5 in. long, of a beautiful brick-red colour. Stamens 5, on long filaments. Ovary stipitate, 1-celled, 1-ovuled. Style long and hairy towards the base, ending above in 5 definite stigmas. Fruit a capsule contained in the glandular persistent calyx.

Fig. 103.—*Plumbago rosea* Linn. × ½

Flowers.—Cold season, but also at other times of the year.

Distribution.—Native of Sikkim and Khasia. Extensively cultivated in the tropics of both hemispheres.

Gardening.—A small shrub with pretty red flowers. It is an excellent subject for winter blooming in pots. Propagated by offsets.

Medicinal uses.—It possesses medicinal properties similar to those of P. zeylanica.

Plumbago rosea Linn. var. *coccinea* is a form with larger scarlet flowers and more common under cultivation than the type.

Chapter 11

EUPHORBIACEAE

Pliny relates in his Natural History how this family got its name. It appears that the medicinal plant, now known as *Euphorbia officinarum*, was called *euphorbea* by King Juba of Numidia, in honour of his favourite court physician Euphorbus.

This enormous family, which contains about 200 genera and more than 4,500 species, extends over the whole of the surface of the earth with the exception of the arctic regions. The characters of the family are as follows:

The family includes large and small trees, undershrubs and herbs. Some species are succulent and resemble the genus *Cereus* of the Cactaceae. The juice is often milky, opaline or watery, sometimes acrid. The leaves are alternate, rarely opposite and whorled, petioled or sessile, nearly always simple, rarely trifoliate or pinnate, entire, toothed, lobed, varying considerably in shape, texture and indumentum. Inflorescence various, usually made up of unisexual flowers, often very inconspicuous, gathered into heads or profusely paniculate. Calyx of valvate or imbricate sepals, or completely absent. Petals absent or rarely present and sometimes united (*Jatropha*). Stamens 1-1,000; filaments

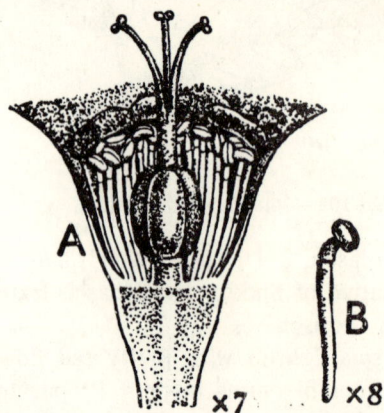

Fig. 104.—*A*. Inflorescence of *E. bojeri*. A typical cyathium showing one female flower surrounded by many male flowers. *B*. A male flower, consisting of a filament with anthers, seated upon a pedicel; there is no floral envelope.

free or connate. Anthers 2-celled, sometimes 3-4 celled, erect or inflexed in the bud, opening lengthwise by slits, rarely by pores; rudimentary ovary usually present in the male flowers. Ovary superior, mostly 3-celled; styles free or united at the base, often again divided

PLATE 19

THE CHRISTMAS FLOWER
Poinsettia pulcherrima Grah.
(⅞ nat. size)

and papillose above; ovules solitary or paired in each cell, pendulous from a funicle which is often enlarged. Disk usually present, circular or of separate glands. Fruit a capsule or drupe. Seeds often decorated with a fleshy appendage called a caruncle.

The family is remarkable for a peculiar type of inflorescence, called a cyathium, which is often, and quite pardonably, mistaken by students for a single flower. In its typical form it consists of an envelope containing inside a number of stamens and a stalked 3-celled ovary. If each stamen is examined carefully with a lens it will be seen that it consists of a short stalk upon which is jointed a filament carrying the anther cells. This structure is to be regarded as a single flower consisting of a single stamen seated upon a pedicel. Calyx and corolla are entirely absent.

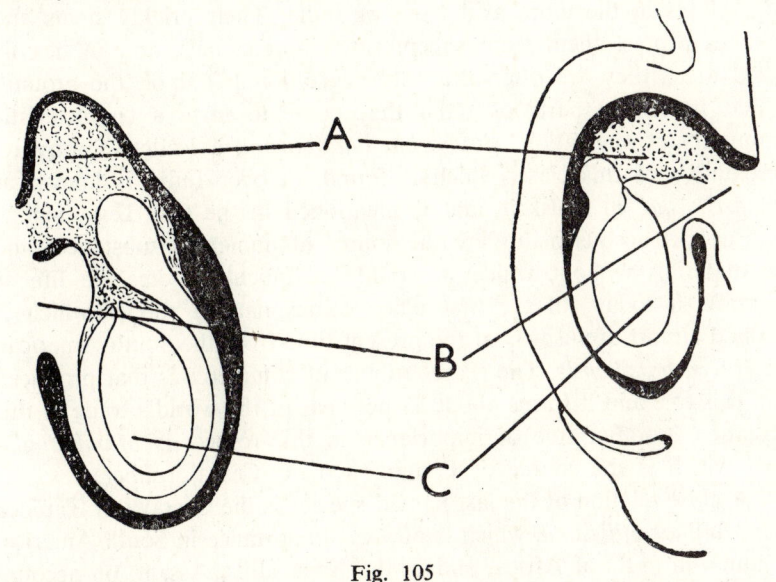

Fig. 105
Longitudinal section of the ovaries of two species of Euphorbiaceae showing:
A. Obturator.
B. Placenta.
C. Ovule.
The obturator is an outgrowth of the placenta and serves to guide the pollen tube to the micropyle.

At the micropyle end of the ovary is to be found a fleshy growth called the obturator, another peculiarity of this family. The obturator is a placental outgrowth whose purpose is to assist in conducting and to nourish the pollen tube. It disappears after fertilisation.

Fertilisation is believed to take place mainly through the agency of Diptera, but there is little doubt that many other insects assist in the process. Some plants are anemophilous; but those that possess nectaries and coloured floral bracts are certainly adapted for cross-fertilisation.

As already stated, this family has representatives over almost the whole of the surface of the earth, but it reaches its greatest luxuriance in the tropics. In the evergreen forests the members of the family occur as lofty trees, shrubs and climbers. In the drier parts of the tropics they occur as out-and-out xeromorphs, resembling some cacti in form. A number of species of *Euphorbia* belong to the latter class and anybody who is familiar with the drier parts of India will remember the arborescent forms of this genus, which are so conspicuous on the arid foothills of the Punjab and United Provinces and in other parts of Central and North-western India. These species of *Euphorbia*, with their prickly fleshy stems full of a white milky juice, are favourite hedge-plants in the more arid parts of India. Their prickly stems and branches protect them from sheep, cattle and camels, and in heavily grazed areas they are often the only vegetation left upon the ground; so that in certain parts of India they come to form a characteristic community of vegetation. *Euphorbia antiquorum*, a leafless thorny representative of this class, which is found all over India is known as 'Maharrik sha' in Sanskrit, and is mentioned in the Old Testament.

Euphorbiaceous plants are the source of numerous substances and raw materials without which it would be difficult to imagine life as we know it today. In the first place comes natural rubber which is obtained almost exclusively at the present time from the South American tree *Hevea brasiliensis*. The 'Para' rubber of commerce is that produced from this tree and it forms about 85 per cent of the world's trade in this substance. Another tree of importance in this respect is *Manihot glaziovii* which is the source of Pernambuco or 'Ceara' rubber.

A close relation of the last named species is the Cassava or Tapioca tree, *Manihot utilissima*, which is of great importance in South America, its home, in parts of Africa, and in the Naga Hills, Assam, on account of the swollen roots which contain a large quantity of starch. The remarkable fact about the swellings is that they contain sufficient hydrocyanic acid to render them extremely poisonous. This poison is liberated by the action of an enzyme on a glucoside, phaseolunatin, particularly when wilting occurs. Thus a fresh root which is free from hydrocyanic acid may become deadly poisonous if allowed to become stale. When a tuber is boiled the enzyme is destroyed and the hydrocyanic acid, being soluble in water, is carried away. As the acid is found chiefly in the cortical layer, peeling the tubers removes the danger of poisoning to a large extent.

Another plant of importance is the tree *Emblica officinalis*, the myrabolan tree. The fruit of this tree is much used in India where the juice of the fruit is taken internally for dyspepsia, dysentery and as a diuretic. It is used externally for conjunctivitis. Recently (1943) it has been discovered that the fruits of this tree are very rich in Vitamin C.

Castor oil is obtained from *Ricinus communis,* another member of the family. It is believed that its native home is in Africa but it is cultivated for its oil in all the warmer parts of the earth. Besides the oil, the castor oil seeds contain a very poisonous substance which is called ricin. Ricin is the substance which causes death when the seeds are eaten, its action being that of a coagulant in the blood. It is known that death has followed the eating of four seeds. The German chemist, Ehrlich, discovered an interesting fact about this poison. When ricin is introduced into the blood stream, an antitoxin (antiricin) is produced in the blood, so that the body can acquire a certain immunity towards the poison. It is now known that several other substances as well as the bacterial toxins act in the same way. Fowls appear to be uneffected by the poison and horses can consume considerable quantities of it. Castor oil is used (apart from its use as a purgative) as an illuminant and as a lubricant for aero engines where its viscosity and low freezing point render it particularly valuable.

Several other species contain poisonous elements in the seed or latex. *Excoecaria agallocha,* a seashore shrub or small tree, contains a substance in the juice which can blister the skin and has been known to cause blindness. The most poisonous of them all is said to be the 'machineel' tree, *Hippomane mancinella,* which is a tall tree of Tropical America. As was the case in other poisonous trees, this particular species was credited with the most extravagant properties by early explorers of the nineteenth century. For example, to sleep in its shade meant death, and even to go near it was certain to cause erysipelas if nothing worse. All these fables were disproved by one brave soul, J. Jacquin, who stood naked under the tree for several hours and took no harm.

A number of species of the *Euphorbiaceae* are cultivated in gardens, either for the striking colour of the floral bracts or for their variegated or handsome foliage.

KEY TO THE GENERA

Flowers contained in a bowl-shaped structure:
  no calyx or corolla .... *Euphorbia*
Flowers with a distinct calyx and conspicuous
  red corolla .... *Jatropha*

Euphorbia Linn.

A very large genus, the majority of the species of which are herbs, but there are a few shrubs, some of which are grown in gardens. All species possess a copious milky juice. The leaves are alternate, opposite or verticillate, stipulate; stipules often glandular or spiny. Inflorescence of cyathia which are almost entirely male, sometimes dioecious. The involucre is cupular or campanulate, often provided with nectar-secreting glands, and containing the closely packed male flowers, each consisting of an antheriferous filament jointed on a pedicel. Interspersed are a number of sterile pedicels usually lacerate or hairy. The stamens are 4-30 or even more in number. The anthers are two in number, 1-celled, one on either side of the tip of the filament, at first depressed, eventually erect at dehiscence. Female flowers seated on a pedicel, naked, 3-lobed, 3-celled; one ovule in each cell; ovules pendulous; styles three in number, connate or not, dividing above, into two stigmatiferous arms. Fruit a capsule consisting of three cocci.

Key to the Species

Stem spiny and prostrate: flowers without
  long red bracts .... *E. bojeri*
Stem spineless, erect; flowers supported by
  long red, pink or white bracts .... *E. pulcherrima*
                                                                      (Poinsettia)

Euphorbia bojeri Hook.

Crown of Thorns

(Named in honour of Wenzel Bojer, 1800-1856, an Austrian botanist, who wrote a flora of Mauritius.)

Description.—A low sprawling shrub with long-spined livid-coloured fleshy branches and small leaves. Spines on the branches in pairs up to 1 in. long or more, very sharp. In between each pair of thorns is a reduced branch which bears pairs of small spines 0.12 in. long. In between each pair of small spines a leaf is developed. The spines may therefore be regarded as stipules. The leaves vary much in size, from 0.25 in. to 1 in. long, obovate-oblong, smooth and glabrous, green with a narrow border or red on the margins, entire, apiculate at the apex, attenuate at the base.

The inflorescence is compound and consists of cyathia grouped dichotomously and seated upon glabrous peduncles which emerge from the axils of the uppermost leaves. The peduncle is divided into two successively three times, and the cyathia are seated upon the ultimate peduncles. Each cyathium is supported by two petaloid crimson orbicular apiculate bracts which have become fused to the involucre in their lower part. The orbicular or broadly elliptic bracts are glabrous and smooth and are a deeper shade of crimson inside than out, and are 0.3 in. broad by 0.25 in. long, often emarginate at the tip instead of apiculate. The involucre itself is about 0.1 in. long, and bears at the top margin five broad rounded glands which are almost parallel to the long axis. Alternate with the glands is a scale, reddish in colour, and laciniate on the margins. Inside the involucre there are many male flowers, in all stages of development, interspersed with hairy organs which look like sterile stamens. The male flower consists of a pair of anthers attached to a filament which is jointed to a pedicel. The ovary is glabrous, sessile, and is wholly concealed within the involucre, 3-celled, 3-lobed, with one ovule in each cell. Styles, three, connate for half their length, each divided at the apex into a pair of short stigmas.

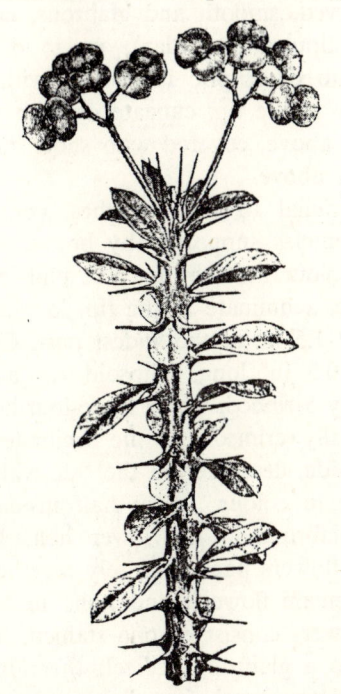

Fig. 106.—*Euphorbia bojeri* Hook.

Flowers.—Practically throughout the year. Does not set fruit in this country.

Distribution.—A native of Madagascar, very common in gardens throughout the country.

Gardening.—A small xerophytic undershrub, armed all over its stem with sharp long spines. The red bracts and the green leaves on the sinuous spiny stems are very striking. It is well suited for a 'rock-garden', in dryish soil and in exposed and sunny situations. It will, however, grow anywhere but thrives best in a mixture of charcoal, leaf-mould and brick rubble. Propagation by cuttings.

Poinsettia pulcherrima (Willd.) Grah. (Euphorbia pulcherrima Willd. ex. Klotzsch)

Christmas flower

(*Pulcherrima* in Latin means most beautiful.)

Description.—A shrub reaching 10 ft in height, with a thick short trunk covered with brown bark and numerous slender unbranched green branches. Sap copious, milky. Branches cylindrical, smooth and glabrous, green, bearing alternate leaves. Leaves petiolate (petiole 2.5 in. long, slender, curved, smooth and glabrous, crimson on the upper surface), stipulate (stipules glandular), up to 6 in. long by 2-3 in. broad, ovate-lanceolate or elliptic in outline with 1-2 blunt teeth on either side, acuminate at the tip, cuneate or even rounded at the base, smooth and glabrous above, covered with short hair on the lower surface; mid-rib reddish above.

Inflorescence terminal to the branches, consisting of groups of cyathia on stout peduncles supported by bracts, which are usually of a beautiful crimson colour but they may be pink or even white. Bracts oblanceolate in shape, acuminate at the tip, long-attenuate at the base, up to 6 in. long by 1-1.5 in. at the widest part. Cyathium seated on a stout peduncle, 0.25-0.5 in. long, ellipsoid-truncate in shape, smooth and glabrous, obtusely 5-ribbed, lower three-fourths green, then banded with yellow and finally crimson at the laciniate mouth, or entirely orange on the gland side, decorated at the side with a large compressed conical gland, orange in colour; upper half inside covered with silky multicellular hairs, glabrous in the lower half. Inside the cyathium there are many male flowers packed closely together, many-sterile male flowers, one or no female flowers, and none to several sterile female flowers. The male flowers consist of one stamen, the glabrous filament of which is jointed to a glabrous pedicel; there is no floral envelope. The two 1-celled anthers are, before dehiscence, depressed on either side of the tip of the filament. After dehiscence they come to stand erect like a pair of disks at each side of the tip of the filament. The sterile male flower consists simply of a filament ending in a reddish tip and covered with multicellular woolly hairs. The female flowers consist merely of an ovary jointed to the tip of a glabrous pedicel; ovary hairy or glabrous, eventually erect and thrust out of the involucre, 3-celled, uniovulate; styles 3, joined almost for the whole of their length and then each divided into two stigmatic lobes. Sterile females on glabrous pedicels.

Flowers.—Cold season.

Distribution.—Native of Mexico and Central America, now one of the most commonly cultivated plants in gardens.

PLATE 60

M. N. Bakshi

THE CHRISTMAS FLOWER
Poinsettia pulcherrima Grah.

PLATE 61

M. N. Bakshi

THE CHRISTMAS FLOWER

Gardening.—A tall unarmed soft-wooded shrub, 8-10 ft high. It bears during the cold weather knobs of insignificant flowers surrounded by deep crimson floral bracts (with variations to shades of pink or even white) which are the chief ornamentation of this plant. It is important among ornamental shrubs as it flowers at a time when practically very few flowers are available in the garden. It is suitable for planting in the open in all areas and is not particular as to its soil requirements. It should be rigorously cut back after flowering, since it is on the current years shoots that the flowers are produced. It is a popular plant for the Christmas season. Readily multiplied by cuttings.

Euphorbia pulcherrima Willd. var. *alba* Hort. is a variety with cream-coloured bracts but comparatively of little beauty.

Jatropha Linn.

This large genus comprises some 160-170 species and contains some ornamental plants which are cultivated in our gardens. In it are to be found trees, shrubs and herbs, the latter frequently with a thick perennial rhizome. The leaves are alternate, petiolate or sometimes sessile, lobed or cut in various ways, glabrous or pubescent, often glandular.

The inflorescence is a terminal cyme. The flowers are unisexual, found on the same, or on different plants, often with calyx and petals. Sepals 5, imbricate, often more or less joined together. Petals 5, imbricate, free or joined together into a five-lobed tube, sometimes absent. The disk may be entire or divided into five glands. Stamens usually about 10, often in two series, the outer series seated upon the petals. The filaments are often joined together. The anthers open by longitudinal slits. In the male flowers there is no rudimentary ovary. The ovary is 2-5-celled with a solitary ovule in each cell. The styles are joined together at the base, shortly divided into 2 branches. Capsule dehiscing into 2-valved cocci. The endocarp is crustaceous and the seeds bear a caruncle.

Many species of the genus are graceful plants with pretty foliage and brightly coloured flowers. A number of them, therefore, are cultivated in gardens in the tropics.

Apart from the striking flowers the species of the genus are remarkable for possessing poisonous substances in the sap or seeds. One or two species protect themselves from browsing animals by the possession of stinging hairs of peculiar construction. Each hair consists of a single large cell expanded at the base and elongated above, ending in a small knob which is bent to one side. The wall of the hair at the bend is extremely thin and the slightest touch is sufficient to break off the head, leaving a sharp oblique point. The sharp oblique point will penetrate the skin of men and animals, and the pressure from the expanded end of the hair injects the contents of the hair into the skin. The principal contents of a hair are formic acid but there are also present some other substances resembling unorganised ferments or enzymes. The formic acid gives the burning sensation but the ferments act as a poison.

Jatropha curcas Linn., a native of America, is not a garden plant but is well known in India where it is frequently grown as a hedge, being easily raised by cuttings or from seed. The sap of this plant is used by Naga children, when diluted with water, for blowing bubbles. The juice also has some repute as a fish-poison. The leaves are used medicinally in a variety of ways, the juice being a styptic and rubifacient.

It is the seed, however, which has made this plant notorious on account of the many cases of poisoning which have occurred after eating it. The seeds contain a toxalbumin which is rich in oil. The poison is an albuminoid called curcin. The absence of any unpleasant taste when eating the fruits, makes the plant a dangerous one when children are about.

The oil which can be extracted from the seeds is put to a variety of uses. It is inflammable and burns without smoke. It can be made into soap and is used in wool-spinning.

KEY TO THE SPECIES

Stem much swollen at the base .... *J. podagrica*
Stem not swollen at the base
   Leaves entire, fiddle-shaped .... *J. panduraefolia*
   Leaves lobed
      Lobes 5; petiole, leaves and stipules glandular .... *J. gossypifolia*
      Lobes 5-11; leaves etc. not glandular .... *J. multifida*

Jatropha podagrica Hook.

(*Podagra* is a Latin word meaning gouty, and refers to the swollen base of the plant.)

Description.—A shrub 2-3 ft tall. The base of the stem is grotesquely swollen as are also the bases of the branches. The branches are soft and succulent, deeply scarred where the leaves have fallen away. Each of the more recent scars has two stipules, one on either side, which persist for some time after the leaves have fallen. These stipules are broad and are deeply cut into setaceous lobes, sometimes glandular. The peltate leaves which are seated on succulent pedicels up to 1 ft long, are 1 ft long or less by 8 in. wide, 3-5-lobed, perfectly glabrous, green above, glaucous beneath; the lobes are entire, somewhat ovate in shape.

FIDDLE-LEAVED JATROPHA
Jatropha pandurifolia Andr.
(nat. size)

The cymose inflorescence is borne on an elongated, succulent, light green peduncle; branches and pedicels red; flowers unisexual; male and female flowers in the same inflorescence. The female flowers are borne in bracts on the branches of the inflorescence. The calyx is cup-shaped, 5-lobed; lobes erect; petals 5, 0.3 in. long, orange-red or scarlet, spathulate, slightly joined together at the base; ovary ovate, seated on a fleshy disk with five glands; style short, divided into several green stigmas. Male flowers much more numerous than the female and above them. Sepals and petals as in the female flowers. Stamens 6-8, seated in a yellow disk, furnished with 5 yellow glands; filaments red, 1.5 in. long; anthers yellow, hastate 0.1 in. long. Fruit a capsule up to 1 in. long, green, 3-celled, with one seed in each cell, ellipsoid in shape, depressed at both ends.

Fig. 107.—*Jatropha podagrica* Hook. × ½

Flowers.—Chiefly during the rains. *Fruits.*—Cold season.

Native country.—Indigenous to Panama, now common in gardens throughout India.

Gardening.—A peculiar gouty-stemmed undershrub having cymes of rather pretty scarlet flowers. Suitable for a rockery. Propagated by division or seed.

Jatropha panduraefolia Andr. (J. hastata Jacq.)

Fiddle-leaved Jatropha

(*Panduraefolia* is Latin for 'with fiddle-shaped leaves'.)

Description.—A shrub up to 6 ft tall with slender graceful branches. Branches greenish brown, with prominent lenticels. Leaves alternate, stipulate, petiolate, fiddle-shaped, shallowly cordate at the base, where there are 3-4 glandular teeth on each side, up to 4 in. long by 2 in. wide, smooth and glabrous, dark green on the upper surface, paler or with a purplish tinge below. Stipules small, subulate. Petioles up to 1.5 in. long.

Inflorescence a small terminal cyme; peduncle purple or purplish green, slender. Male and female flowers on different plants. Male flowers: calyx small, cup-shaped, 5-lobed, purplish red in colour, 1-2 in. long; petals spathulate, 0.25-0.5 in. long, scarlet, twisted in the bud, with a covering of short white hairs inside at the base; stamens 8, often 4 long, 4 short; filaments joined in a column, red; anthers reddish yellow, hastate. Disk present, glandular, 5-lobed. Female flowers: calyx as in the male flower but longer; petals and disk the same as those of the male flower; ovary ovoid, glabrous, attenuate into a column at the summit; styles three, bifid from the middle into filiform branches. Fruit a capsule, purplish green in colour.

Flowers.—Practically all the year round but chiefly during the rains. Fruits ripen in the cold season.

Native country.—Native of Cuba. Commonly cultivated in the Tropics of both hemispheres.

Gardening.—A pretty, rather slenderly branched shrub with fiddle-shaped leaves and bright crimson flowers. It is desirable to prune the plant severely during the cold weather in order to prevent it from becoming scraggy. Readily propagated by cuttings or by seed.

Jatropha gossypifolia Linn.

Bellyache Bush

(*Gossypifolia* is Latin for 'with leaves like the cotton plant.)

Description.—A shrub up to 6 ft tall with rather soft, succulent, brownish green branches, much scarred where the leaves have fallen away. Leaves up to 6 in. long, 6 in. wide, petiolate, stipulate, 3-5 lobed, of a deep purplish red at first, afterwards green; lobes elliptic-acute; petioles up to 4 in. long, colour of dried blood, covered on the upper surface with glandular hairs branched from the base and mixed with simple hairs; stipules of glandular hairs branched from

PLATE 62

M. N. Bakshi

FIDDLE-LEAVED JATROPHA
Jatropha panduraefolia Andr.

PLATE 63

M. N. Bakshi

FIDDLE-LEAVED JATROPHA
Jatropha panduraefolia Andr.

BELLY-ACHE BUSH
Jatropha gossypifolia Linn.

PLATE 65

M. B. Raizada

BELLY-ACHE BUSH
Jatropha gossypifolia Linn.

the base. Margins of the leaves ciliate with simple, white hairs and also furnished with gland-tipped hairs.

Inflorescence a terminal cyme seated upon a short thick dark red peduncle. Male and female flowers on the same plant or on different plants. Calyx 0.25 in. long, consisting of a short cup with 5 lanceolate gland-margined lobes. Petals hardly longer than the calyx lobes, dark red, crimson or purplish in colour, pale at the base, broadly obovate in shape; tip rounded. Ovary globular-ellipsoid, smooth, seated on a glandular cup-shaped disk; style-column short; stigmas three, each dividing into two rugose plates. Stamens 6-8; anthers horse-shoe shaped, crimson; filaments pale, connate in a central column issuing from a glandular disk. Fruit a capsule 0.4 in. long, 3-furrowed, truncate at both ends. Seed greyish red, marbled with black, bearing a caruncle.

Fig. 108.—*Jatropha gossypifolia* Linn. × 2/3

Flowers and Fruits.—Chiefly rainy season.

Native country.—Native of Brazil, cultivated or naturalized in various parts of India and Burma.

Gardening.—A shrub with 3-5-lobed leaves; easily recognised by the stipitate, yellow, viscid glands which cover the leaf margins, petioles and stipules, and by the small red flowers in glandular corymbose cymes. Easily raised from seed. It is deciduous in the cold season.

Economic and Medicinal uses.—The seeds of this plant are said to be eaten by doves and fowls, although they contain much oil which is a drastic purgative and emetic. A decoction of the leaves is used as a blood purifier and for venereal diseases. The root has some repute as an antidote for snake bite (Standley).

Jatropha multifida Linn.

The Coral Plant or Physic Nut

(*Multifida* is Latin for 'much divided' and refers to the palmately divided leaves, the lobes of which are often again divided.)

Fig. 109.—*Jatropha multifida* Linn.× ½

Description.—A shrub or small tree with thick soft branches. Leaves long-petiolate, deeply divided in 5-11 lanceolate-acute linear or elliptic-acute lobes, the lobes often again lobed, green above, often pale-pink below, green at maturity. Stipules short, subulate, branched. Petioles a foot long or more, smooth and glabrous.

Inflorescence a many-flowered long pedunculate terminal cyme; branches slender, reddish in colour. Male and female flowers on the same plant. Calyx gamosepalous, shortly 5-lobed; lobes short, rounded at the tips. Petals five, oblong or broader above, rounded at the tips, red or purplish, 0.2 in. long. Stamens 8-10; filaments connate for a short distance; anthers linear. Disk cup-shaped; glands prominent. Ovary seated on the disk, three-celled. Fruit a capsule, obovate in shape, smooth, yellowish, about 1 in. long.

Flowers and Fruits.—Chiefly during the rains.

Native country.—Native of South America, commonly cultivated in gardens throughout India.

Gardening.—A handsome garden plant with rather pretty foliage and coral red flowers. Propagated by seed.

Economic and Medicinal uses.—The leaves are said to be cooked in Mexico as a vegetable. The yellow sap is used in Brazil for the treatment of wounds, and the roasted seeds for fevers and venereal diseases. The seeds are purgative like those of many other species.

Chapter 12
COMBRETACEAE

This family takes its name from *Combretum*, a name used by Pliny for an, as yet unidentified, climbing plant. It is a tropical family and contains trees, shrubs or undershrubs, often climbing. The leaves are opposite or alternate. The flowers are arranged in spikes which are often panicled, usually inconspicuous, occasionally spectacular. The calyx is superior and often tubular, carrying the small valvate petals, or the petals may be absent. The ovary is inferior, one-celled, with pendulous ovules. Fruit angled or winged.

Some of the tree species will be known to readers. Everybody has seen or heard of the Arjan, *Terminalia arjuna*, a favourite tree in gardens and for the roadside. The Terminalias are not particularly valued for their wood, though *T. tomentosa* gives an excellent timber, but the fruits of some of them, the myrabolans, are greatly valued. The fleshy part of the fruit contains a valuable tanning material. Another genus, *Anogeissus*, has one of the toughest known timbers and is therefore greatly sought after as a wood for hammer handles.

Quisqualis Linn.

The name *Quisqualis* means literally *who? what?* and was first given to the plant by Rumphius (George Eberhard Rumpf), the Dutch botanist, and expresses his astonishment at what he calls the curious behaviour of the species *Quisqualis indica* Linn. which we are about to describe. According to Rumphius the young plant grows up into an erect shrub with scattered leaves and irregular branches. After six months it sends out a runner from the root which is much stouter than the original stem, and this runner then proceeds to climb up the neighbouring trees, not by twining round them, but by means of the petioles which become transformed into stout spines after the fall of the leaves. The name of this climber remains as an exclamation of astonishment but the plant as regards its growth is uninterestingly normal.

Quisqualis indica Linn.
The Rangoon Creeper

A very large woody creeper which is indigenous in Malaysia, south-eastern Asia and west tropical Africa. It is considered not to be indigenous in peninsular India, but it has become so popular as a cultivated plant that it is now one of the commonest creepers to be found in Indian gardens.

PLATE 66

THE RANGOON CREEPER
Quisqualis indica Linn.

M. B. Raizada

Plate 21

THE RANGOON CREEPER
Quisqualis indica Linn.
(nat. size)

Description.—A very large scandent deciduous shrub with a cylindrical stem, green when young, covered with soft brown or golden hairs. Leaves opposite or nearly so, without stipules, oblong or elliptic in shape, slightly cordate at the base, tapering to a blunt or notched apex, 3 in. long; margins entire. The upper surface of the leaf is smooth and glabrous except for the nerves which are hairy: the under surface is softly hairy on the prominent main nerves and on the reticulation. The petioles are short, up to 0.3 in., softly hairy, and are often persistent after the fall of the leaf becoming transformed into stout curved spines which assist the plant to climb.

The flowers are borne in axillary or terminal pendulous racemes. Each individual flower is seated on a short pedicel which emerges from the axil of a small leaf or bract on the main axis of the raceme. The flower itself consists of three parts which can be easily made out: a basal 5-angled ellipsoid portion, 0.25 in. long; a tubular green section, 2.75 in. long, and five-coloured petals. The basal five-angled portion contains the ovary. A longitudinal section through the ovary will reveal that it consists of a single cell from the apex of which hang 2-4 ovules attached to long strap-shaped funicles. The inner surfaces of the funicles are papillose. The filiform style is as long as the calyx tube and ends in a sticky knob-shaped stigma. The calyx tube or hypanthium is slightly constricted above the ovary and is narrowly funnel-shaped, hairy on the outside, dividing at the top into five triangular teeth. The ten stamens are arranged in two rows. One row of five is attached to the tube just inside the mouth, and the stamens are alternative with the calyx lobes. The other row is seated about 0.25 in. lower down on the tube and alternates with the stamens of the upper row. As the filaments are 0.25 in. long, one row of stamens is exserted while the other five only just reach the mouth of the tube. The petals, which are softly hairy outside, are imbricate in the bud and are attached to the calyx tube just at the mouth and alternate to the lobes. The petals are obovate in shape, nearly 0.5 in. long, rounded at both ends, and are attached to the calyx by a mere point. The outer surface of the petals is slightly flushed with pink, the inner surface after the bud has opened is a pure white, which slowly turns pink and finally a rich red. The flowers are beautifully scented in the evening. As the inflorescence is racemose, that is, the flowers open in succession, there are to be found flowers of all ages on the plant, hence the effect of the numerous blossoms, some white, some pink, some red, is very striking. The fruit is narrowly elliptic, five-angled, about 1 in. long.

Economic uses.—The fruits have a great repute for their anthelmintic properties. The bitter liquid which results from pulping the unripe fruit in winter, is often used for this purpose. The ripe fruits although they can be used as a vermifuge, are not so efficacious. The ripe nuts are pleasant to the taste and are eaten but only in strict moderation. In this connection the remarks in Curtis' Botanical Magazine (sub. tab. 2033) are interesting: 'It is observed that, though some persons, and among these are Rumphius himself, could eat these kernels with pleasure and impunity, in others they soon produced nausea, followed by a troublesome hickuping'.

Flowers.—Practically all the year round, but fruit is rarely found.

Distribution.—Indigenous in the Malaya Peninsula, the Philippines and Western tropical Africa. Now commonly planted in all tropical and subtropical countries.

Gardening.—A rapid-growing deciduous scandent shrub requiring a strong trellis for its support. In rich soil its growth is rampant and unmanageable. Consequently it is advisable to cut it back during the dry season. It is constantly in bloom all the year round. The flowers are white, sweet scented, open at night but turning pink at day-break. This mixture of pink and white gives the plant an unique and charming appearance when in flower. Easily raised from layers, cuttings or divisions of the root. It rarely fruits in Northern India.

Chapter 13
MALPIGHIACEAE

A family of flowering plants which commemorates the name of Marcello Malpighi, 1628-93, a distinguished Italian botanist, who wrote on the anatomy of plants. The family comprises trees, shrubs and climbers, some of which are grown in Indian gardens. The leaves are opposite and glands are often present either on the petiole, on the margins or on the under surface of the leaves. Stipules are either present or absent, sometimes large and connate. The flowers are usually hermaphrodite. Sepals five in number, often with two large glands outside. The petals are also five, clawed. Stamens usually 10. Ovary sessile on an obscure disk, usually of three carpels, free or united with 1 ovule in each cell. Styles usually 3, mostly free. Fruit sometimes winged.

The hairs on the shoots of many species are very peculiar and if found on a leafless twig are quite sufficient to place the plant in this family. These hairs are one-celled and branched and are found in three forms (*a*) the magnet-needle type, (*b*) the forked type and (*c*) the morning-star type. The first type has a short pedicel upon which is attached at right angles the two arms of the hair; these may be straight or curved. The foot may be very short in which case the hairs appear to be attached to the under surface by the centre. This

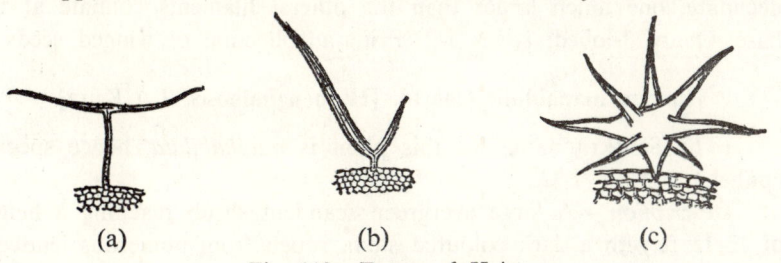

Fig. 110.—Forms of Hairs.
(*a*) Magnet-needle type, (*b*) forked type, (*c*) morning-star type.

type of hair gives a silvery or metallic sheen to the shoot. The second type of hair, in which the foot is long or short, supporting two arms, gives a felty or woolly appearance to the shoot. The third kind of hair, the morning-star type, consists of a foot upon which is found a globular many-branched head. Hairs of this kind give a mealy appearance to the shoot, which recalls the indumentum of many species of *Chenopodiaceae*.

The growth of the stems of certain climbing species resembles that of some species of the *Bignoniaceae* already mentioned in an earlier chapter. Localised growth often leads to deformed stems in which deep clefts are succeeded by protruding woody portions.

Characteristic of the *Malpighiaceae* are the glands which are found so often on stem, under surface of the leaves and upon the outer surface of the sepals. What purpose these glands serve is not known.

The following climbing genera of *Malpighiaceae* are cultivated in India: *Stigmaphyllon, Hiptage* and *Banisteria. Galphimia* is a common shrub.

The climbing species may be distinguished from one another as follows:

Style 1 .... *Hiptage*
Styles 3
   All the stamens fertile; style tops obtuse .... *Banisteria*
   Four of the stamens sterile; style tops leaf like .... *Stigmaphyllon*

Hiptage Gaertn.

(The generic name is derived from the Greek verb *hiptamai*, meaning 'to fly', and refers to the winged fruits of the genus.)

A genus of erect or climbing shrubs. Leaves opposite, entire, exstipulate, sometimes glandular within the margins. Flowers in axillary or terminal recemes, rarely in congested leafy panicles; peduncles bracteate, articulated with the 2-bracteate pedicels. Calyx 5-lobed with one large oblong or linear gland outside and partly on the pedicel. Petals five; four equal in size, white, the fifth yellow. Stamens 10, declinate, one much larger than the others; filaments connate at the base. Ovary 3-lobed; styles 1-2. Fruit a collection of winged seeds.

Hiptage madablota Gaertn. [H. benghalensis (L.) Kurz].

(The Sanskrit name for this plant is *madhabilata*, hence specific epithet.)

Description.—A large evergreen scandent shrub reaching a height of 12-15 ft with a dark coloured stem, rough from numerous lenticels. Leaves opposite, usually without stipules, 4-6 in. long, up to 3 in. broad, coriaceous, dark green and shining above; petiole 2-5 in. long, glabrous or finely hairy below; nervation prominent below. Flowers showy, fragrant, in large terminal and smaller axillary panicles; peduncles and pedicels continuous, 5 in. long with a pair of bracteoles at the centre, covered with dense, appressed, short, silky hairs. Calyx 5-lobed, covered with a dense, silky pubescence; lobes obtuse. A large linear or oblong gland will be found outside partly on the calyx and partly on the pedicel. Petals 5, clawed with an obovate or sub-orbicular limb, silky outside, glabrous within, fimbriate on the margin, four equal in size, white in colour, the fifth smaller and pale yellow, all reflected in the open flower. Stamens 10, one much larger than the others; filaments connate at the base; anthers ovate. Ovary pubescent, 3-locular;

style 1-3 in. long, filiform, circinate in the bud. Fruit of 1-3, 3-winged samaras.

Flowers.—February-April. *Fruits.*—May-June.

Distribution.—Native of India and Malaya, cultivated in gardens in the plains throughout the country.

Gardening.—A large evergreen rampant climbing shrub. It is attractive when in full bloom, with its profuse trusses of white and yellow fragrant flowers, borne on short spikes and resembling somewhat those of the Horse-chestnut. It is rather a heavy climber and

Fig. 111.—*Hiptage madablota* Gaertn.

needs a great deal of space. It would look well on a strong trellis or on a pergola. Propagated by seed.

Medicinal uses.—The leaves are considered medicinal and are useful in chronic rheumatism and skin diseases. It is also said to possess insecticidal properties.

Banisteria Linn.

(The genus *Banisteria* was erected by Linnaeus in honour of John Baptist-Banister, an English traveller and botanist.)

A genus of erect or climbing shrubs with opposite entire, petiolate, exstipulate leaves. The inflorescence is usually a terminal panicle with pedicellate flowers; pedicels bracteate and bracteolate. Calyx of five sepals, each of which (or only four) bears on its back two roll-shaped yellow glands. Petals five, usually pink, sometimes yellow, clawed. Stamens 10, filaments often of different lengths. Ovary of three connate carpels each with a separate style, truncate at the apex; ovule solitary in each loculus. Fruit winged.

Banisteria laevifolia Juss.

(*Laevifolia* is Latin for smooth-leaved.)

Description.—This species is an extensive climbing shrub with rather slender dark brown stems and branches. Branches and branchlets terete, covered with a whitish matted felty tomentum when young, but finally glabrous. The terminal branchlets droop. The leaves are opposite, petiolate (petiole 0.25 in. long, curved, covered with similar tomentum to that on the shoot), cordate at the base, rather variable in shape, being elliptic, elliptic-obovate, ovate-lanceolate or lanceolate, thinly coriaceous in texture, acute or acuminate at the tip, dark olive-green above and smooth below, covered when young with a silvery white silky tomentum which becomes shaggy and dark coloured with age, 4 in. long by 2-5 in. broad; margins entire; venation impressed above, prominent below. On the nerves below one or two sessile globular yellow glands are to be found.

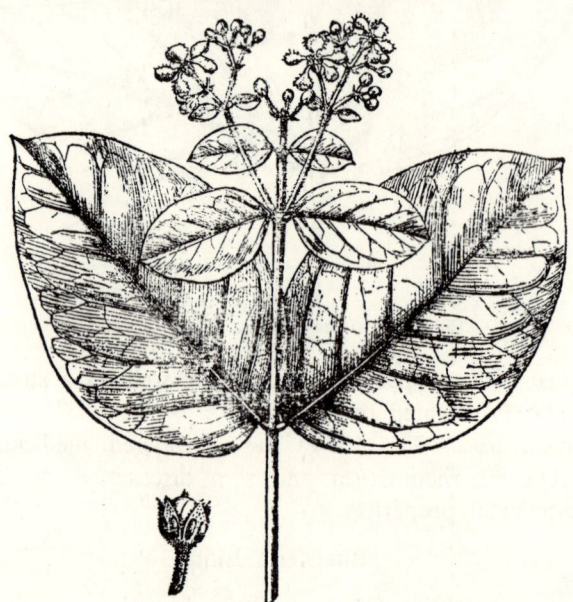

Fig. 112.—*Banisteria laevifolia* Juss.

Inflorescence of trichotomous umbellate panicles, i.e. the flowering shoot divides into three peduncles, each of which is surmounted by a false unmbel. Common peduncle of the false unbel under 0.5 in. long. Flowers 0.5 in. in diameter, yellow, pedicellate, not all arising at the same point but arranged racemosely along a very short axis; pedicels

Stigmaphyllon ciliatum (Lamk.) A. Juss.
(nat. size)

supported by bracts and bracteoles, 0.25 in. long, covered with a yellowish silky tomentum. Sepals 5, about 0.08 in. long, ovate or obovate in shape, acute at the apex, covered with a yellowish silky tomentum, each sepal bearing on its back two yellowish fat sausage-shaped glands, or glands absent from one sepal. Petals 5, alternate with the sepals, orbicular or elliptic in shape, depressed in the centre, almost bowl-shaped, arising from a short stout claw, 0.3 in. long, margin toothed, fimbriate or lacerate. Stamens 10; filaments short and squat, those of the stamens opposite the sepals shorter than the others; anthers 2-celled, finally at right angles to the filament, opening by slits. Ovary ovoid, covered with a silky tomentum; styles three, truncate at the stigmatic tip. In the fruit 1-2 carpels develop a wing on the back which eventually reaches a length of 0.5 in., reddish in colour and covered with a silky tomentum.

Flowers.—Hot and rainy season. *Fruits.*—Cold season.

Distribution.—Indigenous to Brazil, now commonly cultivated in all tropical and subtropical parts of the world.

Gardening.—A rather extensive climber with lanceolate rigid dark olive-green leaves. The large sprays of yellow flowers make it a very handsome object when in bloom. Propagated by layers or seeds which it produces abundantly.

Stigmaphyllon Juss.

(The generic name is derived from two Greek words meaning *stigma* and *leaf* and refers to the leaf-like appendages of the stigmas.)

A genus of woody twiners with opposite leaves. Two glands are visible, usually near the top of the petiole. The flowers are arranged in short, dense, sessile corymbiform racemes. Pedicels usually with 2 bracteoles above the middle. Sepals 5, the four lateral with two glands each. Petals 5, clawed, glabrous, yellow, unequal, the four lateral concave; margin toothed or fringed. Stamens 10, 4 opposite the lateral petals more or less sterile; filaments various. Ovary 3-lobed, 8-locular; styles 3, short with the top dilated into an appendage. Fruit consisting of 3 samaras.

A genus of about 60 species, the great majority of which are twiners, indigenous in the tropics of America. Niedenzu, who monographed the family *Malpighiaceae*, spells the generic name *Stigmatophyllum* A. Juss. Actually it was Spach who spelled the name thus in 1834 in one of his publications. Jussieu published the name, *Stigmaphyllon*, in 1832, and it has priority.

Key to the Species

Leaves cordate .... *S. ciliatum*
Leaves elliptic, oblong or linear .... *S. periplocifolium*

Stigmaphyllon ciliatum (Lamk.) A. Juss.

(*Ciliatum* means hairy on the margins and refers to the fringes of glands on the margins of the leaf; derived from the Latin *cilium*, meaning 'eyelid'.)

Description.—A slender twiner with cylindrical stem, covered in youth with dense white medifixed hairs. Leaves opposite, up to 3 in. long by 3 in. wide, petioled, ovate, obtuse or round in shape, deeply cordate and lobed at the base, palmately nerved, the side nerves being produced beyond the margins and ending in short red glandular processes, covered when young with white appressed medifixed hairs, glabrescent in age; petioles up to 1.25 in. long, furnished with two green elliptic glands at the apex; stipules minute.

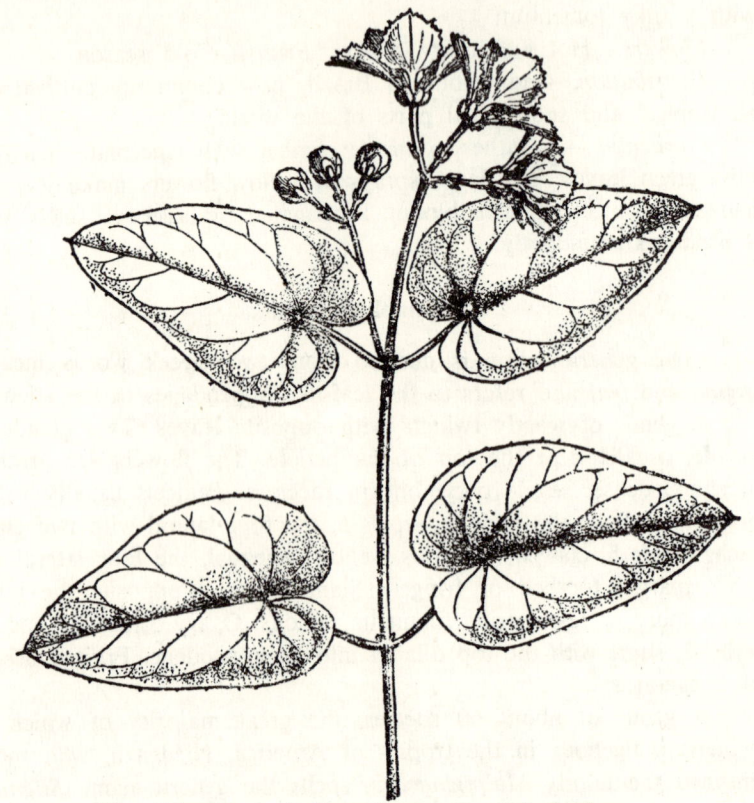

Fig. 113. *Stigmaphyllon ciliatum* (Lamk.) A. Juss.

Flowers yellow, up to 1.3 in. diameter, arranged in umbel-like corymbs which are peduncled and axillary. Peduncles up to 1 in. long, slightly swollen at the apex and sparsely covered with blackish hairs. Umbel of 3-4 pedicellate flowers with two bracts at the base. Bracts lanceolate, glandular-laciniate, with 2 large greenish glands at the base.

PLATE 67

M. N. Bakshi

FRINGED STIGMAPHYLLON
Stigmaphyllon ciliatum (Lamk.) A. Juss.

PLATE 68

M. N. Bakshi

FRINGED STIGMAPHYLLON
Stigmaphyllon ciliatum (Lamk.) A. Juss.

Pedicels clavate with 2 bracteoles at or just above the base. Calyx 5-partite with 8 glands outside; lobes obtuse. Petals 5, four equal in size, one much smaller, all definitely clawed; limbs orbicular or rounded-oblong, concave, irregularly fringed, yellow. Stamens 10; filaments of three, thick and long, the remainder short; anthers oblong, blunt, opening by pores and containing a sticky mass of brown pollen grains. Ovary sunk in or seated upon an obscure disk, 3-celled; styles 3, expanded at the top into leaf-like appendages, which arch over and cover the anthers of the three large stamens; stigmatic areas on the lower surface.

Flowers.—Rainy season. *Fruits.*—Cold season.

Distribution.—A native of tropical America, now widely cultivated throughout the tropical and subtropical parts of the globe.

Gardening.—A medium-sized 'vine' with dark green, attractive foliage almost ivy-like in form. The large clusters of extremely attractive golden-yellow flowers which appear during the rains are very effective. It is suitable for a small trellis or archway and is not particular in its soil requirements. Propagation by layers and seed.

Stigmaphyllon periplocifolium (Desf.) A. Juss.

(The specific name refers to the similarity of the leaves of this species to those of *Periploca gracca* L., a genus of the *Asclepiadaceae.*)

Description.—A twining shrub. Stems cylindrical, lenticellate, dark red or brown, glabrescent or slightly hairy. Young parts covered with silky medifixed hairs, becoming glabrous with age. Leaves opposite or subopposite, elliptic-oblong or linear in shape, shallow, cordate or entire at the base, emarginate or apiculate at the apex, 1-5 in. long, glabrous or glabrescent on both surfaces, coriaceous; nerves prominent below; petiole 0.25-1 in. long, with a pair of stalked glands at the apex.

Flowers yellow, up to 0.75 in. in diameter, arranged either in more or less elongate racemes or in more or less subumbellate corymbs; peduncles up to 0.75 in. long; pedicels 0.6 in. long; peduncles and pedicels covered with appressed silky hairs. Calyx 5-lobed; lobes reflexed at the margins, ovate acute in shape, supported below by 5 oblong glands. Petals 5 in number, yellow, clawed, four subequal in size, the fifth smaller; limbs suborbicular, entire or crenulate on the margin. Stamens 10, of which four opposite the lateral stamens are more or less sterile, the remaining 6, i.e. those opposite the petals and the remaining sepal are fertile. Ovary of three combined carpels, very tomentose with three styles; apex of the styles expanded into a small leaf-like triangular appendage.

Flowers.—Rainy season. Does not fruit in Dehra Dun.

Distribution.—Native of tropical America, now common in cultivation throughout the country.

Fig. 114.—*Stigmaphyllon periplocifolium* (Desf.) A. Juss.

Gardening.—A handsome scandent shrub which produces fine yellow flowers during the rains. Propagated by layers.

Galphimia Cav.

(The generic name is an anagram of *Malpighia*, another, and closely allied, genus of this family.)

The genus consists of shrubs or undershrubs. Leaves opposite, petioled, with linear stipules often glandular on the margins or on the petiole. Flowers terminal, yellow or red; calyx without glands; petals clawed, stamens 10. Fruit a capsule. A small genus of about 10 species, all tropical American.

Galphimia gracilis Bartl.

(*Gracilis* is Latin for slender and refers to the long slender branches of the plant.)

Description.—A handsome shrub reaching 6 ft in height. Old branches a shining brown, slightly fissured, sparsely covered with dark-

Galphimia gracilis Bartl.
(nat. size.)

PLATE 69

M. N. Bakshi

THE SLENDER GALPHIMIA
Galphimia gracilis Bartl.

PLATE 70

M. B. Raizada

THE SLENDER GALPHIMIA
Galphimia gracilis Bartl.

red short hairs, young parts densely covered with dark-red hairs. Leaves opposite, up to 2 in. long by 1 in. wide, petioled, stipulate, ovate or ovate-oblong in shape, unequally cuneate at the base, obtuse at the apex or sometimes apiculate, covered on both surfaces with rufo-sericeous pubescence when young, afterwards glabrescent, though it is usual to find hairs on the mid-rib beneath, with two glands, one on either side, on the margins just above the base; petiole channelled above, up to 4 in. long, puberulous; stipules linear, hairy, persistent.

Flowers arranged in more or less dense, erect terminal racemes, 3-4 in. long, 10-30-flowered. Individual flowers seated on stalks from the main axis; each stalk supported by a bract and bearing two bracteoles at the centre. Calyx 5-partite, without glands; lobes obovate or oblong-ovate in shape, green, 0.25 in. long. Petals 5, yellow, clawed, 0.25-0.5 in. long; limb subcordate in shape or ovate, minutely fimbriate on the margin; claw 0.1 in. long in four petals, that of the fifth 0.2 in. long. Stamens 10; filaments very unequal, 0.1-0.2 in. long, reddish at the base; anthers yellow, large, oblong-obtuse. Ovary ellipsoid, 3-locular, smooth and glabrous. Styles 3-4. Fruit a spherical capsule, 0.2 in. in diameter.

Flowers.—Almost all the year round but profusely during July-November. *Fruits.*—Cold season.

Distribution.—Native of tropical America, now extensively cultivated throughout the plains of India.

Gardening.—A hardy evergreen handsome shrub about 4-6 ft high. It is covered most of the year with small golden-yellow flowers which, against the dark green foliage of the plant, are very effective. The plant will stand wind and poor soil but prefers a protected site and full sun. It has been recommended as a hedge but shows itself best when planted in clumps. Propagation is by seed.

Chapter 14

APOCYNACEAE

The Dogbane Family

The family gets its name from the Greek name *apokynon*, used by Dioscorides in his Materia Medica to describe a plant, *Cynanchum erectum*, whose leaves were said to be deadly to dogs. Incidentally the Greek for dog is *kyon, kynos*.

This family contains many shrubs and climbers with a few species which reach the status of small trees. The vascular bundles in this family are bicollateral so that phloeum is found on the outer edge of the pith. Moreover, most species have a milky juice which is contained in non-articulated lactiferous vessels situated in the pith and bark and also in the veins in the leaves. The wood of many species is normal, but again others possess wedges of soft bast in the normal wood. The leaves are simple, opposite and decussate, seldom whorled, entire, not unusually with many parallel side nerves. Stipules are rarely present, sometimes interpetiolar. Flowers mostly showy, 4-5-merous, with bracts and bracteoles, hermaphrodite, usually produced in cymose inflorescences. Calyx tubular or on free sepals often glandular within. Corolla gamopetalous, funnel- or salver-shaped, sometimes campanulate, or urn-shaped, with the lobes twisted in the bud, often furnished with scales in the throat. Stamens as many as the lobes of the corolla and seated upon it; anthers free, often connate and adnate to the stigma. A glandular disk is present in most of the species and takes the form of scales or is shaped like a disk. Ovary of two carpels, joined together or separate, containing many hanging ovules. Style slender. Fruit of two separate follicles, seldom drupaceous.

Carissa carandas Linn., the Karaunda, is a spiny shrub which is said to be wild and also cultivated in India. The pinkish white fruits are very acid but are capable of being made into a very pleasant conserve with the aid of plenty of sugar.

The family contains about 1,100 species, most of which are found in the tropics. The magnificent flowers and handsome foliage of numerous species make them valuable additions to the garden and they are widely cultivated.

A reference has been made above to the plant *Cynanchum erectum* from which the family gets its name. This plant, however, is not a member of the *Apocynaceae* but of the neighbouring family *Asclepiadaceae*. Both families are well marked, but in one section of *Apocynaceae* the manner in which the stamens are connivent around and applied

THE MOONBEAM
Tabernaemontana coronaria R. Br.
(× ¾)

to the stigma recalls a very definite character of the *Asclepiadaceae*.

In nearly all species a milky juice is found. In certain South African lianes, for example, species of the genera *Landolphia* and *Kickxia*, quantities of latex are found, and this is tapped and marketed under the name 'silk rubber'. Other genera in India (*Alstonia* sp., *Willughbeia* sp.), Brazil (*Hancornia* sp.), Jamaica (*Forsteronia* sp.) can give a valuable rubber-yielding latex.

A large number of the species have an evil reputation on account of their poisonous properties. Many contain glucosides and alkaloids which are valuable drugs in medicine. A few species contain powerful glucosidal heart poisons which belong to the digitalis group of drugs. To this section belongs the genus *Strophanthus* which is mainly confined to Africa, though a few species are found in this country.

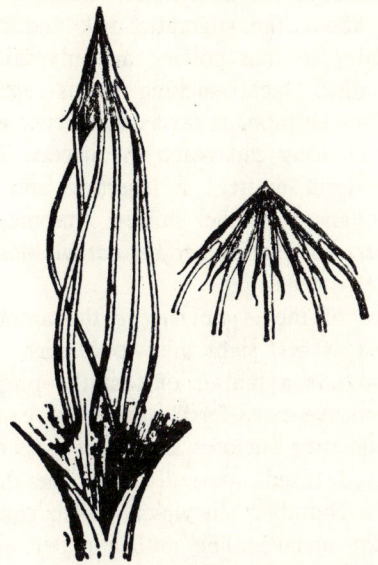

Fig. 115.—*Beaumontia grandiflora* showing stamens connivent into a cone and adnate to the stigma.

Strophanthin is a nitrogen-free glucoside which is used by certain of the African tribes as an arrow poison. The drug has a very rapid and powerful action on the heart and is sometimes used in conjunction with digitalis, whose action is not so rapid but is more lasting. Quebracho bark (from *Cortex quebracho*, which has the hardest wood known) possesses alkaloids which are valued in the treatment of asthma and fevers. Many other species yield drugs which are put to various purposes.

Alstonia scholaris R. Br. is a common tree of the Indian jungles. It gets its specific name from the fact that in bygone days boards of the wood were used as school slates, the writing being rubbed off by the use of the rough leaves of *Delima*. The tree is called 'Devil Tree' in many places and it is an object of devotion in Malay.

The mechanism of cross pollination has been worked out in several species of *Apocynaceae*.[1] In species of *Vinca*, the Periwinkle, for example, the nectar is secreted by two yellow glands near the ovary and is stored up in the corolla tube which is 0.5-0.75 in. long. The entrance to the tube has a wall of hairs as a protection against rain. About the middle of the tube the style thickens conically and terminates in a short cylindrical horizontal plate, the edge of which functions as a stigma and is covered with a sticky secretion. The plate bears a tuft of hairs which takes up the pollen as it is shed from the anthers. The filaments, which spring from the middle of the corolla tube, are bent into 'knees' and are beset with hairs internally. The anthers are situated immediately above the stigmatic disk and dehisce introrsely. Their margins are hairy so that pollen can only fall on the terminal brush of the sigmatic disk. Nectar-sucking insects can insert their heads for a short distance into the tube, as far as the brush, so that a proboscis between 0.3 and 0.5 in. long can reach the nectar. The proboscis becomes covered with viscid matter on insertion and when withdrawn carries away some pollen to another flower. Automatic self-pollination is excluded. As a matter of fact in certain periwinkles the flowers have been proved to be self-sterile.

In certain species of the *Apocynaceae* the anthers are connivent around the stigma and at first sight this would seem to be a device to ensure self-fertilisation. As a matter of fact the purpose is the exact opposite, that is, to ensure cross-fertilisation. For example, in *Nerium odorum* Sol. the corolla tube encloses a cone of anthers covered outside with woody plates, and fused internally with the dilated end of the style to form a pollen chamber (in which pollen can collect) beneath which is the stigmatic surface. The anther plates are produced into points below and covered dorsally with hair. Each stamen is drawn out into a long terminal appendage, which is filiform at its base and then becomes broader and feather-like. These fine appendages are twisted together into a loose, woolly, whitish ball, which with the corona blocks the entrance of the flower in such a way that only long-tongued Lepidoptera are able to penetrate to the nectar. The woody plates on the back of the anthers prevent these being gnawed away in order to open a passage to the nectar. Actually the proboscis of the insect must be inserted in the chinks between the plates. In its withdrawal the proboscis passes over the stigma, becomes sticky, and carries out some pollen on it. This is transferred to the next flower visited. Very often

---

[1] See Knuth's *Handbook of Flower Pollination*.

a number of insects which are not adapted to carry out the process are caught by the anther plates and are killed.

KEY TO THE GENERA

Flowers yellow
  Leaves very narrow .... *Thevetia*
  Leaves broad .... *Allemanda* sp.

Flowers blue or violet
  Flowers blue .... *Vinca major*
  Flowers violet or purplish .... *Allemanda violacea*

Flowers red
  Leaves narrow .... *Nerium*
  Leaves broad .... *Vinca rosea*

Flowers white or pinkish white
  Leaves whorled .... *Nerium*
  Leaves opposite:

Shrubs
  Flowers not over 0.5 in. wide: leaves very
    coriaceous, tip sharp .... *Acocanthera*
  Flowers well over 0.5 in. broad: leaves soft
    A small herbaceous shrub .... *Vinca rosea*
    A large shrub .... *Tabernaemontana*

Climbers
  Flowers large, up to 4 in. long
    Flowers campanulate .... *Beaumontia*
    Flowers salver-shaped .... *Chonemorpha*
  Flowers less than 2 in. long
    Corolla-lobes orbicular or obovate in shape
      Petals orbicular; tube 0.3 in. long;
      flowers creamy white; no corona .... *Vallaris*
      Petals obovate; tube 1.5 in. long;
      flowers white tinted with rose; corona
      present, rose-coloured .... *Strophanthus*
    Corolla-lobes ovate-lanceolate or oblong truncate
      Tube slender below, abruptly inflated .... *Trachelospermum*
      Tube gradually swelling above .... *Melodinus*

Thevetia Linn.

(This generic name commemorates Andre Thevet, a French monk, 1502-1590 who travelled in Brazil and Guiana.)

    This genus contains a few glabrous shrubs and trees with alternate leaves. The large yellow flowers are borne in terminal cymes. Sepals five, spreading, glandular. Petals five, twisted, combined below into a campanulate tube. Stamens five, inserted at the top of the tube; anthers short, lanceolate, without appendages. Disk usually absent. Ovary 2-celled with a filiform style, ending above in a thick two-lobed stigma; ovules two in each cell seated on a conspicuous placenta. Berry fleshy.

Thevetia neriifolia Juss. [**T. peruviana** (Pers.) K. Schum.]
Yellow Oleander; Trumpet Flower

(*Neriifolia* means in Latin, having leaves like Nerium, the Oleander.)

Description.—A shrub or small tree, sometimes reaching 12 ft in height. Stems and shoots quite glabrous, exuding a copious milky juice when cut. Leaves alternate up to 6 in. long by 0.3-0.4 in. broad, narrowly linear in shape, shortly acuminate and obtuse at the apex, decurrent at the base, shining green above, paler and dull below, coriaceous, glabrous; margins revolute; central nerve very prominent below; secondary nervation invisible.

Flowers yellow, arranged in subterminal few-flowered cymes, seated on pedicels from 0.3-1.25 in. long. Sepals 5, about 0.3 in. long, long-acuminate, sharp, spreading, thin, joined below into a short tube.

Fig. 116.—*Thevetia neriifolia* Juss. × $\frac{3}{4}$

Petals 5, 1 in. or longer in length, oblong-acute, joined below into a tube which is about 1 in. long, cylindrical at the base and broadly campanulate at its junction with the petals, decorated with two hairy wings in the cylindrical portion and with hairy scales at the base of the campanulate portion. Stamens five, inserted on the throat of the corolla below the scales; filaments short; anthers minute, oblong, shortly acuminate, with a lamellate appendage. Disk present, thick, fleshy, as high as the ovary; ovary bicarpellary; style slender, 0.5 in. long. Fruit drupaceous, fleshy, usually four-angled; seeds four.

Flowers and fruits.—Practically throughout the year.

Distribution.—Native of tropical America, cultivated or naturalised throughout India.

Gardening.—A large evergreen shrub or small tree with alternate 1-nerved linear leaves and large fragrant yellow, or even creamy-white, funnel-shaped flowers about 2 in. across. It is a great favourite with Hindus and is frequently cultivated near temples, the flowers being offered to the God Shiva. As the plant requires little attention, is easily grown from seed, transplants well, and is not touched by cattle or goats, it is useful for growing in the compounds of rest-houses, etc. It was brought into cultivation in Europe in 1735 and from there distributed to the tropics in general, as a showy and ornamental plant. Owing to its immunity from damage by browsing and the ease with which it is propagated, it might be useful for afforestation work in moister parts of this country. It is also suitable to be grown as a hedge.

Medicinal and economic uses.—The wood is moderately hard and the seeds yield an oil used for burning as well as in medicine. All parts of the plant are poisonous, owing to the presence in the latex of a glucoside which acts as a heart-poison. Accidental cases of poisoning by eating its seeds are not infrequent. According to Heyne its wood is a fish-poison. The seeds are sometimes used to poison wild animals.

Allemanda Linn.

(The genus was erected in honour of Dr. Allemand of Leyden.)

A genus of shrubs with handsome flowers. Leaves verticillate with interpetiolar glands. Flowers arranged in axillary cymes, showy, yellow or purple. Calyx-lobes acute, conspicuous, without glands. Corolla funnel-shaped, 5-lobed; lobes spreading. Stamens inserted at the upper end of the cylindric basal tube, covered by the hairy appendages of the throat. Ovary of two connate carpels, 1-celled, with numerous seeds on two parietal placentas, surrounded by a small annular disk; style filiform; stigma capitate. Fruit a globose prickly capsule. Seeds winged, flat.

Key to the Species

Flowers yellow:
    Base of corolla swollen .... *A. neriifolia*
    Base of corolla not swollen .... *A. cathartica*
Flowers violet or purplish .... *A. violacea*

Allemanda neriifolia Hook.

(*Neriifolia* means having leaves like the Oleander, but in the case of this plant the leaves are very different from those of the Oleander.)

Description.—An evergreen shrub or half climber. Stems terete, woody below, herbaceous above, smooth and glabrous, but covered with down when young. Leaves in whorls of two to five, elliptic or ovate-acuminate in shape, petiolate, dark green above, pale below, shortly hairy on the midrib and principal veins, otherwise glabrous; petiole very short.

Inflorescence a terminal panicle becoming lateral owing to the production of new shoots which continue the growth. Flowers large, conspicuous, showy, gloden-yellow, seated on short terete pedicels up to 0.5 in. long; bracts 0.12 in. long, green. Calyx of five unequal sepals, elliptic-ovate in shape, light green in colour, 0.6 in. long. Corolla-tube 2 in. long; base swollen, angled, bulbous, green, swelling into the golden-yellow tube, which divides above into five orbicular or oval spreading obtuse lobes, 0.5 in. long. The tube is striped inside with reddish brown or orange. Stamens five, situated at the bottom of the tubular portion of the corolla, included. Ovary 1-celled, with many ovules on two parietal placentas. Fruit unknown.

Flowers.—Hot and rainy season. Does not fruit in this country.

Distribution.—A native of Brazil, commonly cultivated in all tropical and subtropical parts of the globe.

Gardening.—A dwarf bush or semi-climber. It differs from all other Allemandas here described, in its habit as also in the form of the corolla, with its singularly short contracted base of the tube, swollen and angled at the base, and very elongated upper portion; the colour is a deep almost golden-yellow, and it is streaked with orange. It is well adapted for planting against a back wall or for training up pillars. It also flowers freely when cultivated as a pot plant, the branches being supported either by stakes or a wire trellis. A mixture of light loam and leaf mould suits it very well and during the season of growth it needs a free supply of water. It is readily increased by cuttings and layers.

Allemanda cathartica Linn.

Description.—An evergreen glabrous shrub which is capable of assuming a climbing or rambling habit. Branches smooth, green, hairy or not. Leaves opposite in whorls of four, obovate or oblanceolate in shape, cuneate at the base, shortly acuminate at the tips, papery to sub-coriaceous in texture, glabrous or more or less hispid or pubescent especially on the nerves, intramarginal nerve conspicuous; petiole very short.

PLATE 71

M. B. Raizada

THE ALLEMANDA
Allemanda neriifolia Hook.

PLATE 72

M. B. Raizada

THE ALLEMANDA
Allemanda neriifolia Hook.

Flowers large and showy, yellow, in terminal (afterwards axillary) cymose panicles; bracts deciduous. Calyx of five sepals, glabrous or hairy without, lanceolate in shape, 0.3 in. long. Corolla of two distinct portions: a lower, which is tubular, about 1 in. long and very narrow, and an upper which is campanulate and about 0.5 in. in diameter, ending above in five orbicular lobes, 1-1.5 in. long. Stamens five, included. Ovary 1-celled, with numerous ovules on parietal placen-

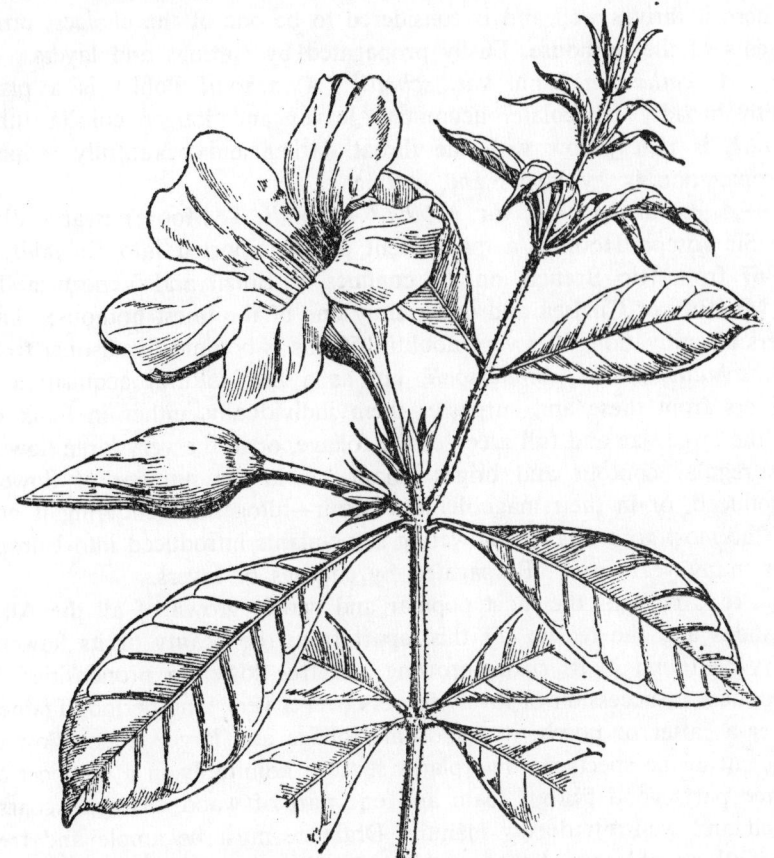

Fig. 117.—*Allemanda cathartica* Linn. × ¼

tas. Fruit a globose prickly capsule; prickles soft about 0.3 in. long. Seeds many, obovate, flat, winged.

Flowers.—Hot and rainy seasons. *Fruits.*—Cold season but only rarely.

Distribution.—Native of tropical America now extensively grown throughout the tropical and subtropical parts of the world. In recent years it has run wild near Travancore.

Gardening.—A rather large shrub of scandent and rambling habit. It is one of the commonest plants in all tropical gardens and one of the showiest and ornamental. The large, pure, bright yellow flowers are borne in great profusion during the hot and rainy seasons and produce a most magnificent effect against the rich deep green foliage. Like all other species of this genus it should be fed liberally with natural or artificial manures during its growing season as it is a gross feeder. It should be cut well back during the cold season to keep it within bounds. It was introduced into England in 1785 by Baron Hake where it thrives well and is considered to be one of the choicest ornaments of the hothouse. Easily propagated by cuttings and layers.

A. cathartica Linn. var. *schottii* (*A. schottii* Pohl.) is a plant with broadly lanceolate, acuminate leaves and longer corolla tube, which is rich yellow with the throat darker and beautifully striped. Propagation is by layers and cuttings.

A. cathartica Linn. var. *nobilis* (*A. nobilis* T. Moore) is according to Sir Joseph Hooker 'a magnificent plant, imported into England in 1867 from Rio Branco, on the confines of Brazil and Venezuela, by a Mr. Bull of Chelsea and is certainly one of the finest hothouse climbers in cultivation. It is very doubtful if this is botanically distinct from *A. schottii* and *A. hendersonii*, but as a horticultural acquisition it differs from these and surpasses them individually, either in habit or in the large size and full green of the foliage, or in the very large flower, its regular contour and bright colour, or in the number of flowers produced, or in their magnolia-like odour—altogether rendering it one of the most gorgeous free-flowering stove-plants introduced into Europe for many years past'. Propagated by cuttings or layers.

It is perhaps the most popular and widely grown of all the Allemandas and the reason for this, apart from the beauty of its flowers, may be found in its strong-growing and free-flowering propensities. It produces a succession of lovely flowers over a very long period. Trained over a rafter or porch, its handsome foliage and large, golden flowers present a fine spectacle. The plants succeed admirably in a compost of three parts good fibrous loam and one part of wood charcoal, coarse sand and well-rotted cow manure. Drainage must be ample and free and all possible sun light must be provided.

Allemanda violacea Gardn.

(The specific name refers to the colour of the flowers.)

Description.—An erect, sometimes scrambing shrub, with terete branches which are glabrous when old but hirsute when young. Leaves verticillate, 3-5 in. by 1.5 in., very shortly petioled, oblanceolate, elliptic or oblong in shape, cuneate at the base, abruptly acuminate at the apex, entire on the margins, rough on both surfaces with short, stiff, stout hairs; hairs on the midrib below and on the margins particularly stiff and sharp; nervation evident but not particularly prominent; stipules small.

Inflorescence a few-flowered cyme, terminal. Calyx of five separate lobes, the two other elliptic in shape, the three interior lanceolate, bristly hirsute. Flowers large and showy, violet, purplish brown or purplish. Corolla-tube up to 2.5 in. long, of two distinct parts, the lower tubular narrow and the upper campanulate up to 1 in. in diameter, ending above in five orbicular or broadly ovate lobes, 0.75 in. long, glabrous without. Stamens five, inserted at the base of the campanulate portion of the corolla, included, covered with the hairy appendages of the corolla; filaments very short. Ovary of two connate carpels, containing numerous ovules on two parietal placentas. Style filiform ending above in a capitate stigma.

Flowers.—Hot and rainy season. Does not set fruit in this country.

Distribution.—A native of Brazil, now cultivated in all tropical and subtropical countries of the world.

Gardening.—A fine, handsome, slender growing climber, quite distinct from all other species and varieties in the colour of its flowers. It is a poor grower on its own roots, but thrives well when grafted on *A. cathartica* var. *schottii*. It prefers a well-drained soil and a sunny site like most other species of this genus.

Medicinal uses.—The root is said to be a powerful cathartic and is used in malignant fevers.

Nerium Linn.

(Dioscorides in his Materia Medica gave the Greek name *nerion* to the plant *Nerium odorum* Sol.)

A genus of erect glabrous shrubs. The leaves are arranged in whorls of threes, and are coriaceous, narrow, with obscure nervation. The inflorescence is of terminal branchy panicles. Calyx divided nearly to the base into 5 lobes, glandular inside. Corolla tubular below, campanulate above, ending in five twisted petals. Stamens five, included, fixed to the base of the campanulate portion of the corolla; anthers connivent round the stigma and adherent to it. Ovary of two distinct carpels; ovules numerous. Fruit dry, of two connate carpels. Seeds oblong, hairy.

Nerium odorum Sol. (Nerium indicum Mill.)

The Oleander

Description.—A glabrous erect shrub. Young shoots greenish, thin, brown, emitting quantities of milky juice when cut. Leaves mostly in threes, sometimes in twos, opposite, 4-6 in. long by 0.3-0.9 in. wide, linear-lanceolate in shape, tapering at both ends, thick, coriaceous, with a thick midrib, decurrent at the base; petiole 0.2-0.3 in. long, thick.

Flowers arranged in terminal panicles, pink or white, single or double, fragrant; peduncles and pedicels minutely pubescent; bracts small, 0.3 in. long. Calyx 0.25 in. long divided nearly to the base into five linear, acute, pubescent lobes. Corolla-tube 0.7 in. long, lower half tubular, hairy within, upper half campanulate, ending above in

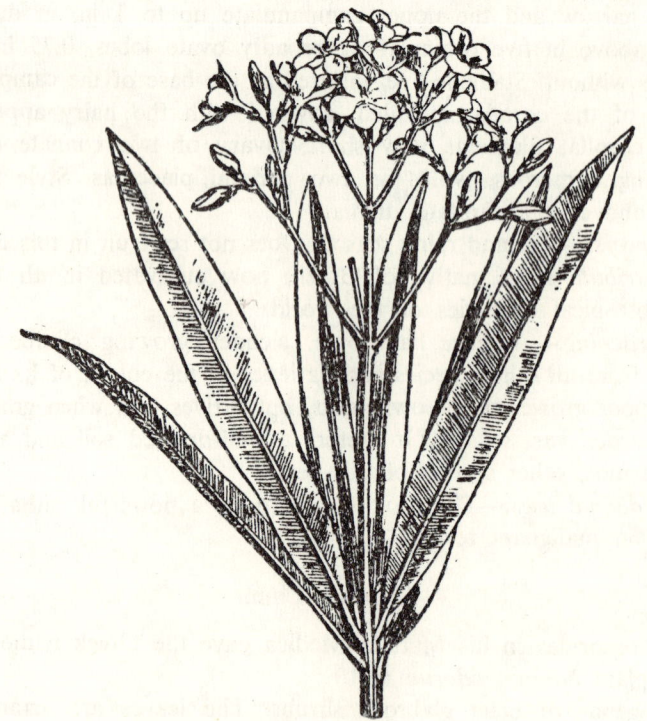

Fig. 118.—*Nerium odorum* Sol. × ¾

five rounded overlapping petals. In the throat of the corolla can be found a corona of five scales, each cleft into 3-7 segments. Stamens five, included; filaments short; anthers connivent into a cone and adherent to the stigma; connective with two long thread-like hairy appendages. Disk absent. Carpels 2, distinct; style filiform, thickened upwards; stigma 2-lobed. Follicles connate, 5-8 in. long by 0.3 in. wide. Seeds small, 0.2 in. long, each topped by a tuft of brown hairs, 0.5 in. long.

Flowers and fruits most of the year in cultivation.

Distribution.—Found from Persia to China and Japan; frequently grown in gardens throughout the country.

Gardening.—A large evergreen hardy, beautiful shrub with narrow deep green tapering leaves, and deep rose or white, fragrant flowers. The plant is very poisonous and is not touched by cattle or goats. It

PLATE 73

M. B. Raizada

MADAGASCAR PERIWINKLE
Vinca rosea Linn.

is commonly grown in gardens with single or double flowers, the double pink variety being perhaps the favourite. Easily multiplied by cuttings or layers during the rains. It is suitable for a screen or as a hedge plant. This shrub is considered the glory of the gardens of Northern India where during the hot season it thrives vigorously and being always in bloom, scents the whole air around with its perfume.

Medicinal and economic uses.—Every part of the plant is poisonous and contains a glucoside allied to digitalin which acts as a heart poison. There is 2½ times as much poison in oleander leaves as in those of digitalis. The roots if taken internally are highly poisonous but a paste is reputed to be useful in skin diseases. The leaves boiled in oil yield a medicated ointment which is also used in skin diseases.

Vinca Linn.

Periwinkle

(In his Natural History, Plinius called the periwinkle by the name *vinca pervinca*, by which it is known in Italy even to the present day.)

This genus comprises upright or prostrate herbaceous shrubs with opposite leaves and a milky juice. Inflorescence of axillary flowers. Calyx of five narrow acuminate sepals. Corolla-tube cylindrical somewhat dilated at the level of the stamens; lobes five, spreading. Stamens five, included; anthers without an appendage at the base. Disk of two scales. Ovary of two distinct carpels; style filiform; stigma annular; ovules 6-8, in two rows in each carpel. Fruit of two follicles.

Two species of this genus are common garden plants in India and can be separated easily by the shape of the leaf.

Key to the Species

Leaf oblong, obtuse; petioles very short 0.1 in. long .... *V. rosea*
Leaf heart-shaped, acute, petioles over 0.25 in. long. .... *V. major*

Vinca rosea Linn. [Catharanthus roseus (L.) G. Don.]
Madagascar Periwinkle

Description.—The species is a herbaceous shrub reaching 2.5 ft in height with numerous, erect branches; young parts hairy. Leaves opposite, oblong, occasionally oblanceolate, obtuse, cuneate at the base, membranous, up to 3 in. long by 1 in. broad, smooth and glabrous, shining green above, paler below; nervation not marked; petiole up to 0.5 in. long.

Flowers solitary in the axils of the upper leaves, pink or white, very fragrant, seated on very short pedicels. Calyx-tube very short, divided above into five sepals, which are linear acute in shape, hairy on the back and about 0.2 in. long. Corolla-tube 1-1.25 in. long, hairy outside, somewhat inflated above, with a narrow throat, having a hairy ring inside below the stamens and with hairy rugosities above; lobes five, oblong-rounded, about 1 in. long, spreading. Stamens five, situated on the swollen portion of the tube; filaments very short. Disk present, higher than the ovary, consisting of two narrow, obtuse, fleshy scales. Ovary of two separate and distinct carpels, hairy at the top; style about 1 in. long; ovules many. Fruit of two follicles.

Fig. 119.—*Vinca rosea* Linn. × ¾

Flowers and fruits all the year round.

Distribution.—Probably a native of the West Indies, commonly grown in gardens all over the country where it springs up readily self-sown.

Gardening.—A beautiful herbaceous or somewhat woody shrub 2-3 ft high, with deep green polished oval or oblong leaves and pure white or deep rose-coloured flowers in axillary pairs. When in full bloom, as it nearly always is, it is a lovely plant. Very readily raised from seed which it produces abundantly.

Vinca major Linn.

Greek Periwinkle

(*Major* is a Latin word meaning larger.)

Description.—A perennial woody herb up to 3 ft tall. Stem prostrate, rooting at the nodes and creeping over the soil, the vertical shoots producing flowers. Leaves opposite, petiolate, broadly ovate in shape, obtuse or acute at the tip, rounded or almost heart-shaped at the base, membranous or somewhat coriaceous in texture, up to 2 in. long by 1.5 in. broad, finely hairy on the margins, glabrous and smooth

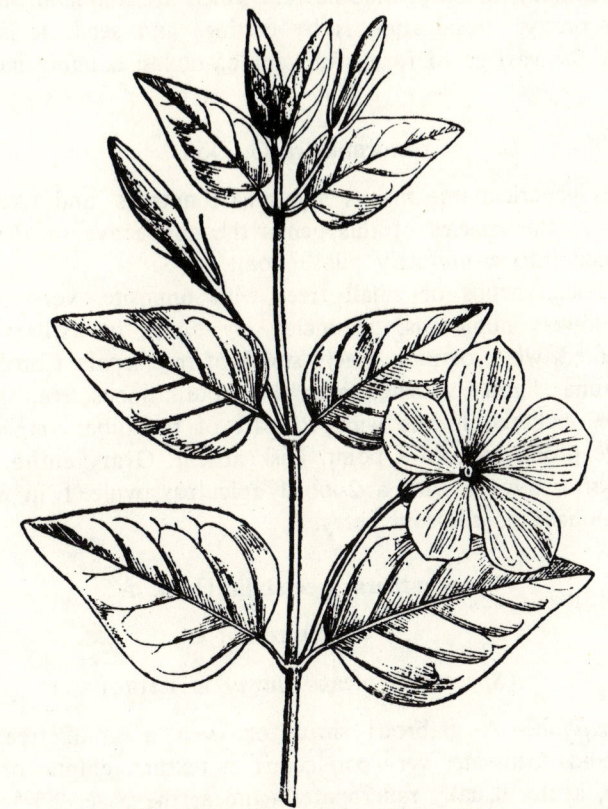

Fig. 120.—*Vinca major* Linn. × ¾

elsewhere; petiole up to 0.5 in. long, grooved above, margins hairy.

Flowers arising singly from the axils of the upper leaves, bright blue or violet, rarely white; pedicel up to 1.5 in. long. Calyx-tube campanulate, very short, ending above in five narrow almost setaceous lobes which are hairy along the margins, 0.5-0.75 in. long; towards the base of the lobes are marginal glands on either side. Corolla tube as long as the calyx-lobes; tubular portion short, dilated towards the mouth where the stamens are included; lobes five; stamens five; filaments short, curved, arising from a tubercle, hairy below, anthers oblong, ending above in a hairy hooded appendage. Ovary of two distinct carpels, style about 0.5 in. long; stigma capitate. Disk of two almost orbicular, fleshy, glandular scales.

Flowers.—December-March. Does not fruit in this country.

Distribution.—A native of Europe, often grown in gardens in the hills and in the plains of this country. Recently it has run wild in Barlowganj, below Mussoorie, and in Simla.

Gardening.—A small evergreen undershrub, trailing over banks, etc. The solitary, axillary, blue flowers which are about an inch across are very pretty. Propagation is by cuttings and seed. It is suitable, especially the variegated form, for veranda-boxes, hanging baskets and the like.

Acocanthera G. Don.

(This generic name means 'mucronate anthers' and refers to the fact that in the species of this genus the connective of the stamens is produced into a minutely pilose point.)

Unarmed shrubs or small trees with opposite, very coriaceous leaves. Flowers numerous, peduncled or sessile, in axillary clusters, sweet-scented, white or pink. Calyx small, of free sepals. Corolla salver-shaped, tube slightly widened near the mouth; lobes five, short. Stamens five, included in the widened part of the tube; anthers ovate; connective produced into a point. Disk absent. Ovary entire, 2-celled; style filiform; stigma with a 2-lobed apiculus; ovule 1 in each cell. Fruit a globose or ellipsoid berry.

Acocanthera spectabilis Hook. f.

Wintergreen

(*Spectabilis* means showy in Latin.)

Description.—A glabrous shrub or even a small tree. Leaves opposite and decussate, very coriaceous in texture, elliptic or oblong-lanceolate, acute, usually mucronate, acute at the base, 2.5-5 in. long, 1-2 in. broad, dark green above, paler beneath, glabrous; petiole short, stout.

Flowers in short many-flowered dense subsessile clusters or corymbs. Calyx of five pubescent ovate-lanceolate green or whitish sepals, 0.15 in. long. Corolla white, tinged with pink, fragrant; tube 0.7-0.9 in. long, pubescent or almost glabrous without, hairy within. Lobes five, elliptic-acute, up to 0.25 in. long; stamens five, alternate with the lobes, seated on short, hairy filaments and included; anthers ovate; connective produced into a short hairy point. Ovary 2-celled; style filiform; stigma short, cylindric, obtusely apiculate. Fruit a small, purplish black berry.

Flowers.—April. Does not fruit in Dehra Dun.

Fig. 121.—*Acocanthera spectabilis* Hook. f. $\times \frac{3}{4}$

Distribution.—A native of western districts of South Africa, now cultivated throughout the tropical and subtropical parts of the world.

Gardening.—An evergreen shrub with dark green leaves. The masses of white flowers which are produced at the commencement of the hot weather are very fragrant and make this plant very ornamental. According to Sir J. D. Hooker it was introduced into England by Mr. B. S. Williams about 1872. Propagation is by cuttings or seed.

Tabernaemontana Linn.

(Linnaeus named this genus in honour of James Theodore Tabernaemontanus of Heidelberg, a physician and botanist of the 16th century.)

This genus consists of glabrous shrubs and small trees with thin or coriaceous, opposite leaves. Inflorescence of few-flowered cymes seated on a short peduncle. Calyx of five imbricate glandular sepals. Corolla tubular, ending in five petals, hypocrateriform; tube swollen at insertion of stamens. Stamens five, inserted towards the middle of the tube, bearing sagittate anthers. Disk absent. Ovary of two distinct or separable carpels; ovules numerous. Fruit often of two follicles.

Tabernaemontana coronaria R. Br.
[Ervatamia divaricata (L.) Burkill]

The Moonbeam; Chandnee; Wax-flower

(*Coronarius* means used for or pertaining to garlands in Latin, and refers to one use to which the flowers are put.)

Description.—A glabrous shrub reaching 6 ft in height, exuding a copious milky juice when cut. The young shoots are bright green in colour and lenticellate. The leaves are 3-6 in. long by 1-2.5 in. broad, elliptic-oblong, abovate or oblanceolate in shape, attenuate more or less caudate at the tip, acute or obtuse, cuneate at the base, membranous in texture, dark green; petiole 0.5 in. long or less.

Inflorescence of geminate or solitary few-flowered cymes in the axils and also terminal. Flowers often double, white, fragrant. Calyx gamosepalous; tube campanulate, 1 in. long, ending in 5 oblong-acute lobes about 0.35 in. long, glandular inside at their base. Corolla-tube 0.75 in. long, cylindrical, slightly dilated below the top, where it divides into five obliquely ovate lobes which are horizontal, and about 0.5 in. long. Stamens five, inserted in the dilated portion of the tube; filaments short; anthers 0.15 in. long. Ovary of two distinct carpels, with a long style ending in an obovoid stigma. Ovules inserted in two series of 3-4. The fruit consists of two recurved fleshy follicles 1-3 in. long, green outside, orange or bright red within. When ripe the follicle splits and turns inside out.

Flowers.—Hot and rainy season. *Fruits.*—Cold season.

Distribution.—Undoubtedly wild and very commonly cultivated in gardens throughout India.

Gardening.—A handsome evergreen shrub, 4-6 ft high. The large pure white flowers, which are usually double in cultivation, are found on the plant at all seasons and look very lovely against the dark green shining foliage. The flowers are inodorous by day but sweet-scented at night and are used for making garlands. It is suitable for a hedge as well as for growing on lawns. Propagated easily by layers or cuttings.

Economic and medicinal uses.—According to Watt, in the Himalayas the pulp of the seeds is used as a red dye, and the wood medicinally as a refrigerant and is burnt for incense and used in perfumery.

Beaumontia Wall.

(This generic name commemorates the name of Mrs. Beaumont of Bretton Hall, Yorkshire, England.)

A genus of tall climbing shrubs. Leaves rather large, opposite, coriaceous. Inflorescence of terminal cymes; flowers large, white, fragrant; bracts often foliaceous. Calyx of five large, foliaceous sepals, glandular or not within. Corolla white, shortly tubular below, widely

Beaumontia grandiflora (Roxb.) Wall.
(× ½)

campanulate above, 5-lobed, lobes twisted. Stamens 5, inserted at the base of the campanulate portion of the corolla; anthers sagittate at the base, connivent around the stigma. Disk present. Ovary of two carpels; style filiform. Fruit of two follicles; seeds compressed, ending in a tuft of hair.

Beaumontia grandiflora (Roxb.) Wall.

(*Grandiflora* means large-flowered in Latin.)

Description.—This species is a huge evergreen climber which ascends to the tops of the tallest trees in its home, Assam, and covers the crown of its support with its beautiful foliage and flowers. The stem is woody with rusty pubescent shoots, and, in old plants, reaches a considerable thickness. Copious milky juice flows from cut stems. The leaves are opposite, exstipulate, membranous to coriaceous in texture, ovate to ovate-oblong in shape, smooth and glabrous, rarely sparsely hairy beneath; nerves 7-16 pairs, arching and forming loops within the margin; reticulation prominent beneath; petiole 0.5-1 in. long, sometimes, especially in young leaves, rusty pubescent. Inflorescence a terminal cyme, few- to many-flowered. Bracts large and leafy. Calyx of five oblanceolate, obovate or lanceolate, acuminate segments, often glandular within, 1.5 in. long, dark red, or reddish brown in colour.

Corolla 3-5 in. long, campanulate from the very short tubular base, with five large rounded acute lobes about 1 in. long, of a beautiful, pure white or cream colour, greenish towards the base without, glabrous or hairy. Stamens five, inserted on the corolla at the top of the tube, alternate with the lobes, anthers sagittate 0.5-0.7 in. long, connivent to the stigma by their inner surfaces forming a five-sided cone, the stamens and style lying along the inner surface of the corolla. Ovary superior, seated on a five-lobed disk, 2-celled, with many ovules in each cell. Style filiform, expanding above into a fusiform stigma to which the anthers adhere. Fruit long, thick, woody, eventually dividing into two horizontally spreading follicles. Follicles turgid, fleshy, green, with a thick, hard, spongy, yellowish endocarp. Seeds many, compressed, ovoid or oblong, contracted at the top, 0.75 in. long, crowded, with a coma of hairs 1.5 in. long.

Flowers.—March-April. *Fruits.*—Cold season.

Distribution.—Native of the Eastern Himalayas, now frequently cultivated in gardens throughout India.

Gardening.—A huge and truly magnificent climbing shrub with large white fragrant trumpet-shaped flowers. It is of very rapid growth and ascends to the height of a lofty tree in no time. Propagated by seed, cuttings or layers.

Economic uses.—The young branches are sometimes used for making coarse ropes.

Chonemorpha G. Don.

A genus of high-climbing shrubs. Leaves opposite, broad, with distant nerves. Inflorescence paniculate, terminal or axillary. Calyx-tube cylindrical with an annular glandular area at the base inside. Corolla-tube cylindrical, slightly narrower at the base, ending above in five large twisted petals, horizontally spreading. Stamens five, epipetalous, connivent into a cone and adhering to the stigma, sagittate at the base. Carpels two, distinct; style filiform; stigma thick. Fruit of 2 follicles.

Chonemorpha macrophylla G. Don. [C. fragrans (Moon) Altston]
(*Macrophylla* means large-leaved.)

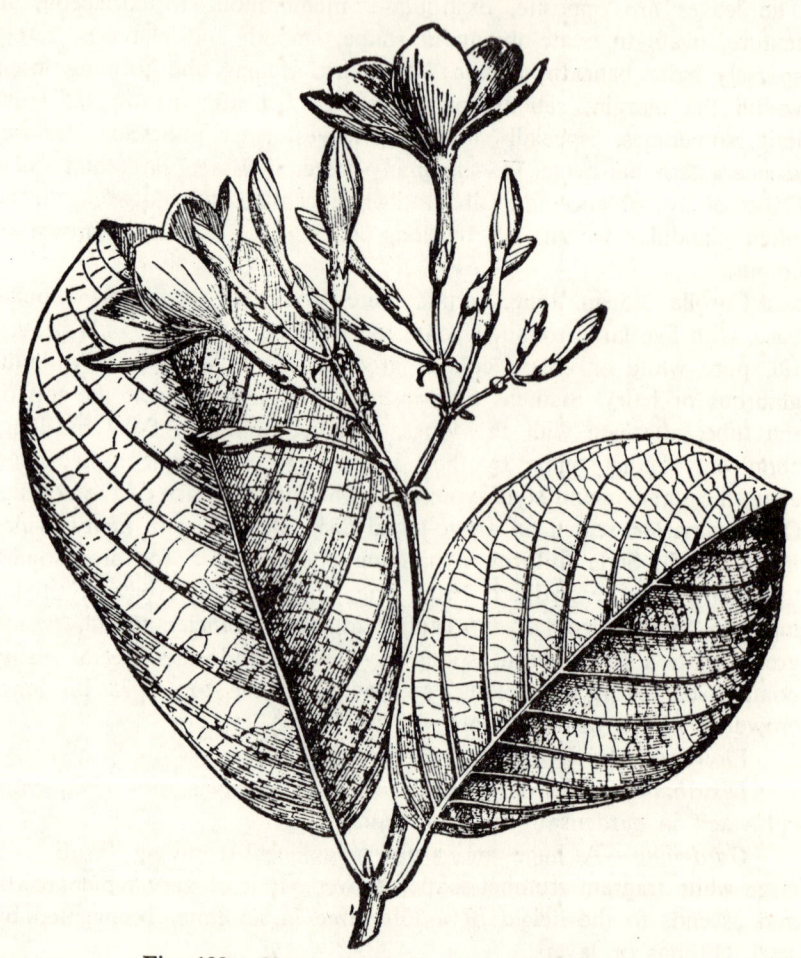

Fig. 122.—*Chonemorpha macrophylla* G. Don. × ¾

Description.—A scandent shrub capable of overtopping the largest trees; stems more or less pubescent, emitting quantities of a white latex if wounded. Leaves opposite, up to 10 in. long and as much broad, ovate, obovate or even orbicular in shape, rounded or cordate or even cuneate at the base, dark green above, paler beneath, almost hirsute below, sparsely hairy above, entire on the margins; petiole short, cylindrical.

Flowers borne in sub-terminal short-pedunculate cymes; branches cylindrical, green, speckled with red. Calyx-tube 0.75 in. long, cylindrical, shortly 5-lobed, at length withering red-brown and tightly embracing the base of the corolla tube. Corolla-tube white, narrow at the base, swelling above into a throat, about 0.75 in. long, ending above in five white petals which spread at right angles to the tube; lobes broadly and obliquely triangular or trapezoid from a very narrow base, white, yellowish in the throat. Stamens five, inserted at the bottom of the throat. Disk 5-lobed. Ovary of two distinct carpels. Fruit of two long straight hard trigonous follicles, 12-18 in. long. Seeds flat with a long tuft of hair.

Flowers.—May-July. *Fruits.*—Cold season.

Distribution.—Throughout India 2,000-6,000 ft extending to Burma, Malaya and Ceylon.

Gardening.—A large powerful climber with copious milky juice and very large leaves. The fragrant, pure white flowers, about 4 in. across, appear in great abundance during the hot weather. Propagation is by seed which is produced abundantly. It is suitable for cultivation in the open on a long trellis or on trees.

Vallaris Burm.

(This name is derived from the Latin word *vallo*, I enclose, owing to the fact that some species are used for screens in Java.)

A genus of climbing shrubs with punctuate, opposite leaves. The white flowers are borne in dichotomous or fascicled cymes issuing from the axils of the leaves. Calyx 5-lobed, from a short tube. Corolla-tube short, with five petals. Stamens five, fixed to the top of the tube seated on short filaments; anthers acuminate and sagittate at the base, connivent at the apex into a cone, with a large gland on the connective. Disk of five scales or lobes. Ovary of two hairy carpels, at first connate, soon free. Fruit of two oblong-acuminate carpels.

Vallaris heynei Spreng. [**Vallaris solanacea** (Linn.) O. Ktze.]

(This plant was named in honour of Frederick Adolf Heyne, a German botanist who lived at the beginning of the nineteenth century.)

Description.—A hoary shrub, climbing or prostrate. Stems covered with a grey lenticellate bark, emitting a copious white latex when wounded. Leaves opposite up to 5 in. long by 2 in. broad, elliptic, oblong, acuminate, sometimes somewhat abruptly narrowed to the acute apex, membranous, cuneate at the base; nervation rather obscure; petiole up to 0.75 in. long, slender, channelled above.

Inflorescence of axillary cymes about 1 in. long, 3-6-flowered; flowers white, fragrant, seated on slender pedicels. Calyx-tube very short, 5-lobed; lobes up to 0.1 in. long, acuminate, thin, shortly ciliate on the edges. Corolla-tube about 0.3 in. long, narrow and cylindrical and hairy within below, campanulate above, ending in five petals, which are orbicular in shape, 0.25 in. long. Stamens five, inserted at the top of the narrow portion of the tube; anthers acuminate above, sagitate at the base; carrying a large horseshoe-shaped appendage on

Fig. 123.—*Vallaris heynei* Spreng. × $\frac{3}{4}$

the connective. Disk annular, 5-toothed. Ovary of two connate carpels. Fruit of two follicles, straight.

Flowers.—December-April. *Fruit.*—January-April of the following year.

Distribution.—Native of India and Burma, often grown for ornament in gardens.

PLATE 74

M. B. Raizada

CREAM FRUIT
Strophanthus gratus (Benth.) Baill.

CREAM FRUIT

Gardening.—An extensive vigorous climber, evergreen and drought-resistant, with small elliptic-ovate leaves. The creamy white-scented flowers are produced in profusion during December-April. It is an excellent plant for cultivation in the open on a trellis, or on arches or trees. Propagated easily by layering, cuttings, suckers or seed. It was brought into cultivation in Europe in 1818.

Medicinal and economic uses.—The bark is bitter and astringent and according to Haines is chewed by Kols for fixing loose teeth. The latex from the stem is applied to sores and wounds. According to one story it is believed to keep away snakes during the rains when it is suspended from the roof on a certain day in June.

Strophanthus DC. (Roupellia Wall. et Hook.)

(This generic name comes from two Greek words which mean *twisted cord* and *flower* and refer to the long, caudate petals of some species. The other generic name *Roupellia* was selected by Wallich and Hooker to perpetuate the name of Charles Roupell of Charlestown, South Carolina, 'commemorated in many of the pages of Sir James E. Smith's Correspondence of Linnaeus'.)

This genus contains a number of shrubs, some of them are climbers. The leaves are opposite and penni-nerved. The inflorescence is terminal and consists of few-flowered cymes, compact, or many-flowered corymbs. Sepals five in number with 5-20 glands at the base. Corolla-tube short, campanulate in the throat, ending above in five petals which are more or less long-caudate; throat with 10 scales inserted in pairs. Stamens five, attached to the top of the tube, included; filament short; anthers sagittate, more or less acuminate, connivent. Disk absent. Ovary of two distinct carpels; style filiform. Fruit of two follicles.

The seeds of the species of this genus are well known for the presence of an active chemical substance called strophanthin. It raises the blood pressure, acts as a diuretic and is a powerful cardiac poison. Because of these properties different species have been used as arrow poisons in different parts of the world.

Strophanthus gratus (Benth.) Baill. (**Roupellia** grata Wall. & Hook. ex Benth.)

Cream-fruit

(*Gratus* is a Latin word meaning pleasing.)

Description.—A handsome climbing shrubby plant. Shoots-green, terete, very glabrous. Leaves opposite, petiolate, stipulate, up to 6 in. long by 2 in. broad, oblong-elliptic or elliptic in shape, glabrous, entire on the margins, acute at the tip, cuneate at the base, dark green on the upper surface, paler below; nerves somewhat impressed on the upper surface, slightly prominent below, joined by an intramarginal vein; stipules short, awl-shaped; bases of the petioles joined by a stipular line.

Inflorescence terminal, of 6-8 white, rose-tinted flowers, crowded into a cyme. Bracts ovate-lanceolate, acute or acuminate, keeled on the back, 0.15-0.2 in. long. Pedicels longer than the bracts but shorter than the calyx. Calyx of five lobes, each 0.6-0.8 in. long, obovate in shape, obtuse, greenish but reddish at the tip. Corolla-tube swelling above, glabrous without and within, 1.5 in. long, ending above in five petals; petals broadly obovate in shape, crisped on the margins, under 1 in. long. In the mouth of the flower is a corona of 10 linear-lanceolate scales, connate at the base, erect, of a beautiful rose colour, 0.4-0.5 in.

Fig. 124.—*Strophanthus gratus* (Benth.) Baill. × ¾

long. Stamens five, inserted at the beginning of the swollen portion; filaments short, thick, slightly papillose; anthers produced above into a subulate process protruding from the corolla tube. Ovary surrounded by a nectariferous disk. Carpels distinct, joined by a filiform style; ovules numerous, on axile placentas.

Flowers.—Hot and rainy season. *Fruits.*—Cold season, but very rarely in India.

Distribution.—Native of Sierra Leone, tropical Africa, now in cultivation in all tropical and subtropical countries of the world.

Gardening.—A handsome climbing shrubby plant or a rambling shrub requiring considerable space for its full growth but it can be kept within bounds by judicious pruning. The large leathery bell-shaped flowers are white, tinged with rose-purple, and are attractive just as they are opening. As it is a fast grower it requires plenty of water during the summer. Good fresh loam, with a little leaf-mould suits it best. It is well adapted for a trellis. Propagated easily by cuttings which strike root readily during the rains. It is commonly known under the name of *Roupellia grata* in Indian gardens.

Trachelospermum Lemaire.

(This generic name comes from two Greek words, *trachelos*, a neck, and *sperma*, a seed. The combination refers to the long seed of the species.)

A genus of climbing shrubs with opposite leaves. Inflorescence of terminal or axillary cymes. Flowers white or purple. Sepals 5, small, glandular or scaly at the base inside. Corolla-tube cylindrical, dilated at the insertion of the stamens, ending above in five spreading lobes. Stamens five, included in the tube, connate at the apex and connivent around the stigma. Disk annular, truncate, or 5-lobed. Ovary of two distinct carpels. Fruit of two dry follicles.

KEY TO THE SPECIES

Calyx-lobes erect .... *T. fragrans*
Calyx-lobes reflexed .... *T. jasminoides*

Trachelospermum fragrans Hook. f. [**T. lucidum** (Don.) K. Schum.]

Description.—An evergreen, twining shrub. Young shoots slightly hairy, emitting a milky juice when cut. Leaves opposite, up to 6 in. long by 2 in. broad, elliptic-lanceolate, acuminate, glabrous, bright green above, paler below; petiole about 0.2-0.4 in. long, minutely hairy.

Inflorescence a lax, terminal or axillary, trichotomous panicle. Flowers white, seated on glabrous pedicels decorated by minute bracts. Calyx about 0.1 in. long. cleft almost to the base, 5-lobed; lobes ovate, obtuse, ciliate. Corolla-tube about 0.4 in. long, very slender below, inflated at the top, glabrous within and without; mouth thickened, hairy, 5-lobed; lobes spreading, cuneate below, obliquely truncate at the apex, overlapping to the right, twisted to the left. Stamens 5, included; anthers connivent to the top of the style. Disk of 5 erect lobes. Carpels two, glabrous, distinct; style filiform; stigma columnar; ovules many in each cell. Follicles up to 12 in. long by 0.2 in. broad, incurved, cylindrical. Seeds 0.7 in. long, linear, flattened, dull-brown; coma copious, 1 in. long.

Flowers.—April-June. *Fruits.*—October-December.

Distribution.—North-West Himalayas 3,000-7,000 ft, Assam, Cachar, and Upper Burma.

Fig. 125.—*Trachelospermum fragrans* Hook. f. × ¾

Gardening.—A tall climber with dark green leaves and white fragrant flowers which are produced in great abundance during the hot season. It is suitable for covering embankments and the like and prefers moist shady places. Propagation is by cuttings of half-ripened wood during the rains or by seed.

Trachelospermum jasminoides Lem.

[**T. divaricatum** (Thunb.) K. Schum.]

Star Jessamine

(The specific name *jasminoides* means jessamine-like in Latin.)

Description.—A pretty evergreen climbing shrub; young parts puberulous. Leaves petioled, opposite, elliptic, elliptic-lanceolate or even oblanceolate in shape, glabrous and smooth, somewhat coriaceous in texture, entire on the margins, up to 2.5 in. long by 1 in. wide; petiole very short, grooved above.

Inflorescence a few-flowered cyme seated on a peduncle much longer than the subtending leaf. Flowers pure white, very fragrant. Calyx divided almost to the base into five reflected lanceolate ciliate segments 0.1-0.2 in. long. Inside the calyx at the base of the corolla are several jagged scales. Corolla-tube 0.25-0.3 in. long, abruptly contracted below the middle, glabrous outside, hairy within at the mouth, ending above in five lobes; lobes oblique, obovate-spatulate, spreading, waved, with reflexed margins. Stamens five, sessile on the corolla; anthers lanceolate; connnective produced above into a spur, the whole five stamens connivent into a cone round and adherent to the stigma. Ovary of two carpels. At the base of the ovary are five large glands, two united, three free.

Flowers.—March-April. *Fruits.*—Cold season.

Distribution.—Indigenous to China and Japan, commonly grown in gardens throughout the country.

Fig. 126.—*Trachelospermum jasminoides* Lem. × 1

Gardening.—A pretty white-flowered evergreen climbing shrub with deep green, smooth foliage. The delightfully fragrant flowers are produced in great profusion during the early part of the hot weather. It was collected by a Mr. Fortune from Shanghai and introduced by him into European gardens. Propagation is by cuttings or layers during the rains. It is best suited for pergolas, arches and the like.

Melodinus Forst.

Climbing shrubs. Leaves opposite, penni-nerved; nerves parallel. Inflorescence terminal, of trichotomous panicles, many-flowered or axillary and then few-flowered; flowers white, often fragrant. Calyx of 5 sepals, without glandular scales. Corolla-tube dilated at the insertion of the stamens, ending above in five twisted petals; throat of the corolla furnished with scales. Stamens five, inserted towards the base or in the middle of the tube. Disk absent. Ovary entire, bilocular, style short; stigma thick; ovules numerous. Fruit globular, pulpy, containing many seeds.

Melodinus monogynus Roxb.

(*Monogynus* means with a single ovary.)

Description.—A large glabrous scandent shrub. Branches and branchlets at first green, afterwards turning brown, containing copious

Fig. 127.—*Melodinus monogynus* Roxb. $\times \frac{3}{4}$

quantities of latex. Leaves opposite, petiolate, up to 6 in. long; elliptic, elliptic-oblong or oblong-lanceolate, acuminate, glabrous and smooth, chartaceous in texture; margins incurved; petiole 0.2-0.4 in. long.

Inflorescence close, terminal, of trichotomously branched panicles; branches puberulous, rather thick. Flowers white, fragrant. Calyx of five sepals, elliptic-obtuse, elliptic-oblong or even ovate, ciliate on the margins, 0.1-0.15 in. long, imbricate. Corolla-tube 0.7 in. long, tubular below becoming funnel-shaped above, glabrous without but hairy within, ending above in 5 lobes; lobes spreading, up to 0.5 in. long, oblong or ovate, obtuse; corona of scales present, five in number, villous, bifid at the apex. Stamens five, filaments very short; anthers without basal appendages. Disk absent. Ovary of two connate carpels. Fruit a globose berry, smooth, yellow or orange in colour, up to 3 in. in diameter.

Flowers.—April. *Fruits.*—Cold season.

Distribution.—Native of Sylhet and Assam extending to Malaya Peninsular and China. Occasionally cultivated in gardens throughout the plains and up to 4,000 ft in the hills.

Gardening.—A large climber with milky juice and bright dark-green lanceolate leaves. The pure white very fragrant flowers are produced during spring and make this plant very ornamental. The fruit which is the size and colour of an orange is said to be edible. Propagation is by seed, cuttings or layers during the rains. It was introduced into England by Capt. Craigie as a present from Dr. Wallich of the Calcutta garden. It is well suited for growing over arches, pergolas and the like.

Chapter 15

OLEACEAE

THE OLIVE FAMILY

This family takes its name from one of its genera, *Olea*. *Olea* is itself derived from the Greek word, *elaia*, a name for the Olive. Olive oil was called *elaion* in Greek.

Oleaceae is a large family of 22 genera and about 400 species, distributed throughout the temperate and warmer regions of the earth. Included in the family are deciduous and evergreen trees and shrubs with opposite leaves. The leaves are simple or compound, exstipulate. Inflorescence various, axillary or terminal; flowers hermaphrodite, rarely unisexual, regular. Calyx nearly always present, small, 4- or occasionally more-lobed, usually bell-shaped. Corolla gamopetalous, sometimes of four petals, often 4-lobed. Stamens 2; anthers apiculate, often back to back, opening lengthwise; filaments short. Ovary superior, 2-celled; ovules usually 2 in each cell. Fruit a capsule, berry or drupe.

The flowers of the species of *Oleaceae* are often sweetly scented and numbers of them secrete a nectar at the base of the corolla, characteristics which indicate fertilisation through the agency of insects. An interesting fact about some species, e.g. *Nyctanthes arbor-tristis* (the well-known *Harsingar*) and various species of *Jasminum*, etc. is that their flowers are fragrant after sunset. In these cases nocturnal Lepidoptera act as the unconscious agent of cross-fertilisation. In certain cases, however, for example in the ash, *Fraxinus*, where there is no corolla, the blossoms are pollinated by the wind.

The fruits of the ash are winged but a good many species depend upon an edible fruit for the distribution of their seeds.

One species, the olive tree, is of importance economically. In all countries surrounding the Mediterranean, the tree *Olea europaea* Linn. is cultivated for its fruit. The well-known olive oil may be pressed out of the fruits, which are also used for pickling. Its original home is in Asia Minor but it has been introduced with great success into California. Efforts have been made in the past to introduce this tree into India but with little success so far, though reports from Kashmir are encouraging. In India either the tree does not bloom in which case there is no fruit, or if it does bloom, the fruits drop off before they are ripe. It may, however, be possible to achieve success by grafting the European olive on to one of the indigenous olives.

The largest genus in the family is *Jasminum*, of which a large

PLATE 76

HILL JASMINE
Trachelospermum fragrans Hook. f.

N. L. Bor

PLATE 77

HILL JASMINE
Trachelospermum fragrans Hook. f.

N. L. Bor

number of species occur in India. The genus is well represented in our gardens and is valued for its pretty sweet-scented flowers and glossy foliage. *Syringa* and *Ligustrum* are sometimes grown in the hills. *Syringa vulgaris* Linn. is the well-known common Lilac of England.

KEY TO THE GENERA

Leaves simple or compound; if simple articulated on the petiole .. *Jasminum*
Leaves simple, continuous with the petiole .... *Osmanthus*

Jasminum Linn.

The Jessamin, Jasmin or Jasmine

(This generic name is said to be derived from *ysmym*, an Arabic word. There is, however, no certainty about this and others derive it from two Greek words, *ia* flower and *osme* scent.)

An important genus of shrubby or climbing plants, comprising about 200 species. Many of these are cultivated in the open in the warmer, and under glass in the colder, regions of the world, not only for their pretty flowers and handsome foliage but also for the delicate perfume of the blooms. Leaves simple, 3-foliate or odd-pinnate; petioles articulated. Inflorescence usually terminal, rarely axillary, cymose; flowers bracteate, hermaphrodite, yellow, red or white in colour. Calyx usually bell-shaped, sometimes cylindrical with 4-9 lobes. Corolla-tube slender, 4-10-lobed; lobes spreading. Stamens 2, included within the tube on short filaments. Anthers oblong, connective shortly produced. Ovary 2-celled; ovules 2 in each cell, basal, erect, style slender; stigma linear. Fruit a berry.

The well-known perfume, Jasmine, is extracted from the flowers by the process known as 'enfleurage', first developed in France. The flowers are lightly spread over a layer of solid fat. Every 24 hours or so the old flowers are replaced by fresh ones so that the fat eventually becomes saturated with the sweet smelling substances. These are subsequently extracted with the aid of alcohol, acetone and other solvents. Another method in use in France to extract the perfume is to spread the flowers on blankets which have been soaked in olive oil, from which it is recovered in the usual way. The principal ingredient of the perfume is a pale yellow oil with small quantities of benzoyl acetate, benzyl alcohol, indole and certain esters. The trade in Jasmine oil is very large. In France alone, about 600 tons of flowers are used for this purpose and to this must be added further large quantities produced in Tunis, Algeria and in other countries of the world. Some of the species, used as medicine, bear a high reputation for the treatment of various diseases. The medicinal uses of the various plants will be outlined under the species about to be described.

Key to the Species

Leaves opposite
  Leaves 1-foliate
    Calyx-lobes short; plant hairy .... *J. sambac*
    Calyx-lobes very long; plant very hairy .... *J. pubescens*
  Leaves 3- or more-foliate
    Flowers white
      Terminal leaflets much larger than the others; distal pair not with broad connate bases; leaflets 3-7, lateral acute .... *J. officinale*
      Terminal leaflets not or scarcely larger than the others; distal pair with broad connate bases; leaflets 7-11, lateral usually very obtuse .... *J. grandiflorum*
    Flowers yellow .... *J. primulinum*
  Leaves alternate; flowers yellow .... *J. humile*

Jasminum sambac (Linn.) Ait..

The Arabian Jasmine

(*Sambac* is the Arabic name for the shrub; *zambak* in Persian.)

Description.—A scandent or suberect shrub with pubescent branches. Leaves opposite, exstipulate, petiolate, ovate in shape, 1.5-3.5

Fig. 128.—*Jasminum sambac* (Linn.) Ait. × ¾

in. long by 0.8-2.5 in. wide, thin, glabrous, obtuse, acute or acuminate at the apex, rounded at the base; lateral nerves 4-6 pairs, prominent beneath and looping within the margin; petioles 0.2 in. long, pubescent.

THE YELLOW JASMINE
Jasminum humile Linn.
(× ¾)

Inflorescence of few-flowered terminal cymes or, occasionally, of solitary flowers. Flowers white, fragrant, pedicellate or not. Bracts linear-subulate, hairy. Calyx-tube campanulate, tubular, 0.5 in. long, 5-9-toothed; teeth linear-subulate, longer than the tube. Corolla-tube cylindrical, 1 in. long, 5-9-lobed; lobes acute or obtuse, as long as the tube. Stamens 2, included; filaments short. Ovary 2-celled. Fruit of 1-2 globose berries, each 0.5 in. in diameter, black, surrounded by the erect, persistent calyx teeth.

Flowers.—Hot and rainy season. Does not set fruit in this country.

Distribution.—Believed to be indigenous to South India, much cultivated in the tropics.

Gardening.—A straggling shrub much prized for its exquisitely fragrant flowers. The white, sweet scented flowers are considered sacred to Vishnu and are largely used by Hindus for making into garlands. In the tropics the plant is almost invariably attacked by scale insects, usually resulting in a black fungus growth on the leaves. The shrub is consequently best allotted a place in the background in an unfrequented spot. The plant prefers a dry habitat and water must be applied to the roots and not on the leaves and blossoms. If water touches the flowers they usually turn black and fall. It flowers best and most profusely when grown in direct sun. Stripping off the leaves is a method commonly used to induce more blossoms. Propagation is by cuttings. It is commonly known by the name of Bela, Motiya or Mogra and has been under cultivation since very early times.

Economic and medicinal uses.—This plant has been cultivated since very early times. Double-flowered races are common. The flowers are used to give an aroma to Chinese teas. The perfume is extracted in India by the method known as enfleurage, but instead of fat or oil, crushed sesamum seeds are used. The leaves are used in India as a lactifuge, and are said to be at least as efficacious as belladonna. A decoction of the leaves and root is used for sore eyes.

Jasminum pubescens Willd. [**J. multiflorum** (Burm f.) Andr.]

(*Pubescens* means hairy in Latin.)

Description.—A scandent shrub; young parts velvety-tomentose, often rusty. Leaves opposite, simple, entire, petioled, exstipulate, ovate in shape, 1-3 in. long, up to 1.5 in. wide, acute or acuminate at the tip, rounded or cordate at the base; upper surface pubescent becoming glabrous; under surface tomentose or pubescent, especially on the nerves; petiole stout, 0.2-0.4 in. long, densely tomentose.

Inflorescence of dense capitate cymes, terminal at the tips of dwarf lateral branches. Flowers white, fragrant, 0.7-1.5 in. across, sessile or nearly so. Calyx-tube 0.5-0.6 in. long, densely rusty tomentose, 6-9-toothed; teeth subulate, twice as long as the tube or more Corolla-tube 0.7-0.8 in. long, slender, 6-9-lobed; lobes oblong-lanceolate, acute, shorter than the tube; stamens 2, included. Ovary 2-celled; ovules 2 in each cell. Style slender. Fruit ellipsoid, about 0.5 in. long, black when ripe, surrounded by the long hairy calyx-teeth.

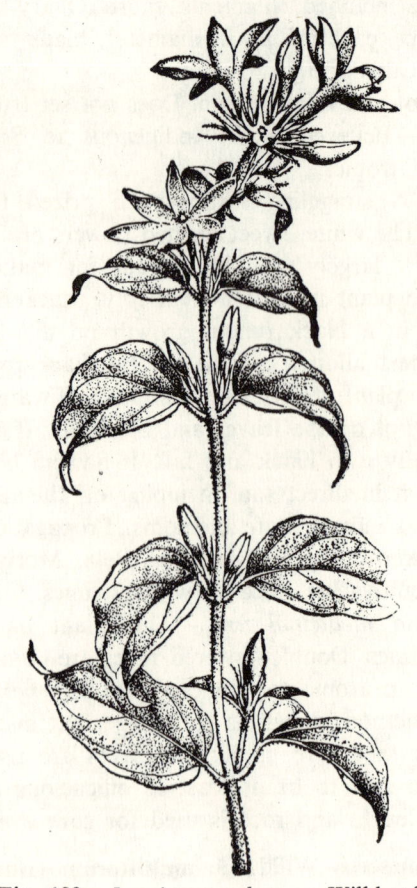

Fig. 129.—*Jasminum pubescens* Willd. × ¾

Flowers.—December-April and also during the rains. Fruits May-July.

Distribution.—Throughout the greater part of India ascending to 4,000 ft in the Himalayas, also in Burma and China.

Gardening.—A scandent shrub with all parts covered with pubescence. The star-like slightly scented flowers appear practically throughout the year and make this plant one of the most useful plants for covering a trellis or as a ground cover and low shrub. It is not particular as to soil and sun requirements.

PLATE 78

N. L. Bor

THE HAIRY JASMINE
Jasminum pubescens Willd.

PLATE 79

PRIMROSE JASMINE

M. B. Raizada

PLATE 80

M. B. Raizada

THE YELLOW JASMINE
Jasminum humile Linn.

PLATE 81

M. B. Raizada

THE SPANISH JASMINE
Jasminum grandiflorum Linn.

Medicinal uses.—A decoction of the root has some repute as an antidote to cobra venom, and that of the leaves is said to be of use in stimulating static ulcers.

Jasminum primulinum Hemsley

(*Primulinum* means primrose-like in Latin.)

Description.—An evergreen twiggy shrub with 4-angled glabrous stiff branches. Leaves opposite, 3-foliate, exstipulate, petiolate, up to 4 in. long. Leaflets almost sessile or with a petiolule up to 0.1 in. long, glabrous, narrowly elliptic or oblong-lanceolate, entire, 1-2 in. long, rather thick in texture, apiculate, wedge-shaped at the base, dark green and shining above, paler beneath.

Flowers solitary on axillary peduncles, primrose yellow in colour, orange in the throat; bracts and bracteoles foliaceous, often scale-like.

Fig. 130.—*Jasminum primulinum* Hemsley × ¾

Calyx bowl-shaped, 0.1 in. long, 6-lobed; lobes lanceolate, sparsely pubescent, 0.2 in. long. Corolla-tube stout, increasing slightly in diameter towards the top, usually 6-lobed; lobes obovate; rounded, about 1 in. long.

Stamens 2, included. Ovary glabrous, 2-celled; ovules 2 in each cell. Style slender, exserted in single flowers, glabrous; stigma capitate, deeply and acutely 2-lobed.

Flowers—March-May. Does not set seed in this country.

Distribution.—Native of Yunan and China apparently as an escape from cultivation, now extensively cultivated throughout the tropical and subtropical parts of the world.

Gardening.—A rambling, evergreen shrub of recent introduction. The scentless blossoms of a rich golden yellow characterize this plant. It will thrive in poor soil and under adverse conditions. It has pretty foliage and being semi-trailing it makes a successful ground cover. The plant spreads by layering itself. Propagation also by cuttings or root suckers. The double-flowered form is the one usually seen in cultivation.

Jasminum humile Linn.

Yellow Jasmine

(*Humilis* is a Latin adjective meaning low as opposed to high.

Fig. 131.—*Jasminum humile* Linn. × ¾

It is hardly appropriate when applied to this species.)

Description.—A diffuse shrub reaching 6 ft at Dehra Dun, evergreen, glabrous. Branches green, angular. Leaves alternate, exstipulate, petioled, imparipinnate, very variable in size, 2-3 in. long, often 6 in. long in cultivated bushes; petiole and rhachis channelled above. Leaflets 3-5, very variable in size, rather thick, dark green, paler below, elliptic, ovate or lanceolate in shape, sessile or subsessile, acute or obtuse, wedge-shaped at the base, the terminal up to 4 in. long in cultivated shrubs, the lateral smaller.

Inflorescence of terminal corymbose panicles. Flowers bright yellow, seated on pedicels, 0.2-0.6 in. long, furnished with linear bracts. Calyx-tube 0.1-0.15 in. long, 5-lobed; lobes 0.05 in. long. Corolla-tube 0.5-1 in. long, 5-lobed; lobes broadly ovate-obtuse or round, usually reflexed when the flower is fully open. Stamens 2, included. Ovary 2-celled, 4-ovuled. Fruit of 1-2 ripe carpels, ellipsoid, 0.3-0.4 in. long, black when ripe, full of crimson juice.

Flowers.—April-June. *Fruit.*—September-December.

Distribution.—Native of the North-West Himalaya, up to 9,000 ft, also on the Salt Range, Mount Abu and Nilgiris; widely cultivated throughout the country.

Gardening.—An erect rigid shrub with bright yellow flowers. It is advisable to prune it hard after flowering so as to keep the bush within bounds. Easily propagated by cuttings or seed. It is locally known as 'shanjoi'.

Medicinal uses.—This plant is not of much repute medicinally but the root is said to be useful in curing ringworm.

Jasminum officinale Linn.

The White Jasmine

(*Officinalis* means, medicinal, officially recognised as a drug.)

Description.—A twiny shrub with striate branches, sparsely hairy when young. Leaves opposite, exstipulate, imparipinnate, 2-4 in. long; petiole and rhachis narrowly margined. Leaflets 3-7, the terminal 1-3 in. long by 0.4-1 in. wide, ovate or lanceolate, acuminate, usually larger than the lateral leaflets which are shorter and relatively broader, acute, sessile or shortly petiolulate, the distal pair sometimes with broad connate bases.

Inflorescence of terminal few-flowered corymbs or cymes and axillary pedunculate few-flowered cymes shorter than the leaves, or the cymes often reduced to single flower, pedicel of the cyme-flowers 0.3-0.7 in. long, those of the solitary and corymb-flowers often much longer;

bracts up to 0.5 in. long, linear-subulate or narrow-linear. Calyx-tube 0.1-0.15 in. long, puberulous, 5-lobed; lobes subulate, 0.2-0.6 in. long. Corolla-tube 0.5-0.7 in. long, cylindrical, 5-lobed; lobes ovate or elliptic. Stamens 2, included. Ovary 2-celled; ovules 2 in each cell. Fruit black when ripe, elliptic or globose, 0.3-0.4 in. long, full of crimson juice.

Flowers.—May-June. *Fruit.*—October-November.

Distribution.—Native of Persia and Kashmir now widely distributed throughout India, wild or cultivated.

Gardening.—This plant, whose native home is in Persia and Kash-

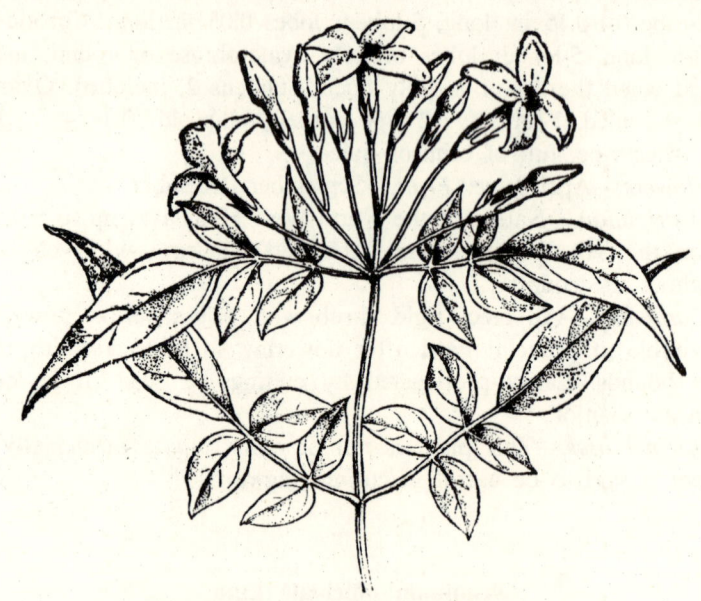

Fig. 132.—*Jasminum officinale* Linn. × $\frac{3}{4}$

mir, has been in cultivation in India and China since very early times. It has been introduced into the milder parts of Europe and has become established. It is a loose climbing 'vine' requiring a support but scarcely self-climbing. The glossy foliage and fragrant white flowers which appear during the hot weather render the plant very attractive. Like *J. grandiflorum* it is of vigorous growth and hardy and requires periodic pruning. Easily propagated by cuttings.

Medicinal uses.—It is mentioned in Chinese medical books dated about the 17th century, as a valuable aphrodisiac. A decoction of the root is said to be of use in ringworm cases. The fruits are reputed to be narcotic and sedative. The fragrant oil from the flowers is mixed with sesamum oil and rubbed on head as a nerve-sedative.

Jasminum grandiflorum Linn. [**J. officinale** Linn. var. **grandiflorum** (L.) Kobuski]

The Spanish Jasmine

(*Grandiflorum* means large flowered.)

Description.—A large shrub with striate, glabrous, almost angled branches. Leaves opposite, exstipulate, petiolate, 2-5 in. long; petiole and rhachis margined. Leaflets 7-11 in number, the terminal somewhat larger than the lateral but not markedly so, glabrous, dark green, entire, the upper lateral pair with broad flat base, often confluent with the terminal, the lowest pair with short petiolules, the intermediate pairs sessile, the terminal acuminate at the tip, wedge-shaped at the base, the other apiculate at the tip and rounded, often obliquely, at the base.

Fig. 133.—*Jasminum grandiflorum* Linn. × ¾

Inflorescence in lax axillary or terminal cymes longer than the leaves. Flowers white, very fragrant, often tinged with red outside, pedicellate; pedicels 0.5-1 in. long; bracts ovate to spathulate-oblong, foliaceous; bracteoles small, linear. Calyx-tube campanulate, 0.1 in. long or less, glabrous, 5-toothed; teeth subulate, two to three times as long as the tube. Corolla-tube cylindrical, 0.7-1 in. long, glabrous 5-lobed; lobes elliptic or obovate, obtuse. Stamens 2, included. Ovary 2-celled, 4-ovuled.

Flowers.—Hot and rainy season. *Fruit.*—Cold season.

Distribution.—Native of the North-West Himalayas up to 7,000 ft; extensively cultivated in gardens both in the plains and hills.

Gardening.—A large twining or scandent shrub of vigorous growth, hardy and drought resistant. The leaves are imparipinnate and the flowers, which are pure white and fragrant, are bigger than those of *J. officinale*. On account of the great demand for the buds of this species it is extensively cultivated. It is rather a troublesome plant to keep within bounds as it spreads over a large space sending forth roots from its stems wherever they touch the ground. It is suitable for a light trellis and is quite attractive because of its graceful deep green foliage and scented flowers. To induce profuse flowering it is best to prune in November-December and manure in March-April. Easily propagated by cuttings. It is popularly known as 'chameli'.

Economic and medicinal uses.—The leaves and flowers have long been known in Hindu medicine. The leaves contain a resin, salicylic acid, an alkaloid, jasminine, and an astringent principle. The leaves are astringent in action. The whole plant is considered to be anthelmintic, dicretic and emmenagogue. The perfume from the flowers is extremely valued.

This is the plant commonly cultivated in Europe for the perfumery trade. Up to the present the chemists have not been able to copy it exactly in the laboratory, as the synthesis of a ketone found in the oil of the flowers, which gives a distinctive scent, has not yet been accomplished. The juice is said to be anthelmintic and an antidote for scorpion sting.

Osmanthus Lour.

(The generic name means fragrant flower in Greek.)

A small genus of evergreen trees and shrubs with serrate or entire, opposite or alternate short-petioled leaves. Flowers fascicled or in very short racemes, perfect, polygamous or dioecious. Calyx short, 4-toothed. Corolla short- or long-tubular, 4-lobed; lobes 4, obtuse. Stamens 2, rarely 4, inserted on the tube. Ovary 2-celled; ovules 2 in each cell pendulous from the apex; style short, 2-lobed or entire. Fruit an ovoid or globose drupe with a one-seeded stone.

Osmanthus fragrans Lour.

Description.—A shrub or small evergreen tree. Leaves opposite, petiolate, exstipulate, glabrous, coriaceous, 7 in. long by 2 in. wide, entire in wild but serrate in cultivated plants, elliptic to oblong-lanceolate in shape, acute or acuminate at the tip, wedge-shaped at the base; reticulation prominent beneath; petiole 0.6 in. long.

Flowers yellowish, very fragrant, 0.25-0.75 in. long, densely fascicled in the upper axils, rarely terminal, shortly pedicellate. Calyx minute, 0.03 in. long, 4-toothed. Corolla-tube very short, 4-lobed; lobes

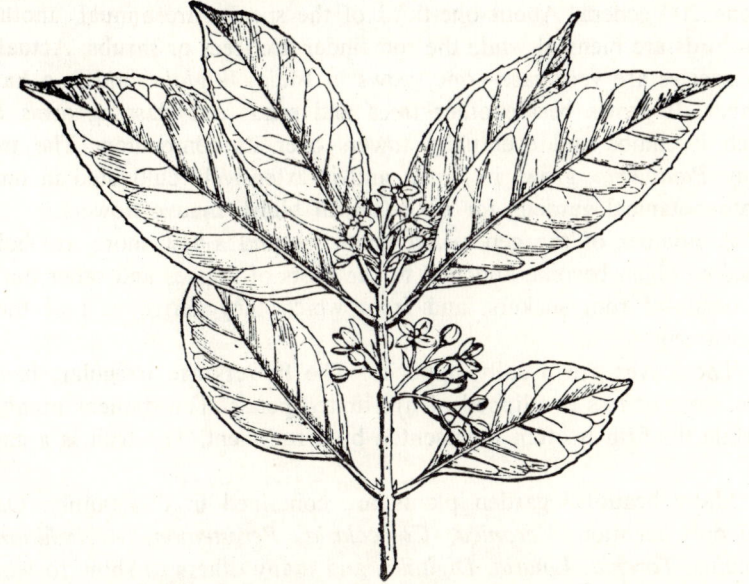

Fig. 134.—*Osmanthus fragrans* Lour. × ¾

oblong, about 0.3 in. long. Stamens 2, inserted in the corolla tube; anthers exserted; filaments short. Ovary 2-celled; ovules 2 in each cell, pendulous. Drupe 1-seeded.

The flowers of this species are extremely fragrant and will scent the air for a considerable distance round the plant. The sweet-smelling flowers are said to be used by the Chinese for scenting their finer qualities of tea.

Flowers.—October. Also at other time of the year. *Fruits.*—April.

Distribution.—Native of the Himalayas extending to China and Japan.

Gardening.—An attractive, evergreen shrub with dark green glossy foliage. The tiny pale yellow flowers have a strong smell very like apricots and it is for the love of the fragrance that it is extensively cultivated. It prefers partial shade and is propagated by cuttings or 'gooties'. It is popularly known as *Olea fragrans* in gardens.

Chapter 16

SCROPHULARIACEAE

This cosmopolitan family consists of about 2,600 species divided among 200 genera. About one-third of the species are annual, another two-thirds are biennial while the remainder are trees or shrubs. Actually tree species are very rare—one grows in India, *Wightia*, and is a half-climber. It grows beside other trees and sends out clasping arms by which it climbs and eventually towers over its companion. The tree genus, *Paulownia*, grows in Japan and is extensively cultivated in temperate botanical gardens for its beautiful bluish mauve flowers.

A number of the genera are parasites and a few more are half-parasites which become attached to the roots of grasses and other herbs by means of root suckers, and from which they derive part of their nourishment.

The leaves are usually opposite. The flowers are irregular, being often more or less two-lipped of five united petals. The stamens number 4, while the fifth is often represented by a rudiment. The fruit is a capsule.

Many beautiful garden plants are contained in this family. One need only mention *Veronica, Calceolaria, Pentstemon, Antirrhinum, Mimulus, Torenia, Linaria, Digitalis,* and many others to show to what an extent our gardens are indebted to this family.

Pollination is normally carried out by insects. Honey is secreted by a disk surrounding the base of the ovary, and there are various devices to ensure pollination by the insects which visit the flowers.

Russelia Jacq.

(A genus named in honour of Alexander Russel, an English physician and traveller, who died in 1768.)

A genus of shrubby species, Mexican in origin, which are favourite plants in Indian gardens on account of their showy flowers. The leaves are opposite or whorled, often reduced to scales. The flowers are red and are arranged in many-flowered cymes; calyx gamosepalous, cleft into five lobes; corolla-tube cylindrical with five lobes spreading at the mouth; stamens, four; fruit a capsule.

KEY TO THE SPECIES

Leaves scale-like or very small; corolla 0.75 in. long .... *R. juncea*
Leaves well developed; corolla less than 0.5 in. long .... *R. sarmentosa*

THE CORAL PLANT
Russelia juncea Zucc.

M. B. Raizada

THE CORAL PLANT
Russelia juncea Zucc.

M. B. Raizada

TWIGGY RUSSELIA
Russelia sarmentosa Jacq.

M. B. Raizada

PLATE 85

M. B. Raizada

TWIGGY RUSSELIA
Russelia sarmentosa Jacq.

Russelia juncea Zucc. (**R. equisetiformis** Cham. et Schl.)

The Coral, Fountain or Fire-cracker Plant

(*Juncea* means rush-like in Latin.)

Description.—A much-branched shrub up to 6 ft tall with glabrous, noded stems. Branches whorled, noded, smooth and glabrous, angled and grooved, noding at the tips. The leaves are opposite or whorled, petioled, very small, ovate-lanceolate or linear-obtuse, sometimes spathulate, often crenate, up to 0.5 in. long, but usually reduced to mere scales at the nodes. In the latter case the whole task of photosynthesis is undertaken by the stems and branches.

The inflorescences are produced at the tips of whorled branches. The flowers are arranged on 2-3 flowered peduncles and are produced in profusion all the year round. Pedicels very slender, about 0.25 in. long. Calyx 0.1 in. long, bell shaped, cleft into five, ovate-acute, imbricating lobes, greenish brown or reddish in colour. Corolla tubular in shape, of a beautiful clear red colour, 0.75 in. long, ending above in five rounded imbricating lobes, glandular inside at the base. Stamens four, perfect, one rudimentary, inserted on the corolla near the base, reaching to the mouth. Anther cells divergent but at length confluent. Ovary seated on a fleshy disk, 2-celled, with many ovules on central placentas. Fruit a globose capsule.

Flowers.—Throughout the year. Does not set seed in this country.

Distribution.—Indigenous to Mexico. Very commonly grown in gardens in the plains throughout India.

Gardening.—An exceedingly common and at the same time a very graceful plant, with long rush-like ribbed green stems and scarlet flowers. It seems to grow readily almost anywhere and is propagated by cuttings or division of the roots during the rains. It is quite suitable for growing in a rock garden and in hanging baskets.

The brightly coloured flowers and generous production of nectar are indications that the flowers are adapted to cross-pollination by insects or birds. In this country, however, the *Russelia*s apparently never set seed and are always propagated by other means.

In Dehra Dun the sunbird, *Nectarinia asiatica,* is a very frequent visitor to these flowers. Instead of doing its job in the proper way, this sunbird bores a hole through the base of the corolla and steals the nectar and the transference of pollen from one shrub to another is avoided.

Russelia sarmentosa Jacq. [**R. coccinea** (Linn.) Wettst.]

(*Sarmentosus* means twiggy or branchy in Latin.)

Description.—An erect twiggy shrub reaching 5 ft in height.

Branches angular (often octangular) in cross-section, smooth and glabrous. Leaves well developed, often in fours at the nodes, petioled, ovate-acuminate in shape, 2-4 in. long coarsely toothed, with nerves deeply impressed on the upper surface, prominent beneath. Upper surface and petiole more of less covered with coarse short hairs, under surface hairy on the nerves only. The under surface between the nerves is deeply pock-marked with glandular pits.

Inflorescence in congested bracteate dichotomous cymes. Bracts, bracteoles, peduncles and pedicels sparsely covered with coarse hairs. Calyx divided almost to the base into 5 lanceolate-acuminate lobes, dark red in colour, sparsely hairy, 0.2 in. long. Corolla red, tubular, divided above into 5 lanceolate-obtuse lobes, 0.3-0.5 in. long. Stamens four, with widely divaricate anthers; filaments arising from the base of the corolla. Ovary seated on a disk. At the insertion of the stamens towards the base of the corolla is a narrow ring of club-shaped yellow glands. There is another line of glandular rod-shaped hairs running down from the mouth towards the base.

Flowers.—Most part of the year. Does not set seed in this country.

Distribution.—Indigenous to Mexico, ascending to 8,000 ft, now grown in gardens throughout the tropics.

Gardening.—A handsome erect plant with deep scarlet flowers borne in crowded bunches along the stem. Propagated by division of the root, as cuttings are less successful. This species is also known as *R. multiflora* Sims.

PLATE 27

THE CORAL PLANT
A. *Russelia juncea* Zucc.; B. *Russelia sarmentosa*
(¾ nat. size)

Chapter 17

MALVACEAE

This family, which takes its name from one of its genera, *Malva*, the Mallow, contains 42 genera with over 900 species and is confined, with two exceptions, to the tropics of both hemispheres.

The family comprises herbs, shrubs and trees. Mucilage cells either singly or in rows, occur in the bark and pith. The leaves are alternate and often more or less palmately divided, but it frequently happens that both entire and palmately divided leaves are found on the same plant. Stipules are present, but they fall off early and leave a scar at the base of the petiole. Young parts are usually covered, sparsely or thickly, with hair. The pubescence in *Malvaceae* is termed stellate because the individual hairs are not single and simple but a number are attached to one point and radiate from it like the rays of a conventional star.

The flowers are usually large and showy and are most often solitary in the leaf-axils. Below the calyx is a structure known as an epicalyx, which in this family consists of a whorl of bracteoles. The epicalyx functions as an extra protection for the young parts in the bud. The five, usually large and brightly coloured petals are twisted in the bud. They are free to the base where they are attached to the staminal-tube and fall with it when the flower withers. The stamens are numerous. The filaments are joined together into a complete tube which arises from the base of the petals and surrounds the filiform style. The top of the staminal-tube is usually produced upwards above the anthers and 1-celled. The pollen-grains are large and spherical and adhere to one another in masses. The extine of the pollen grain is covered with spines. The ovary is superior and generally 5-celled, with many ovules attached to the inner angle of the cells. The style is long and filiform; stigmas the same number as or double the number of the carpels, capitate. The fruit is usually a dry capsule, very rarely fleshy, or a berry. The seeds are sometimes hairy.

The flowers of the *Malvaceae* are termed protandrous, i.e. the anthers mature earlier than the stigmas. When the petals untwist themselves the stigmas are hidden within the staminal-tube and the anthers mature and burst before the stigmas emerge. By the time the stigmas have emerged the anthers have turned downwards but the slimy spiny pollen-grains adhere for sometime to the dehisced anther-cells. There are five nectar pits on the inner surface of the calyx-tube, one between each pair of petals. The nectar is protected from the rain by the fringes at the base of the petals. Insects which come to take the nectar carry away some of the pollen with them and transfer it to the stigma of an older flower. Some American species of *Hibiscus* are said to be cross-fertilised by birds. Although cross-fertilisation is the rule in these brightly coloured plants, self-fertilisation is quite a common phenomenon.

Many shrubby plants belonging to this family are cultivated in Indian gardens for the sake of their gorgeous flowers. *Hibiscus* is perhaps the best known genus but *Pavonia, Malvaviscus, Malvastrum, Thespesia* and *Gossypium*, as ornamental plants are by no means to be despised.

It has been mentioned above that the seeds of some species are enveloped in hairs. This is a device to secure wide dissemination of the seed. This characteristic of certain genera is of very great importance to mankind. For it is not too much to say that, if cotton, the hairy covering of the seeds of *Gossypium*, was not available to man, life as we know it, would be very different. These long 1-celled hairs, which can be spun into thread, enter into a host of indispensable articles. It can also, as if to counteract its usefulness, be used to produce a devastating explosive.

Many of the species growing wild in India are well known for their possession of a very tough fibre which is developed in the bast. *Urena lobata*, a pretty pink-flowered untidy undershrub, is very common all over India in plains and hills. The fibre of this plant, which resembles flax, can be made into cloth, string-bags, fishing-lines, twine and so forth. Several species of *Abutilon*, another genus of *Malvaceae*, yield a fibre which has some repute but for which no commercial demand exists as yet.

Mucilage is found in large quantities in the roots, stems and leaves of certain species and this property led to their extended use as medicines in ancient times. A very considerable number of malvaceous species are mentioned by Dioscorides and by the herbalists of the seventeenth century. These species were prescribed in various ways as tonics, aperients, aphrodisiacs, palliatives and curatives for all sorts of diseases. The followers of Pythagoras considered that magical formulae, written upon the leaves of *Malva*, were far more potent than when written upon any other surface. This particular plant also has the dubious honour of being mentioned by the Latin Poet Martial who in the following lines—

> *Exoneraturus ventrem mihi vilica malvas*
> *Attulit et varias, quas habet hortus, opes*

sings of its virtues as a purgative.

Although members of the *Malvaceae* have been known in Europe since very early times, competent botanists consider that no malvaceous plant can be deemed beyond all doubt to be indigenous to Europe. As stated above the real home of the family is in the warmer regions of the earth.

Hibiscus Linn.

(This name is derived from *ebiskos, ibiskos,* Greek words used by Dioscorides to designate *Althaea officinalis,* the Marsh Mallow.)

Herbs, shrubs, climbers or trees. Leaves alternate, usually palmately lobed or cut; stipules early caducous. Flowers large, showy, axillary or rarely in a terminal raceme. Bracteoles below the calyx, 4-12 or rarely 0, usually free from one another and from the calyx; calyx bell-shaped, 5-lobed; lobes valvate. Petals 5, connate at the base and adnate to the staminal-tube. Staminal-tube 5-lobed or 5-toothed at the top, giving off the free ends of the stamens at different levels. Ovary 5-celled; ovules 3 or more in each cell; styles 5, connate below; stigmas capitate. Fruit a 5-valved or 5-celled capsule. Seeds reniform, globose or obovoid, glabrous, velvety, cottony or scaly.

This genus contains about 200 species, of which a fair number are indigenous to India. Some of the indigenous species are used as food, some as medicines, while others are valued for the fibre which they yield. A few of these which are valued for some reason or another but not cultivated in gardens for show, are the following:

Hibiscus abelmoschus Linn.

The Musk Mallow

(Arabic *Hab-ul-mushk,* from which the specific name is derived. The Sanskrit name is *Gandapura.*)

An annual or biennial plant found in the hotter parts of India. The seeds are musk-scented and contain an oil which is extensively used in native medicine; it is said to be antispasmodic, stomachic, diuretic and to be of value in venereal diseases.

Hibiscus cannabinus Linn.

Deccan Hemp, Ambari Hemp

(Hind.: Ambari)

A small herbaceous shrub which is extensively cultivated as a fibre crop outside the great jute areas. The fibre is said to be stronger than jute. The seeds contain an oil which is a good lubricant and illuminant. Burkill says it is suitable for the manufacture of linoleum paints.

Hibiscus esculentus Linn. [**Abelmoschus esculentus** (Linn.) Moench.]

Ladies' Fingers

(Hind.: Bhendi)

A tall herb cultivated everywhere in India for the sake of its young capsules which are largely eaten as a vegetable. Opinion as to the palatability of Ladies' Fingers vary, some liking them, while others consider them to be absolutely revolting in their insipidity. The latter opinion is due to the amount of mucilage they contain which, when the vegetable is boiled, appears in large quantities. This unpleasant aspect of the vegetable can be got rid of by boiling the capsules in vinegar. The plant is also valued for its medicinal virtues. A decoction of various parts of the plant is used in venereal diseases.

Hibiscus sabdariffa Linn.

The Rozella or Red Sorrel

(Hind.: Patwa)

A shrub which is very extensively cultivated in India. It has red stems and a succulent, red, fleshy, edible calyx. The calyx can be made into jellies or preserves and has a very delicate flavour. The stem yields a fairly strong fibre.

Hibiscus in the garden

A number of species of *Hibiscus* are cultivated in gardens for their very showy flowers. As always happens with popular plants, the horticulturists have bred, crossed, rebred and recrossed the various species until we have a bewildering set of names and colours among which the original species are almost lost. In the *Journal of the Bombay Natural History Society*, Vol. 20, 1906, p. 892, et seq. Millard gives a list of the varieties of *Hibiscus* as known to him at that time. He lists 7 double and 24 single varieties.

More recently Mrs. Robinson (*J. Bombay nat. Hist. Soc.*, Vol. 40, 1938, pp. 1-7), published a paper on the 'Raising of Hibiscus from seed' in which she discourses upon the many varieties of *Hibiscus*, their pollination, and other interesting themes.

It may interest readers to know that in Hawaii up to the year 1913 over 1000 crosses of *Hibiscus* were made. It is not too much to conclude that in the intervening years several thousand more have been accomplished.

When making a cross it is advisable that the unopened anthers of the female parent be cut away. This is accomplished by removing the petals of a bud that is going to open the next day. One can easily judge this by the size of the bud. Having removed the unopened anthers the staminal-column is placed in a pollen bag, i.e. a bag which prevents the transference of pollen to the stigmas through the agency of birds or insects. Dry pollen from the male parent should be deposited upon the receptive stigmas and the pollen bag replaced until the stigmas wither. As has already been mentioned the stigmas of certain species do not appear until the pollen has been shed, several hours after the flower has opened. In such cases the anthers may be removed after the flower opens.

Seeds resulting from natural or artificial pollination are ripe in about 6 weeks. These seeds may be planted as soon as dry or they may be kept for some time as their viability is high. All seeds resulting from a cross must be sown as the amount of variation resulting from a cross is considerable and all seedlings obtained must be raised in order to get material for propagation and for further crossing.

In Hawaii the seeds of *Hibiscus* are usually planted in pots 0.25 in. deep in a 1 in. deep layer of coral sand on top of the soil. Seedlings are transplanted after 1 month when they are 2 in. high. When they reach a height of 5 in. they are again transplanted. The seedlings are said to flower from 9-12 months after the seed has been sown.

In order to perpetuate the desirable characters of the hybrid as revealed by the flowering of the seedling, these hybrids must be propagated by grafts or cuttings. Cuttings of well-matured wood 0.5-1 in. diameter and 5 in. long do best. Cuttings should be planted 3 in. deep in sand and if the latter is kept well watered roots should form in 6 weeks.

With few exceptions all varieties are scentless. The flowers usually open in the early morning and stay fresh for about twelve hours or so. When picked, all varieties stay just as crisp and fresh with or without water. They are, therefore, suitable for table decoration.

Most hibiscus plants flower best in the hot months, although a few blooms appear in the cold weather. Their cultivation is easy and demands but little care and plants can easily be multiplied by cuttings, although the hybrids are often slow in striking. The plants lend themselves to a variety of uses: they will grow as shrubbery masses, either tall or short; they can be trimmed into hedges for which purpose the common single red, *H. rosa-sinensis* is the best; they will cover arbors and will even form standard trees.

Hibiscus are sun-loving plants and should not be planted in shade. They grow so rapidly that much pruning is necessary and it is the lack of such care that makes them so often look ragged and unsightly. The blooms occur on the new wood, so that heavy pruning also induces extra flowers. The plants prefer a deep rich soil and a good unfailing supply of moisture.

In this chapter we shall only deal with the 'undiluted', as it were, species to be found in our gardens. It is not possible to trace the innumerable hybrids which have been produced at various horticultural centres.

Key to the species of Hibiscus

Flowers double, changing in colour from white to red .... *H. mutabilis*
Flowers single or double, not changing as above
   Petals cut and lobed .... *H. schizopetalus*
   Petals not cut, entire
     Stamens prominently exserted; flowers never lilac
       or purple .... *H. rosa-sinensis*
     Stamens not exserted; flowers lilac, purple or blue .... *H. syriacus*

Hibiscus mutabilis Linn.

Changeable Rose

(*Mutabilis* means changing in Latin, and refers to the flowers of

Fig. 135.—*Hibiscus mutabilis* Linn. × ½

the species which change colour from white to red during the course of the day.)

PLATE 28

THE CHANGEABLE ROSE
Hibiscus mutabilis Linn.
(nearly nat. size)

Description.—A deciduous shrub or small tree with brownish bark on the old stems; younger parts greenish covered with a stellate tomentum, among which are to be found erect, simple, golden glandular hairs. Leaves alternate, petiolate, stipulate, 4-9 in. long and as broad, deeply cordate, 3-5-lobed, crenate on the lobes, stellate-tomentose on both surfaces, but much more thickly on the lower surface, yellow glandural hairs present; middle lobe long, caudate, acute; petiole up to 9 in. long, terete, tomentose; stipules linear-subulate.

Flowers large, single or double, 3-4 in. across, pedunculate, axillary; peduncles 2-3 in. long, terete, stellate-tomentose, glandular-hairy, articulate about 0.3 in. below the flower. Epicalyx of 6-9 lanceolate, tomentose lobes; calyx bowl-shaped, yellowish green, glandular-hairy, 5-lobed; lobes triangular up to 1 in. long, acute, valvate. Petals 5 or several times that number, orbicular, obovate, shortly clawed, white at first, fading to pink, 2-2.5 in. long, glabrous, somewhat hairy below. In cases where the corolla is double the extra petals arise from the staminal-tube. Stamens numerous; anthers 1-celled. Ovary covered with a dense mat of short silver hairs. Capsule subglobose, 0.8 in. in diameter, hirsute, endocarp with dense white hairs. Seeds brown, densely bearded on one side.

Flowers.—Sept.-Oct. *Fruits.*—Oct.-Nov.

Distribution.—Roxburgh states that this plant is a native of China. It has, however, been cultivated in this country for a very long time as an ornamental shrub.

Gardening.—A very common large deciduous shrub with large heart-shaped leaves. During September and October it bears, in constant succession, a profusion of large, handsome, usually double flowers, somewhat like an immense double rose, which are white as they open fading to a deep rose tint. Haines, however, states that the sequence of fading, white to red, does not always occur and that occasionally individual flowers are red or pink from the bud. Propagated by cuttings which root readily, or by seed for even the double-flowered form fruits freely in this country. It should be cut back after flowering. The shrub is not particularly ornamental when not in bloom. The flowers are, however, handsome and provide colour in the garden when it is badly needed. It is not particular in its soil requirements but insects sometimes attack it making it rather unsightly.

Hibiscus schizopetalus (Mast.) Hook. f.

Coral Hibiscus

(*Schizopetalus* means split petals and refers to the beautifully cut and laciniate petals of this species.)

Description.—A glabrous shrub sometimes rambling. Leaves

alternate, stipulate, petiolate, ovate or elliptic in shape, shining, crenate-serrate on the margins except at the base, 3-5-nerved from the base, 2-3.5 in. long; stipules minute, subulate, caducous; petiole 0.5-1 in. long.

Flowers axillary, drooping and fuchsia-like; peduncle 6 in. long, jointed in the middle, glabrous. Epicalyx of 7 minute, subulate bracteoles, or absent. Calyx 0.7 in. long, cylindrical-spathaceous, at length split into 2-3 lobes; lobes obtuse. Petals 5, oblanceolate in outline, clawed, the margins beautifully cut or laciniate, of a deep crimson colour, reflexed, 2-3 in. long. Staminal-tube very long, up to 6 in. in length, red, slender divided at the top into an irregular number of lobes. Ovary pear-shaped, minutely hairy; style very long, slender, dividing at the top into 5 long arms, each of which terminates in a capitate stigma. Fruit a long capsule, with smooth seeds.

Fig. 136.—*Hibiscus schizopetalus* (Mast.) Hook. f. × ½

Flowers.—April-Sept. Does not set seed in this country.

Distribution.—A native of tropical Africa; common in gardens throughout the country.

Gardening.—A large, evergreen, shrub with slender drooping branches. The flowers are red or orange-red, drooping and fuchsia-like, with the petals deeply cut and fringed. It was discovered by Dr. Kirk, Consul at Zanzibar who found it first in 1874 on the coast hills at Mombasa where it grows both on dry rocky slopes and in damp mountain glens, in dense shade, amongst bignonias, balsams, and ferns. Propagated by cuttings.

CHANGEABLE ROSE
Hibiscus mutabilis Linn.

M. B. Raizada

PLATE 87

M. B. Raizada

CORAL HIBISCUS
Hibiscus schizopetalus (Mast.) Hook. f.

The coral hibiscus has been frequently crossed with other varieties and thus many of the newer sorts show a longer central column and petals more frilled than are found in ordinary kinds. The colour range also varies except that it is not blue or purple.

Hibiscus rosa-sinensis Linn.

Chinese Rose, Common Garden Hibiscus, Chinese Shoe-flower
(The specific name refers to the origin of the plant.)

Description.—An evergreen shrub but in favourable situation reaching the dimensions of a small tree, glabrous (younger parts slightly pubescent). Leaves alternate, stipulate, petiolate 2.5-5 in. long, ovate-acuminate, coarsely serrate, glabrous and shining, sometimes lobed, 3-nerved at the base; stipules ensiform or subulate; petioles up to 1 in. long.

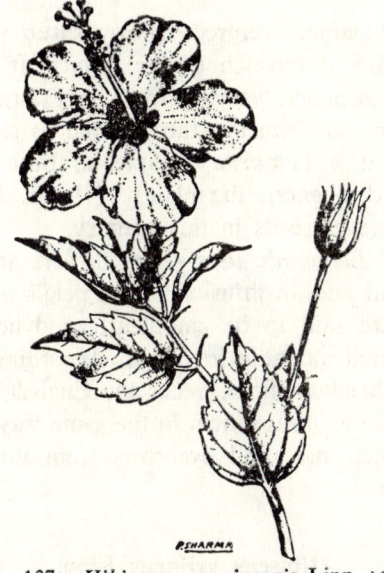

Fig. 137.—*Hibiscus rosa-sinensis* Linn. × ½

Flowers solitary from the upper axils, pedunculate; peduncles longer than the petioles, as long as the leaves, jointed 0.5 in. below the epicalyx. Epicalyx of 6-7 linear bracteoles shorter than the calyx, connate at the base, often with a few hairs along the margins. Calyx campanulate, including the lobes 1.25 in. long, light green in colour with darker nerves, 5-lobed; lobes triangular-acute, 0.5 in. long. Petals 5, crimson, with a darker eye, obovate-obtuse, covered with short white hairs in the bud, afterwards glabrous, clawed, ciliate on the claw, up to 4 in. long by 2.5 in. wide at the broadest part. Staminal-tube 3.5 in. long or more, terminating above in short linear-acute lobes; stamens numerous; anthers 1-celled, yellow. Ovary conical, obtuse, 0.4 in. long, cream-coloured, covered with a very short erect pubescence. Style filiform, dividing into 5 branches 0.25 in. below the capitate red hairy stigmas. The fruit is a capsule up to 0.75 in. long.

The colour of the corolla of this species is extremely variable in cultivation. Double-flowered forms are common and in others the corolla may be magenta, cherry, bright red, yellow or striped white and red. After the pollen has been shed the corolla begins to fade and wraps itself round the stamens and any pollen which may still be sticking to them. The stigmas are, however, still receptive and are guarded against self fertilisation. Sunbirds certainly do aid in cross-fertilisation here.

Flowers.—Practically all the year round but profusely from April-September. Does not set seed in this country

Distribution.—Probably a native of China, now common in all warm countries.

Gardening.—A large, evergreen showy shrub which is commonly cultivated for ornament throughout the tropics. It includes numerous single and double varieties, varying from very large brilliant crimson flowers, often 5.5 in. in diameter, through red to salmon-coloured and yellow flowers. It is a favourite ornamental bush and is commonly grown in gardens throughout the plains of India. Propagated by cuttings as the plant never seeds in this country.

Medicinal and Economic uses.—The flowers are considered refrigerant and emollient and an infusion of the petals is given as a demulcent. The leaves are said to be emollient, anodyne and laxative and the root is considered valuable in cough. According to Roxburgh the petals are used to blacken shoes, hence the English name of the plant. The Chinese are said to utilise them in the same way, and also to make a black dye for their hair and eyebrows from the petals. The bark yields a good fibre.

Hibiscus syriacus Linn.

Shrubby Althaea, Rose of Sharon

(The specific name refers to the alleged native country of the plant.)

Description.—A shrub reaching 9 ft in height, young parts covered with a soft sparse pubescence. Leaves alternate, petiolate, stipulate, triangular, rhomboid, or rhomboid-ovate in shape, 5-nerved at the cuneate base, of which 3 (the midrib and 2 lateral) are strongly marked, glabrous on both surfaces or with a few scattered tufts of stellate pubscence, 2-3 in. long; margins with crenate teeth in the upper two-thirds; petiole up to 1 in. long; stipules minute.

PLATE 88

M. B. Raizada

CHINESE SHOE-FLOWER
Hibiscus rosa-sinensis (Double-flowered form)

PLATE 89

M. B. Raizada

CHINESE SHOE—FLOWER

PLATE 90

M. B. Raizada

ROSE OF SHARON
Hibiscus syriacus Linn.

PLATE 91

M. B. Raizada

ROSE OF SHARON
Hibiscus syriacus Linn.

Flowers axillary, solitary, seated on stout peduncles which are shorter than the petioles. Epicalyx consisting of 6-7 linear narrow lobes, shorter than the calyx. Calyx bowl-shaped with the lobes 0.75 in. long, 5-lobed; lobes triangular-acute, valvate in the bud, stellate tomentose outside, Petals 5, pale purple or white, orbicular, clawed, 2 in. long by 2.5 in. wide at the broadest part, ciliate with white hairs. Staminal-tube short, not exserted, about 1.5 in. long, white or very pale purple. Ovary pubescent; style filiform; stigma white. This species is also extremely variable as regards the colour of its flowers; purple-pink, deep-purple, white, violet-red and pink forms are all met with.

Flowers.—June-August. Does not seed in this country.

Distribution.—Native country uncertain, but probably not Syria as Linnaeus supposed. It is now commonly cultivated in the plains and in hill stations throughout India.

Gardening.—A deciduous shrub which is extremely variable in the character of its flowers, the colour ranging from blue-purple to violet-red, flesh-colour and white; also in full double forms. Like other species of Hibiscus it will grow in any good soil, but thrives best in the hills and cooler plains districts. This species is of slender habit and the growth is thinner than with the majority of other species. Propagated easily by cuttings.

Fig. 138.—*Hibiscus syriacus* Linn. × ½

Chapter 18
PASSIFLORACEAE

This family takes its name from the most famous of its genera, the Passion flower or *Passiflora*, which may be taken as embodying the characteristics of the family in itself.

Passiflora Linn.

This genus, one of twelve which go to form the *Passifloraceae*, is commonly cultivated in Indian gardens on account of its strange flowers and handsome foliage. Some species produce an edible fruit of very delicate flavour. The name of the genus is derived from two Latin words, *passio*, suffering, and *flos*, a flower. The reason for this derivation will be seen later. With some unimportant exceptions the Passion-flowers are inhabitants of tropical South America where there is to be found a very large number of species.

The characteristics of the genus are as follows:

Mostly climbers or scramblers with alternate usually lobed leaves and gland-bearing petioles. Stipules are present, often foliaceous, sometimes cut into filamentous gland-tipped threads. The plants climb by means of lateral simple spiral and elastic tendrils. The flowers, which are often large and showy, are axillary and may be solitary, in pairs or racemose, seated on a short pedicel which is joined to a 3-bracteate peduncle. The calyx is 4-5-lobed with a very short tube to which is attached the same number of petals and a double or triple showy fringe or corona. The ovary is supported on a stalk which also bears the stamens, 3-5 in number. The anthers are elliptic or oblong in shape and dehisce downwards. The ovary is globular or ellipsoid, one-celled with 3-5 parietal placentas, each bearing numerous ovules. The fruit is large or small, fleshy, containing many flat seeds, each surrounded by a fleshy envelope or arillus. The seeds are sculptured or pitted on both surfaces.

L. H. Bailey[1], quoting from Folkard's *Plant Lore, Legends and Lyrics*, gives an excellent account of the reasons why these plants have been named *Passiflora* and we reproduce it verbatim.

'The peculiar charm of these plants lies in the odd flowers, the parts of which were fancied by the early Spanish and Italian travellers to represent the implements of the crucifixion (whence both the technical and popular names). Legend and superstition have attached to these plants from the first. The ten coloured parts of the floral envelope

---

[1] L. H. Bailey, *The Standard Cyclopedia of Horticulture*, p. 2480.

were thought to represent the ten apostles present at crucifixion, Peter and Judas being absent. Inside the corolla is a showy crown or corona of coloured filaments or fringes, taken to present the crown of thorns, or by some thought to be emblematic of the halo. The stamens are

Fig. 139.—Old conception of the Passion-Flower, from Folkard's *Plant Lore*, and there taken from Zahn.

five, to some suggestive of the five wounds, by others thought to be emblematic of the hammers which were used to drive the three nails, the latter being represented by the three styles with capitate stigmas. The long axillary coiling tendrils represent the cords or the scourges. The digitate leaves suggest the hands of the persecutors. The following sketch of the passion-flower legend is from Folkard's *Plant Lore, Legends and Lyrics*. The passion-flower (*Passiflora caerulea*) is a wild flower of the South American forests, and it is said that the Spaniards, when they first saw the lovely bloom of this plant, as it hung in rich festoons from the branches of the forest trees, regarded the magnificent blossoms as a token that the Indians should be converted to Christianity, as they saw in its several parts the emblems of the passion of our Lord. In the year 1610, Jacomo Bosio, the author of an exhaustive treatise on the Cross of Calvary, was busily engaged on this work when there arrived in Rome an Augustinian friar named Emmanuel de Villagas, a Mexican by birth. He brought with him, and showed to Bosio, the drawing of a flower so 'stupendously marvellous', that he hesitated making any mention of it in his book. However, some other drawings and descriptions were sent to him by inhabitants of New Spain, and certain Mexican Jesuits, sojourning at Rome, confirmed all the astonishing reports of this floral marvel; moreover, some Dominicans at Bologna engraved and published a drawing of it, accompanied by poems and descriptive essays. Bosio therefore conceived it to be his duty to present the *Flos passionis* to the world as the most wondrous example of the *Croce trionfante* discovered in forest or field. The flower represents, he tells us, not so directly the Cross of our Lord, as the past mysteries of the Passion. It is a native of the Indies, of Peru, and of New Spain, where the Spaniards call it 'the Flower of the Five Wounds', and it had clearly been designed by the great Creator that it might, in due time, assist in the conversion of the heathen among whom it grows. Alluding to the bell-like shape assumed by the flower during the greater part of its existence (i.e. whilst it is expanding and fading), Bosio remarks: 'And it may well be that, in His infinitive wisdom, it pleased Him to create it thus shut up and protected, as though to indicate that the wonderful mysteries of the Cross and of His Passion were to remain hidden from the heathen people of those countries until the time preordained by His Highest Majesty.' The figure given to the Passion-flower in Bosio's work shows the crown of thorns twisted and plaited, the three nails, and the column of the flagellation just as they appear on ecclesiastical banners, etc. 'The upper petals,' writes Bosio in his description, 'are tawny in Peru, but in New Spain they are white, tinged with rose. The filaments above resemble a blood-coloured fringe, as though suggesting the scourge with which our blessed Lord was tormented. The column rises in the middle. The nails are above it; the crown of thorns encircles the column; and close in the centre of the flower from which the column rises is a portion of a yellow colour, about the size of a reale in which are five spots or stains of the hue of blood, evidently setting forth the five wounds received by our Lord on the Cross. The colour of the column, the crown, and the nails is a clear green. The crown itself is surrounded by a kind of veil, or very fine hair, of a violet colour, the filaments of which number seventy-two answering to the number of thorns with which, according to tradition, our Lord's crown was set; and the leaves of the plant abundant and beautiful, are shaped like the head of a lance or pike, referring, no doubt, to that which pierced the side of our Saviour, whilst they are marked beneath with round spots, signifying the thirty pieces of silver.'

A comparison between the figure from Folkard's book and our artist's drawing of the same plant (*Passiflora caerulea*) will show to what length men, their minds obscured by religious fanaticism and superstition, were prepared to go in the seventeenth century in order to further their religious beliefs. The generic name perpetuates this phantasy and remains as a monument to fanatical self-deception but there is no need to seek a supernatural explanation for the odd shape assumed by the flower. In *Passiflora* the tendency is towards cross-fertilisation and the flowers are adapted to this end, an explanation

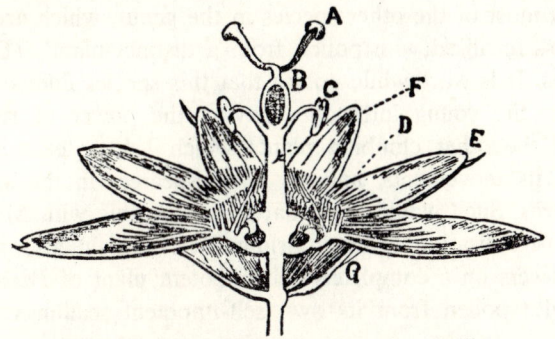

Fig. 140.—*Passiflora caerulea* after Baillon
A. Styles and stigmas; B. Ovary on a gynophore; C. Stamens; D. Corona; E. Perianth lobes; F. Gynophore; G. Bracts.

which, no doubt, will be deemed prosaic when compared with the flights of imagination of the early explorers of South America.

If a bud of a species of *Passiflora* is examined it will be found that the anthers, erect in the bud, face inwards. As the flower expands the filaments assume a horizontal position and the anthers face downwards. A portion of the pollen is discharged upon the corona and portion is retained by the anthers. The corona is often attractively coloured and sometimes the flowers are fragrant. As an additional attraction a ring of nectar secreting glands is developed within the calyx tube.

The flowers of *Passiflora* are protandrous which means that the stamens are mature before the stigmas are receptive. An additional safeguard against self-fertilisation is the position of the styles and stigmas which stand erect, well above the level of the stamens. After the stamens have discharged their pollen the styles curve downwards and the stigmas take up a position which is below the original level of the anthers. Consider, therefore, what happens when an insect visitor comes to a freshly opened flower to obtain nectar. The visitor must circumambulate the corona to get at the nectar and in doing so either picks up pollen from the corona on its ventral surface or is well dusted over the back with pollen from the stamens. Should the visitor fly to an older flower it will almost certainly transfer some of its load of pollen to the stigmas and cross-fertilisation is effected.

Bumble bees and humming birds are the unconscious agents of cross-pollination in the tropics of South America.

Henslow writing in the *Transactions of the Linnaean Society* for 1879, p. 366, has the following passage regarding *Passiflora gracilis*. 'This species, unlike other Passion-flowers, is an annual, a feature characteristic of self-fertilisers and "produces spontaneously numerous fruits when insects are excluded, and behaves in this respect very differently from most of the other species in the genus, which are extremely sterile unless fertilised with pollen from a distinct plant" (Darwin), or even species. It is worthwhile noting that this species differs from other members in the young internodes having the power of revolving. It exceeds all the other climbing plants which I have examined in the rapidity of its movements, and all tendril-bearers in the sensitiveness of the tendrils. Such would seem hardly compatible with Mr. Darwin's idea of self-fertilisation being injurious. Mr. Darwin also records the fact that flowers on a completely self-impotent plant of *Passiflora alata* fertilised with pollen from its own self-impotent seedlings were quite fertile.'

The edible passion-fruit is hardly larger than a hen's egg and contains little pulp. The fruit is, however, prized on account of its delicate flavour, and the pulp is widely used to add zest to puddings, drinks, ices and the like.

Key to the Species

Leaves simple

    Flowers very small less than 0.5 in. across, green .... *P. minima*
    Flowers large, 4 in. or more across, mauve .... *P. quadrangularis*

Leaves lunate

    Lobes of leaf rounded at the tips, upper surface with glands ...... *P. lunata*
    Lobes of leaf sharp, upper surface of leaf without glands .... *P. leschenaultii*

Leaves 5-lobed .... *P. caerulea*

Leaves 3-lobed

    Flowers small, not more than 0.75 in. across
        Leaves with short lobes, glabrous: flowers green, 0.5 in.
            across, bark not corky .... *P. minima*
        Leaves with long lobes, hairy: flowers greenish yellow,
            0.5 in. across; bark corky .... *P. suberosa*
    Flowers more than 0.75 in. in diameter, often much more
        Lobes of leaf closely serrate.
            Petals pale pink; flower 2-2.5 in. across .... *P. incarnata*
            Petals white often tinted with purple; flowers 1.5 in. across .... *P. edulis*
        Lobes of leaf not closely serrate
            Leaves hairy
                Flowers reddish or pink .... *P. ciliata*

PASSIFLORACEAE 267

    Flowers white or whitish
        Lobes of the leaf projecting in a bristle .... *P. holosericea*
        No bristle to the lobe of the leaf
            Plant very glandular; lobes entire .... *P. foetida*
            Plant not glandular; lobes toothed..........*P. morifolia*
    Leaves glabrous
        Flowers red or reddish .... *P. racemosa*
        Flowers whitish or greenish
            Stipules foliaceous; lobes of leaf deep........*P. calcarata*
            Stipules not foliaceous; lobes of leaf shallow....*P. gracilis*.

Passiflora minima Linn. (**P. suberosa** Linn.)

Small Passion-Flower

(*Minimus* in Latin means 'least', referring to the size of the flowers which, although small, actually are not the smallest in the genus.)

Description.—A slender climber with green stems. Shoots covered

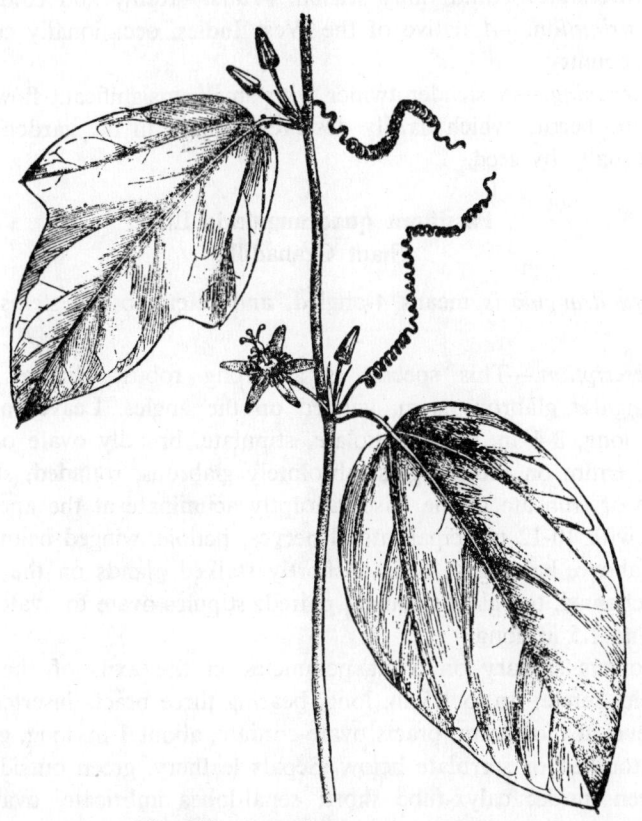

Fig. 141.—*Passiflora minima* Linn.

with an evanescent crisped hairy covering. Leaves up to 4 in. long by 2 in. broad, stipulate, ovate, elliptic-ovate or oblong-entire, or 1-2-lobed, glabrous, petiolate, dark green above, paler below; petiole about 0.75 in.

long, terete but grooved above, covered above with crisped white hairs, decorated with a pair of stalked glands above the middle; tendrils axillary; stipules linear-setaceous.

Flowers small and very inconspicuous, being hidden by the leaves, on the solitary peduncles in the axils of the leaves. Perianth-tube brownish green, flat, very short, 4-5-lobed; lobes triangular or oblong-obtuse, glabrous, green outside, about 0.3 in. long. Petals absent. Corona 4-partite, the outer filaments shorter than the sepals, rather stout, truncate at the tips, yellow; second row very short, club-shaped; third row a closely pleated membrane, very short, purple outside, woolly at the apex; inner row a very short purple continuous membrane. Ovary erect upon a gynophore, 3-celled with 3-parietal placentas; styles three, spreading; stigmas globular, stamens five. Fruit a juicy black or purple berry, 0.3 in. in diameter.

Flowers.—Hot and rainy season. *Fruits.*—Rainy and cold season.

Distribution.—A native of the West Indies, occasionally cultivated in this country.

Gardening.—A slender twiner with small, insignificant flowers and dark blue berries which hardly deserves a place in the garden. Propagated usually by seed.

Passiflora quadrangularis Linn.
Giant Granadilla

(*Quadrangularis* means 4-angled, and refers to the stems of this plant.)

Description.—This species is a strong robust climber with a quadrangular glabrous stem, winged on the angles. Leaves alternate, 4-8 in. long, 3-4 in. wide, petiolate, stipulate, broadly ovate or ovate-oblong, entire on the margins, absolutely glabrous, rounded, shallowly cordate or truncate at the base, abruptly acuminate at the apex, peninerved with 10-12 principal lateral nerves; petiole, winged below, channelled above, bearing 4, 5 or 6 shortly stalked glands on the margins of the channel, the glands usually paired; stipules ovate to ovate-lanceolate up to 1.5 in. long.

Flowers solitary on stout peduncles in the axils of the leaves; peduncles terete, up to 1.5 in. long, bearing three bracts inserted at the same level at the centre; bracts ovate-cordate, about 1 in. long, glabrous, entire, toothed or serrulate below. Sepals leathery, green outside, pinkish green inside; calyx-tube short; sepal-lobes imbricate, ovate, very obtuse at the apex, glabrous, margins incurved. Petals 5, alternate with the sepals, oblong-ovate or oblong-lanceolate, obtuse or truncate at the apex, glabrous, rather fleshy, pink or mauve outside, white, blue, violet or pinkish inside, equal in length to the sepals, 2-2.5 in. long. Corona

PLATE 92

M. N. Bakshi

MAYPOP
Passiflora incarnata Linn.

PLATE 93

M. B. Raizada

PURPLE GRANADILLA
Passiflora edulis Sims.

5-ranked, the outer two ranks thread-like, with the threads up to 2.5 in. long, terete, radiate transversely banded with reddish purple and white at the base, blue in the middle, densely mottled with pinkish blue mauve on the upper half; third rank of deep reddish purple tubercles about 0.1 in. long, the fourth rank thread-like, with white threads banded at equal intervals with reddish purple; the fifth rank membranous, lacerate. Operculum membranous, denticulate, white, reddish purple on the margin. Ovary ovoid; stamens five; stigmas three. Fruit oblong-ovoid, yellow, longitudinally 3-grooved, up to 12 in. long (according to Killip). Seeds broadly obcordate, reticulate at the centre of each face, radially striate at the margins.

Fig. 142.—*Passiflora quadrangularis* Linn.

Flowers and fruits.—Rainy season.

Distribution.—A native of tropical America; widely grown in the tropics and subtropics of both hemispheres.

Gardening.—A strong, woody, quick-growing climber with quadrangular stems which are winged at the angles. It is a superb species with large ornamental flowers which are produced in great abundance during the rains. It is a good climber for a screen or a large trellis work and is frequently planted to cover pergolas and arbours. It is said that the wild prototype of tropical America is not good to eat and the cultivated plant should be called var. *macrocarpa*. Its fruits sometimes weigh several pounds. There is also a form in which the foliage is blotched with yellow.

Economic and medicinal uses.—The fruit is eaten green or ripe. If eaten green, it is boiled as a vegetable. If taken ripe, it may be iced and eaten with sugar. The fruit-wall may be candied.

The leaves are considered dangerous as they produce hydrocyanic acid and the roots are also stated to be poisonous.

Passiflora lunata J.E. Sm. (P. biflora Lamk.)
Crescent-leaved Passion-Flower

(*Lunata* means crescent-shaped in Latin.)

Description.—A wiry climber with angled stems covered with a minute pubescence which disappears with age. Stipules linear-acute, per-

Fig. 143.—*Passiflora lunata* J. E. Sm.

sistent. Tendrils axillary. Leaves alternate, short-petioled, shaped roughly like a quarter-moon, with blunt rounded lobes, shallowly cordate at

PLATE 94

M. B. Raizada
SILKY-LEAVED PASSION-FLOWER
Passiflora holosericea Linn.

PLATE 95

THE CORAL CREEPER
Antigonon leptopus Hook. & Arn.

M. B. Raizada

the base, smooth and glabrous, reticulation prominent on both surfaces, but particularly so on the undersurface; midrib, and sometimes the nerves of the lobes, projecting as a small bristle; a row of glands arranged in a V-shape are often found between the midrib and the central nerves of the lobes; these glands are small pits on the lower surface but their position is quite obvious as raised spots when seen from above.

Flowers solitary or in pairs from the axils. Pedicels jointed half way up, 0.5-0.75 in. long, bearing three filiform bracts at the joint. Hypanthium flat; sepals 5, oblong-obtuse, greenish white, 0.5 in. long. Petals oblong-obtuse, about 1 in. long or less. Corona double; outer spreading, of narrow filaments up to 1 in. long or a little longer, upper half mauve, the inner half banded with purple and white; inner erect, very short, crowded at the base of the gynophore, upper part purple crimson. Gynophore pale green, spotted with purple. Stamens 5, filaments green, spotted with purple; anthers versatile. Ovary ovoid, superior, with three club-shaped styles. Fruit 2 in. long, oblong, yellow when ripe, edible.

Flowers.—Practically throughout the year. Does not fruit in Dehra Dun.

Distribution.—Native of Mexico and Jamaica, now cultivated in all tropical and subtropical parts of the globe.

Gardening.—An extensive climber with 5-angled stems, remarkable for its curious crescent-shaped leaves. The whitish flowers of moderate size are produced in constant succession all the year round. It was introduced into England by William Houston in 1738. Propagation is by layers and cuttings.

Passiflora leschenaultii DC.

(This species was named in honour of Leschenault, a French botanist.)

Description.—A wiry climber with hirsute stems, becoming glabrous with age. Stipules deciduous. Tendrils axillary, rather stout, glabrous. Leaves alternate, 2-3 in. long, 4 in. wide from tip to tip, petioled, semi-circular, attached to the petiole by the centre of the rounded margin with a cuspidate point in the middle of the opposite side, sometime slightly 3-lobed with the two side angles produced, glabrous or rarely hirsute, 5-nerved from the base, the upper three straight from the base to the points of the opposite margin; petiole up to 1.5 in. long, glabrous or hairy, with two scale-like glands in the lower half.

Flowers in pairs in the axils or paniculate, pedicelled. Pedicel jointed 0.5 to 0.3 in. below the flower. Bracteoles 3, linear, hairy. Receptacle flat; sepals 5, 0.75 in. long, oblong-obtuse, greenish. Petals 5, similar in shape to the sepals, slightly shorter and narrower, white. Outer corona of 2 rows of linear filaments, the outer row shorter than the petals. Inner corona a much folded membranous cup; inside this again a small shallow cup round the base of the column. Gynophore 0.25 in. long. Stamens 5. Ovary globose or ellipsoid. Fruit ovoid, 1.5 in. long.

Flowers and fruits.—Cold season.

Distribution.—Native of the Western Ghats, in the Nilgiris and Pulneys, above 5,000 ft.

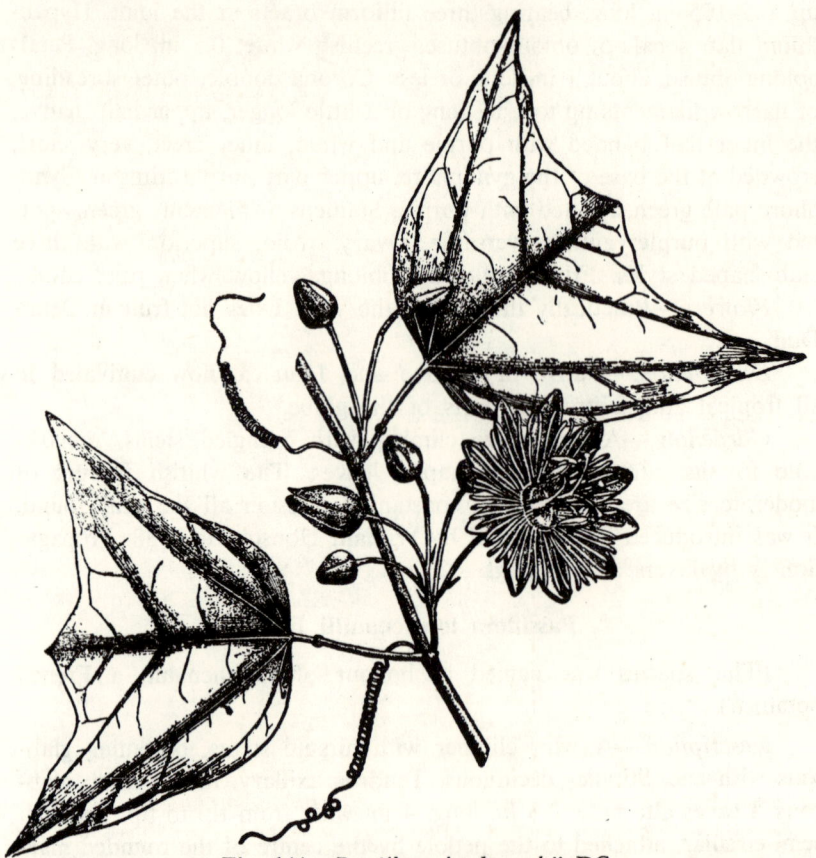

Fig. 144.—*Passiflora leschenaultii* DC.

Gardening.—A wiry climber with half-moon-shaped leaves and white flowers. It can in no way be recommended for the garden and is for all practical purposes a weed. Easily raised from seed.

Passiflora caerulea Linn.

(*Caerulea* means blue in Latin, and refers to the colour of the tips of the coronal filaments.)

Description.—A strong climber. Stems terete, glabrous and smooth. Stipules foliaceous, semi-circular, shallow cordate at the base, margins lobed or toothed, often glandular. Tendrils axillary simple. Petiole smooth and glabrous often with several stalked globular glands scattered irregularly over its length. Leaves divided almost to the base into 5 lanceolate or narrowly elliptic-acute lobes, green above, somewhat glaucous below.

Flowers large, 3-4 in. across, solitary, axillary, pedicellate. Pedicels nearly 1 in. long, bearing three large, ovate-cordate obtuse bracts, about 1 in. in length. Hypanthium saucer-shaped. Sepals 5, oblong, ending in a short awn-shaped process, greenish outside. Petals 5, oblong-obtuse,

Fig. 145.—*Passiflora caerulea* Linn.

pale pink in colour. Rays of the corona in two series, the outer blue at the tip, white in the centre, purple at the base. Gynophore and filaments green, spotted with purple. Styles 3, purple, club-shaped. Fruit the size of a small olive, pale orange-golden in colour.

Flowers.—Rainy season. Does not fruit in Dehra Dun.

Distribution.—A native of Brazil, now very commonly cultivated in all tropical, subtropical and temperate countries of the world.

Gardening.—A hardy vigorous climber with dark green, shining 5-lobed leaves. It is one of the commonest and certainly one of the handsomest of all species of this genus. Flowers profusely during the rainy season. It prefers a sandy loam with admixture of leaf mould and animal manure. It will, however, grow successfully in a variety of soils. It is suitable for covering a great space of walls or trellis. Readily propagated by the numerous young suckers, which it sends forth for a great distance all round, or by seed, layers or cuttings. According to William Curtis it was introduced into England from Brazil about 1775. This species hybridises easily and hybrids with *P. alata*, *P. raddina* and *P. racemosa* are commonly seen in cultivation.

P. caerulea-racemosa—is a hybrid between the two species whose names it bears. The flowers are very large and handsome, though not very brilliant, being of a pale lilac colour, prettily relieved with a pure white crown of rays. When grown in a large pot it continues to bloom. It is preferable to re-pot the plant with fresh soil annually in the cold weather.

Passiflora suberosa Linn.

(*Suberosa* means cork-like and refers to the bark on the old stems.)

Description.—An annual or perennial twiner. Older stems usually corky, whitish. Tendrils axillary, covered with short, appressed, white

Fig. 146.—*Passiflora suberosa* Linn.

hairs. Stipules lanceolate-acuminate, hairy. Leaves alternate, petiolate, slightly cordate at the base, 3-lobed; lobes lanceolate-acute. the two side lobes at an angle of 45° with the central, ciliate on the margins, glabrous above and below, or with short, white, appressed hairs, or densely appressedly hirsute all over; petiole about 0.5 in. long, velvety, with two shortly stalked, round glands at the middle or thereabouts.

Flowers small, about 0.5 in. across, axillary, solitary, pedicellate. Pedicels jointed half way up, with a few hairs at the joint, about 0.5 in. long. Receptacle saucer-shaped; sepals 5, 0.3 in. long, oblong-obtuse in shape, greenish yellow in colour. Petals absent. Corona of several filamentous series, the outermost spreading and half as long as the sepals, followed by shorter threads, the innermost hair-like; the median series plicate, fringed at the apex. Gynophore 0.1 in. long, slender. Stamens 5; filaments filiform; ovary globose, glabrous, ending in 3 slender styles. Fruit ovoid or globose, blue or purple, up to 0.75 in. long.

Flowers.—July to September. *Fruits.*—Cold season.

Distribution.—A native of tropical America now naturalized in damp shady places in Dehra Dun.

Gardening.—A slender twiner with 3-lobed leaves and small greenish white flowers. It is more of a weed than an ornament in any garden. Propagation is by seed.

Passiflora incarnata Linn.

Wild Passion-Flower; Maypop

(*Incarnata* means 'flesh-coloured' in Latin and refers to the colour of the petals.)

Description.—A wide-spreading climbing creeper. Stems wiry, cylindrical, green, shining, almost glabrous. Leaves alternate with small stipules, petiolate; petiole hairy with two glands at the top, dividing at the apex into three nerves which pass out into the three lobes of the leaves; margins of the lobes serrate, those of the side lobes joining the two side nerves about 0.25 in. above the apex of the petiole. Lobes of the blade elliptic-acuminate in shape, dark green above, paler below; nerves hairy above and below. Petioles up to 1 in. long; leaves up to 4 in. long by 4 in. across. Tendrils axillary, long, simple.

Flowers axillary, solitary, seated on peduncles up to 3 in. long, showy, 2-2.5 in. across; peduncle hairy, striate, triangular in section just below the flower. Just below the flower are three glandular leaf-like bracts; bracts small, ovate-acuminate in outline, serrate oblong-obtuse, with a short awn-like projection just below the apex, above green or pale green in colour, rather soft, up to 1 in. long. Petals 5, delicate, pale pink in colour.

Flowers.—Hot and rainy season. *Fruits.*—Cold Season.

Distribution.—Native of the Southern States of North America, now widely cultivated throughout the world.

Gardening.—A tall, perennial climber with broadly cordate, ovate 3-lobed leaves, bearing 2 glands near the top of the petiole. The pretty,

Fig. 147.—*Passiflora incarnata* Linn.

axillary solitary flowers, 2 in. across are white with a purplish corona and are borne in great profusion. It is a fine cover for arbours, verandas, arches and the like. A liberal supply of manure is beneficial to its growth as this plant is a gross feeder and exhausts the soil quickly. Propagation is by seed and layer.

Passiflora edulis Sims

The Edible Passion Flower; Purple Granadilla

(*Edulis* means eatable in Latin.)

Description.—A widely spreading woody climber. Tendrils axillary, long, simple. Stipules linear, small. Leaves alternate, petiolate, up to 6 in. long by 6 in. broad, ovate-cordate in outline, 3-lobed to half their length or more, coarsely serrate on the margins; terminal lobe broadly elliptic, acute or acuminate; side lobes obovate-acute or elliptic. Petiole up to 1 in. long, with two glands at the apex.

Flowers solitary, terminal or axillary, pedunculate. Peduncles up to 1.5 in. long. Below the flower are three bracts, ovate-acuminate in shape, the upper portion glandular serrate, the lower portion with 1-2

Fig. 148.—*Passiflora edulis* Sims

large glands on the margin. Calyx-tube hardly any, shallow, saucer-shaped; lobes 5, oblong-obtuse, greenish, with a short awn-like projection below the apex. Petals 5, oblong, about 0.75 in. long, white, often tinted with purple. Corona in several series, the outer spreading, not as long as the sepals and petals, white in the upper half, violet or purple below. Stamens 5; filaments inserted on the gynophore. **Ovary** ellipsoid, glabrous or woolly, surmounted by the three club-shaped styles and stigmas. Fruit globular-oblong, thickly purple-dotted, when ripe with a hard rind.

Flowers.—Hot and rainy season. *Fruits.*—Cold season.

Distribution.—A native of Brazil, now extensively cultivated in all parts of the world. Run wild below Kodaikanal, near Ootacamund and near Yercaud, in South India.

Gardening.—A widely spreading rather woody 'Vine' with large glossy, deeply 3-lobed and serrate leaves. The white, tinged with purple, flowers, are produced in succession all through the summer and rains and are quite ornamental. It makes an excellent and rapid-growing cover for fences and trellises. It is, however, for the edible fruit that it is so extensively cultivated in many countries. The fruit under cultivation, according to Bailey, is considerably smaller than the Granadilla (*P. quadrangularis*), rarely larger than a hen's egg, and dull purple when ripe. Its pulp is slightly more acid than that of the Granadilla, but of very pleasant flavour and highly esteemed in Queensland and New South Wales, where the plant is cultivated commercially. It is used for flavouring sherbets, for confectionery, for icing cakes, for jams and for a variety of other purposes.

Like most species of *Passiflora* it grows with great rapidity and soon exhausts the soil. It is, therefore, necessary to give it a liberal supply of manure each year.

The fruits produced in Dehra Dun and elsewhere in this country have, however, usually little pulp, the seeds occupying most of the interior. Propagation is by seed, layers or cuttings.

Medicinal and Economic uses.—According to Burkill, hydrocyanic acid is met with in the stem and roots of this plant.

Passiflora ciliata Dryand (**P. foetida** L. var. **ciliata**)

Fringed-leafed Passion-Flower

(*Ciliata* refers to the fringed margins of the leaves.)

Description.—A slender climber with smooth and glabrous terete stems. Stipules laciniate, each segment a glandular hair. Leaves alternate, petiolate, hastate in shape, the two lower lobes widely spreading, the terminal lobe much longer, elliptic-obtuse in shape, 2 in. long, densely hirsute on both surfaces with appressed fulvous hairs, or only slightly hairy above, or almost glabrous with gland-tipped hairs along the margin, densely hirsute on the margins. Petiole little more than 0.5 in. long, covered with golden hairs, with here and there a longer gland-tipped hair and sometimes 1-2 sessile globular glands. Flowers axillary, pedicellate, solitary; pedicel 1 in. long. Involucral bracts 3, bi- and tri-pectinately divided, the ultimate segments capillary and gland tipped. These bracts are foetid when bruised. Calyx-tube a hypanthium; sepals 5, 0.3 in. long, greenish outside, pale on the margins. Petals 5, reddish or pink, oblong-obtuse. Corona several-seriate, the outer filamentous, variegated white and purple, the inner erect and purple-tipped. Gynophore deep purple in colour bearing the 5 purple-

Fig. 149.—*Passiflora ciliata* Dryand

dotted filaments. Ovary purple, ellipsoid, ending in three clavate stigmas. Fruit ellipsoid, 0.5 in. long, enclosed in the involucral bracts.

Flowers and fruits.—Rainy season.

Distribution.—A native of the West Indies, occasionally cultivated in gardens throughout the country.

Gardening.—A slender climber at times attaining a considerable height. The dark green, glossy, 3-lobed leaves are usually glabrous except on the edges where they are beset with glandular hairs. The leaves of this species vary greatly in form according to the vigour and luxuriance of the plant. The rather small flowers are of a purplish colour. Multiplication is by seed or cutting. It prefers a cool situation. According to Curtis it was introduced into England by Mrs. Norman from the West Indies in 1783.

Passiflora holosericea Linn.

Silky-leafed Passion-Flower

(*Holosericea* means covered with silk and refers to the pubescent stems and leaves.)

Fig. 150.—*Passiflora holosericea* Linn.

Description.—A woody climbing shrub with very silky, terete stems. Stipules thread-like hairy. Tendrils axillary, somewhat stout, hairy. Leaves alternate, petiolate, up to 4 in. long by 3 in. wide, ovate in outline, shallowly 3-lobed densely covered on both surfaces with a short, crisped, silky pubescence; lobes rounded or even emarginate, sometimes obtusely-acute; midribs of the lobes projecting as a short bristle; petiole up to 0.5 in. long, short-silky, with two scale-like glands about the middle.

Inflorescence axillary of 2 or more flowers in hairy peduncled umbels or racemes. Individual flowers pedicellate. An involucre of 3 linear bracts arises from the centre of the pedicel. Receptacle flat. Sepals 5, 0.75 in. long, narrowly ovate-obtuse, longitudinally veined, densely hairy. Petals 5, elliptic obtuse, white. Corona in several series, the outer filamentous; filaments yellow in the outer half, reddish or purplish in the inner; median membrane lacerate, purplish-tipped. Gynophore 0.4 in. long, purplish. Stamens 5, filaments greenish, spotted with purple. Ovary ellipsoid, densely hirsute, ending in 3 slender styles, each with a capitate stigma.

Flowers.—Hot and rainy seasons. Does not set fruit in Dehra Dun.

Distribution.—Native of Vera Cruz, in South America, now cultivated in all tropical and subtropical parts of the world.

Gardening.—An extensive, woody climber with ovate, 3-lobed leaves, with pretty dull-red veins. The flowers are not very large, whitish, but are borne in great profusion and strongly sweet-scented. Propagation is by layers and cuttings. It was introduced into the Chelsea gardens by Dr. William Houston from South America before 1733. It is suitable for growing on an arch or a trellis work.

Passiflora foetida Linn.

Stinking Passion-Flower

(*Foetida* is a Latin word meaning evil-smelling and refers to the odour emitted by the plant when crushed.)

Description.—A herbaceous climber, more or less viscous, densely hirsute throughout with yellow-brown hairs. Stipules semi-circular about the stem, deeply cleft into filiform, gland-tipped segments. Leaves alternate, petiolate, ovate-cordate in outline, 3-lobed, membranous, usually glandular-ciliate, sparingly to densely appressed hirsute, 3 in. long by 2 in. wide, the middle lobe lanceolate or ovate lanceolate, acuminate, the lateral lobes suborbicular, often apiculate; petioles slender up to 2.5 in. long, glandular-ciliate but without true petiolar glands.

Flowers solitary, seated on axillary peduncles which are up to 1.5 in. long. Bracts ovate in general outline, 0.75-1.5 in. long, bi- or tri-pinnatisect, the ultimate segments filiform, gland-tipped, closely interwoven in the bud but much less so when the flower opens. Calyx-tube short, campanulate. Sepals and petals oblong or ovate-oblong, 0.7-1 in. long, white, often tinged with lilac. Corona filaments in several series, the outer 2-radiate, filiform, violet and white or purple and white, the other series short, violet. Ovary globose, densely pilose with white or brown hairs. Fruit globose up to 1 in. in diameter, yellowish, densely to sparingly hairy.

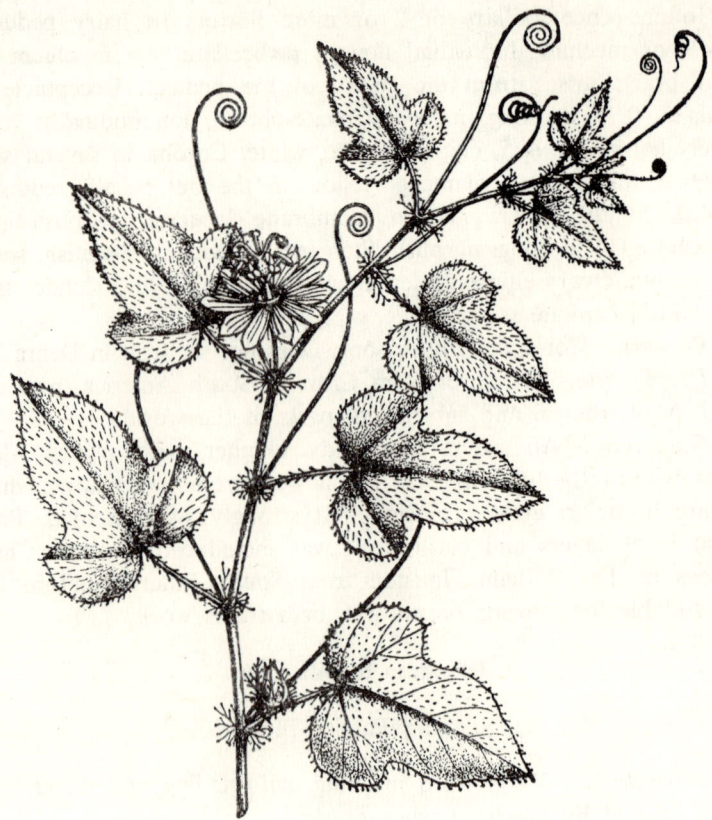

Fig. 151.—*Passiflora foetida* Linn.

Flowers and fruits.—Rainy season but often at other times of the year.

Distribution.—A native of tropical America, now run wild in various parts of India, Sri Lanka, China and Malaya.

Gardening.—A small herbaceous species at once recognised by the moss-like pectinate involucre of the flowers. The leaves when bruised emit a disagreeable smell and are not eaten by cattle. The flowers are small and white and not showy. Bears fruits abundantly. Propagation is by seed. It is more like a weed and is not worth growing in a garden. It was introduced into England as far back as 1731 and in India over a century ago.

Medicinal uses.—The plant is reported to be used in Malaya to cure itches. The hydrocyanic acid which is present in the leaves apparently helping in doing this. The leaves are applied to the head for giddiness and headache and a decoction is given in biliousness and asthma. It is considered to be an emetic.

Passiflora morifolia Mast.

(*Morifolia* means in Latin mulberry-leaved, and the plant is so called from the supposed resemblance of its foliage to that of the mulberry.)

Description.—A slender twiner with hairy angled stems. Stipules foliaceous, 0.25 in. long, ovate cordate, or semi-orbicular, hairy. Tendril

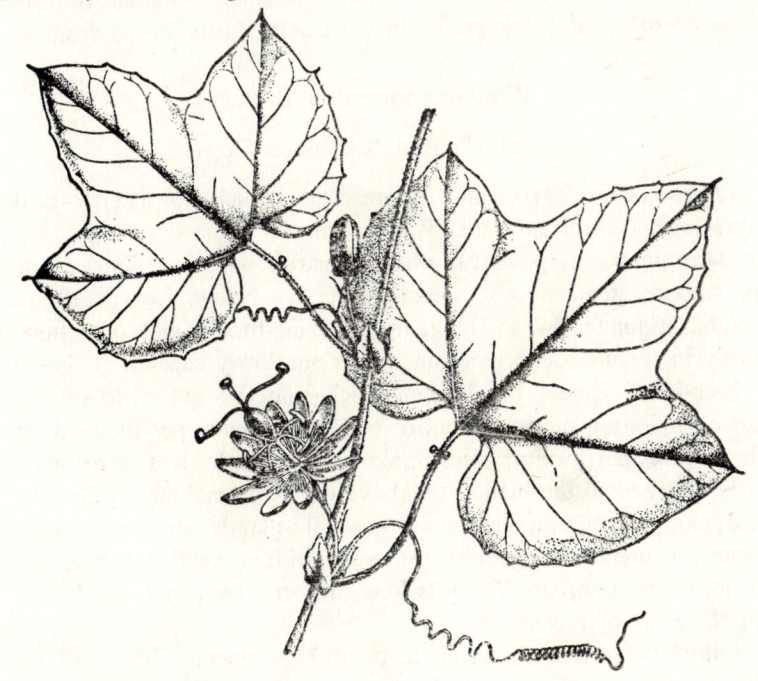

Fig. 152.—*Passiflora morifolia* Mast.

axillary, hairy, slender. Leaves alternate, petiolate, ovate-cordate in outline, glabrous or hairy on the upper and lower surfaces, particularly on the nerves, 3-lobed; lobes ovate-acute, margins of the centre lobe entire or with widely spaced glandular teeth; side lobes dentate on the outer margin, rounded to the insertion of the petiole; the teeth ending in glands; petiole up to 2 in. long, bearing 2 stalked glands above the middle.

Flowers axillary, solitary, pedicellate; pedicels with three filiform bracteoles just below the calyx or scattered. Calyx-tube campanulate, short, pubescent; calyx-lobes lanceolate, 0.75 in. long, green, pale on the margins. Petals absent. Corona in three series, the outer filamentous, half as long as the sepals, the centre series short, curved, the inner a narrow membrane. Gynophore slender; stamens 5; filaments attached to the top of the gynophore. Ovary globular, velvety, ending above in three slender styles. Fruit globose, purple, 1 in. long; seeds pitted on both surfaces.

Flowers and fruits.—Rainy season.

Distribution.—Native of Brazil, occasionally cultivated in this country.

Gardening.—A slender hairy twiner with small, insignificant flowers. It is not worthy of being grown in a garden. Easily raised from seed.

Passiflora racemosa Brot.

Princess Charlotte's Passion-Flower

(*Racemosa* refers to the inflorescence which consists of stalked flowers on a central axis, that is, a raceme.)

Description.—A powerful climbing shrub with a cylindrical, glabrous, smooth stem, woody, sometimes corky below. Leaves alternate, petiolate, stipulate, 3-4 in. long, up to 3 in. broad, glabrous, smooth, leathery in texture, deep green in colour shallowly cordate at the base, all three-lobed except the lowest which may be ovate; lobes ovate-acute or oblong-ovate-acute, entire on the margins, penninerved, each with a principal central vein making the whole leaf three-nerved; petiole attached to the blade just above the base, making it sub-peltate, cylindrical, glabrous, bearing about 4 sessile glands; stipules foliaceous, obliquely ovate-acute, glabrous, smooth, with a median nerve, up to 1 in. long and as broad. Tendrils long, filiform, spirally twisted, arising from the axils of the leaves.

Inflorescence a long, terminal, pendulous raceme, 7-8 in. or more in length, leafless, having stipules in place of leaves; sometimes, especially when it does not support any fruit, it remains alive and produces new branchlets from the axils of the dead peduncles. Flowers solitary (but the peduncles usually two from an axil, with the tendril in the middle between them, becoming racemose on the ends of the shoots), 4 in. or more across, inodorous, the narrow petals deep red and widespreading, the short, upright crown purple. Calyx inferior, glabrous, forming a tube below, limb deeply 5-cleft; sepals not conforming with

PRINCESS CHARLOTTE'S PASSION-FLOWER
Passiflora racemosa Brot.
(about ½ nat. size)

the corolla petals, dirty purple outside, very broadly keeled, the keel curved like a sword, scarlet in colour. Before the flower opens the keels of the calyx give the buds a five-winged appearance. Petals 5, inserted on the throat of the calyx, and a little shorter than the lobes of the calyx, spreading, semi-lanceolate, somewhat obtuse, entire, somewhat flattened, outside slightly keeled, inside channelled in its native ground, purple on both sides. Nectary corona triple, filamentous, all the hairs being whitish on the upper side, blue on the lower side; the innermost corona equal in height to the outer corona, the hairs closing round the cylindric stipes, equal among themselves, simple, joined with a membrane from below, adnate to the elevated margin of the receptacle which goes round the base of the calyx; the two other coronas inserted on the throat of the calyx below the petals, the intermediate one shorter, hairs being scarcely exserted outside the calyx throat, erect, subcapitate at the apex, equal among themselves; the exterior corona with simple hairs which are produced two or three lines outside the calyx throat, patent, unequal, the hairs which are opposite to the calyx-lobes being longer. Staminal filaments 5, inserted on the apex of the stipe below the ovary, slightly united at the base, green, sublinear, erect-spreading. Anthers sublinear, obtuse with a short acumen, leaning, in colour from green to yellowish, unisulcate in the middle and at the sides, bilocular. Ovary superior, oblong, subovoid, obsoletely trisulcate, glabrous, pale green, with the styles prolonged up to one inch beyond the calyx throat, cylindric, green, broader at the base which is pentagonal. Styles 3, somewhat thicker above, pale green, recurved towards the anthers, stigmas capitate, from yellowish to green. Fruit a berry, pedicellate, ovoid, trisulcate, glabrous, pale green, up to 2 in. long, unilocular, many-seeded, fleshy, finally dry.

Flowers.—Hot and rainy season. Has not fruited in Dehra Dun up till now.

Distribution.—A native of Brazil, now commonly grown in all tropical and subtropical parts of the world.

Gardening.—This very choice and handsome climber which is decidedly one of the best red-flowered *Passifloras*, bears scarlet or deep red flowers in great profusion, but unfortunately all the flowers of the raceme do not open out at the same time. A liberal supply of manure and leaf mould is considered beneficial at the beginning of the cold weather. It is a very suitable object for growing over an arch or pergola. Propagation is by seed or layers. It is a fine species and is a parent of various garden hybrids of this genus. According to Firminger it thrives much better if grafted on a strong species but in Dehra Dun plants raised from seed received from Leipzig are doing excellently well on their own roots.

Passiflora calcarata Mast.

Madagascar Passion-Flower

(*Calcarata* is a Latin word meaning spurred, and refers to the long spur on the sepal.)

Description.—A wiry slender climber with glabrous stems. Stipules ovate-lanceolate, foliaceous, 0.75-1 in. long. Tendrils axillary, slender. Leaves alternate, subrotund in shape, petiolate, 2 in. long by 2.5 in. broad, glabrous, smooth, thin in texture, 3-lobed, lobes rounded at the

Fig. 153.—*Passiflora calcarata* Mast.

apex; margins of the lobes with 2 glands just above the obtuse sinus; petiole slender, glabrous with 2 or more glands irregularly distributed.

Flowers solitary, axillary, pedicellate. Bracts 3, ovate-lanceolate up to 0.5 in. long. Hypanthium saucer-shaped. Sepals 5, oblong, obtuse at the tip, greenish, furnished with a short spur just below the tip

0.5-5.75 in. long. Petals 5, white, more delicate than the sepals. The corona consists of an outer set of slender filaments, white with a purple base and blue tips, and an inner set of much smaller filaments which fit closely about the central column and cover a honey-secreting circular channel. Stamens 5; filaments attached to the gynophore. Ovary egg-shaped, with a slight bloom; stylar arms spreading upwards and outwards ending in large 2-lobed stigmas.

Flowers and fruits.—March to May.

Distribution.—A native of Madagascar, naturalized and run wild in the Nilgiris and elsewhere in the hills in South India.

Gardening.—A slender climber (of the section Granadilla) easily distinguished from the rest by its 3-lobed leaves and large stipules 0.75-1 in. long. Propagation is by seeds and layers. The flowers are ornamental.

Passiflora gracilis Jacq.

(*Gracilis* means slender in Latin and refers to the habit of the plant.)

Description.—A slender climber with terete, glabrous stems. Ten-

Fig. 154.—*Passiflora gracilis* Jacq.

drills very slender, axillary. Stipules minute, linear-acuminate. Leaves alternate, petiolate, deltoid in outline, very shallowly and bluntly 3-lobed, shallowly cordate at the base, thinly membranous, green on the upper surface, glaucous beneath, glabrous; petiole up to 1 in. long, with 2 glands at or below the middle.

Flowers solitary, axillary, on peduncles about 1 in. long; the peduncles bear a pair of slender glandular bracts well below the bloom. Calyx-tube a hypanthium, flat; lobes 5, oblong-obtuse, green, tinged with pale violet on the margins; lobes 0.75 in. long, persistent. Petals 5, as long as the sepals and similar in shape, whitish or greenish in colour. Corona in 2 series, the outer filamentous, the inner a lancerated membrane. Gynophore 0.3. in. long, bearing the 5 stamens with versatile anthers; ovary ellipsoid, glabrous, ending in 3 clavate styles. Fruits ellipsoid, globose, purple, 1 in. long.

Flowers and fruits.—August to September.

Distribution.—A native of Brazil, now occasionally cultivated in all tropical and subtropical countries of the world.

Gardening.—A slender climber with broadly deltoid-ovate leaves which are shallowly and bluntly 3-lobed. The solitary, axillary, pale green or whitish flowers about 1 in. across are not very showy. Easily multiplied by seed, layers or cuttings.

PLATE 30

THE CORAL CREEPER
Antigonon Leptopus Hook. & Arn.
(nearly nat. size)

Chapter 19

POLYGONACEAE

The family derives its name from one of its genera, *Polygonum*, which in its turn comes from a Greek word meaning many-angled, presumably from the fact that the nuts are 3-angled.

This family of 32 genera and about 700 species consists for the most part of herbs, and there are very few shrubby arborescent or climbing plants contained within it. The leaves are alternate and the stipules are of a peculiar kind, in that they sheath the stem. The flowers are bisexual with a perianth of 3-6 parts. The stamens are 6-9 in number. The ovary is 1-celled and the fruit a nut.

One climbing genus is very common in Indian gardens.

Antigonon Endl.

The name is derived from the Greek and probably refers to the kneed or angled character of the stem.

Antigonon leptopus Hook. et Arn.

The Coral Creeper; Sandwich-Island Creeper

(*Leptopus* means thin- or slender-stalked.)

Description.—A herbaceous plant ascending from a tuberous root. Stems green, hairy, angled and grooved. Leaves alternate, 3-6 in. long, ovate or triangular in shape, cordate or sagittate at the base, tapering to the acuminate apex which ends in a short spine, entire or undulate on the margins. Nerves flat on the upper surface of the leaves but very prominent beneath, arching several times within the margins, covered with a short pubescence. Petioles 0.5-1.5 in. long, hairy; stipular sheaths inconspicuous.

Inflorescence racemose, terminal or axillary. The rachis of the inflorescence is slender, angled, clothed with a sparse pubescence and bears branched tendrils, by which the plant is enabled to climb. The flowers, rosy red in colour, are borne on pedicels either solitary or several at a node, each pedicel being subtended by a forked bract. The pedicels are up to 0.3 in. long, jointed about the middle, the upper portion more hairy than the lower. The flowers are hermaphrodite. The perianth segments are five in number, the three outer are broadly ovate in shape, rounded or cordate at the base, while the two inner are narrower and not cordate, about 0.5 in. long increasing to 0.9 in. when in fruit. The stamens are 8-9 in number. The filaments are connate into a cup for about half their length, which is extended into teeth which alternate with the stamens. The filaments of the stamens are covered with shortly stipitate glands. The ovary is 1-celled and three-angled. The ovule is solitary and hangs from a long funicle which is attached to the base of the ovary. Styles 3; stigmas capitate, kidney-shaped. The ripe nut is enclosed within the enlarged, free, stiff, perianth segments.

Flowers.—Throughout the rainy and cold season. *Fruits.*—Cold season.

Distribution.—Indigenous to South America. Largely grown in gardens throughout the country.

Gardening.—This lovely climber of moderate growth is best suited for arbours, verandas, screening unsightly objects and the like. Propagated by division of the root, layers, cuttings and seeds during the rains. It grows 30 to 40 feet high in good soil. Too much manure or other fertilizer effects a vigorous growth at the cost of flowering. The plants form large tuberous roots and when killed down by heavy frost sprout readily again during the spring.

Antigonon leptopus Hk. & Arn. var. *albus* Hort.—A variety with white flowers. It is not so luxuriant in growth as the type.

Antigonon guatemalense Meissn.

A very hairy species with broader leaves, more numerous flowers and perianth segments twice as long as in *A. leptopus* Hook. et Arn. It is a native of Guatemala.

Bougainvillea glabra Choisy
(⅔ nat. size)

Chapter 20

NYCTAGINACEAE

This family is important in Indian gardens because one of its genera is the far-famed *Bougainvillea,* so commonly cultivated nowadays. The family comprises herbs, shrubs and trees. The flowers may contain only the male sexual element or the female, or both may be combined in one flower. A brightly coloured involucre of bracts often surrounds the flower or group of flowers and is usually taken for the calyx. The actual calyx is tubular. This again is often taken for the petals which are, however, absent. The stamens are usually many but are sometimes reduced to one, inserted below the ovary, which is 1-celled and ends above in a slender style. The ovule is solitary within the ovary and inverted.

Another common plant belonging to this family which is found in Indian gardens and also running wild, is *Mirabilis jalapa* more usually known under its popular names Marvel of Peru or Four-o'clock Plant. This South American species is now cultivated everywhere in the tropics and its magenta coloured perianth makes a striking patch of colour. The flowers only open in the evening and this fact gives rise to the trivial name. The scientific name of this plant preserves the erroneous belief, at one time widely held, that the tuberous roots were identical with those of the true jalap. The jalap is a renowned purgative, so drastic and powerful in its effects, that it has been suggested that 'jalap' and 'faith' are synonymous. Be that as it may, the purgative effect of the roots of *Mirabilis* are very mild, though a decoction of the roots is said to possess curative powers in a variety of diseases and disorders ranging from dysentery, cholera and diarrhoea on the one hand to constipation on the other!

Bougainvillea Commerson.

This genus was published by Commerson, the French botanist, in 1789, in honour of L. A. de Bougainville, the famous French navigator, based upon specimens collected in Rio de Janerio, Brazil during de Bougainville's voyage round the world in 1766-69. The name as published by Commerson was *Buginvillea* but the correct spelling as adopted in the Index Kewensis Suppl. 9, 1931-35 is *Bougainvillea* Comm. corr. Spach. In between these dates the name has been spelt in half a dozen different ways.

With its spectacular mass of brightly coloured bracts, the Bougainvillea is unrivalled both in beauty and utility, particularly in the gardens of the tropics and subtropics. It is easy to cultivate and remarkably free from diseases and pests. In India it is a very popular garden plant and is appreciated for its brilliantly (brightly) coloured bracts and profuse flowering almost throughout the year.

Around homes the Bougainvillea can be grown as pot plants, or trained over arches, bowers and pergolas, along wires up posts and tree trunks, on high porches, over the roofs of garages and outhouses or as covers of unsightly walls. Architects choose them for providing large sheets of colours for vast landscape, or for relieving monotony created by rows of concrete buildings.

A versatile plant, it has many different uses in the garden and has a wide adaptability to various soil and climatic conditions. It needs very little care for growing.

The leaves of the species of the genus are simple and alternate. The stem is covered with spurs or spines which aid the plant to climb upon and ascend through other vegetation.

An examniation of the 'flower' of *Bougainvillea* will reveal that at the tip of the flowering shoot there are three (usually) brightly coloured leaves or bracts which are heart-shaped when flattened. These bracts touch along their margins and give rise to the common belief that they constitute the floral envelope. Actually the true flowers will be found within the floral bracts. These flowers are rarely solitary and usually three of them are seated together within the bracts in an umbel-like structure. Each individual flower is erect upon a pedicel attached to the midrib of the subtending bract near its base.

The true flowers are tubular in shape, greenish-white or cream-coloured, about 0.75 in. long and 0.3 in. in diameter, pedicelled, glabrous or hairy. The tubular perianth is roughly pentagonal in section and terminates in five equal lobes which often persent a frilled appearance from lateral outgrowths from the lobes. Within the perianth-tube will be found the stamens, seven or eight in number, the anthers being seated upon long flattened, capillary filaments which are unequal in length so that in a mature flower 2 or 3 anthers will be found emerging from the perianth-tube. Within the filaments is a solitary pistil surmounted by a solitary style terminating in a hairy stigma. The pistil contains a single ovule. Some of the bud sport varieties, e.g. var. *Louis Wathen* have imperfect flowers.

All of the known varieties of *Bougainvillea* are derived from the four species *B. spectabilis* Willd., *B. glabra* Choisy, *B. peruviana* Humb. et Bonp. and *B. buttiana* Holttum et Standley, but it must be remembered that there is also a host of hybrids. The flowers of *Bougainvillea* are self-sterile and in the ordinary course of events seed is very rarely produced. A number of hybrids have been produced by

Mr. Percy Lancaster in Calcutta by means of cross-pollination. There is likely to be a great advance in the coming years and we may expect a large number of hybrids with tinted bracts of every imaginable colour. All who are interested in *Bougainvilleas* should read the works of Holttum (Malayan Agri-Horticultural Association Magazine), Parsons (Tropical Agriculturist) and Percy-Lancaster (Agri-Horticultural Society of India) from which most of the information in this article is compiled. It is quite impossible to describe all the cultivated varieties of the species of this genus. Not only are some of the species of later hybrids and sports given varietal rank based upon very subtle differences in colour of the bracts, but also the true name of the parent is not always known. It has happened more than once that a variety has been introduced into several gardens in India under different names, Hybrids have been produced from these and the result has been nomenclatural confusion. Again a name used to denote a colour does not always convey the same impression to the reader as it does to the writer and there is the additional difficulty that most of the varieties exhibit a change in colour in the bracts from youth to age. Hence keen gardeners are informed that while chaos has not yet come upon us in the naming of *Bougainvillea* varieties it cannot be long delayed. With the big gardens producing hybrids and the small nurseries following suit, there is bound to come a time when each valid variety has three or four names. We sympathise with our readers but cannot do anything about it.

As has been said above four main species have given rise to the true varieties known in India at the present time.

Key to the Species

Plant definitely hairy ....*B. buttiana*
Plants glabrous or nearly so.
  Tip of the floral bracts acute;
    bracts magenta ....*B. glabra*
  Tip of the floral bracts obtuse.
    Bracts crimson ....,......*B. spectabilis*
    Bracts delicate mauve pink .... *B. peruviana.*

Bougainvillea spectabilis Willd.

Description.—An arborescent shrub climbing over large trees and through other vegetation by means of the curved spines on the stem and branches. Bark pale and corky, cracking into rectangular plates. Branches and shoots hairy; very hairy when young. Spines woody, axillary, 1-2 inches long, more or less curved, tomentose or sometimes glabrous. Leaves petioled, ovate or even rotundate in shape, obtuse or shortly acuminate, entire, somewhat wavy on the margins, membranous to somewhat leathery in texture, 2.5-3 in. long, 2 in. wide, sparsely hairy above, hairy to hirsute below. Flowerheads terminal or anxillary, seated on peduncles and more or less gathered into branched panicled inflorescence; peduncles simple or sometimes divided. Each flower head consists of three bracts arranged as an involucre and carrying the flower. Bracts purple or rosy purple, ovate in shape, 1.5 in. long, 1.2 in. wide, obtuse at the tip, cordate at the base, memberanous in texture, reticulately nerved. The flowers are adnate to the median nerves of the subtending bracts. Perianth tubular, corolla-like, hairy, 1 in. long, ending above in a 5-lobed narrow mouth. Fruit club-shaped, 5-ribbed, glabrous or pubescent.

Distribution.—This species grows wild in the mountainous parts of eastern and central Brazil.

This was the first species of *Bougainvillea* to be introduced to cultivation. A specimen was brought alive to Paris in the eighteen-twenties and flowered in a tropical greenhouse there. It was cultivated in England in 1844 and the plant was well known in India as far back as 1860. The flowering of this species is always seasonal and depends upon the occurrence of dry weather. It is said that only very large plants, especially those grown over tall trees, flower well.

Varieties of **B. spectabilis** Willd.

Var. *lateritia.* This very fine variety is difficult to propagate and is therefore not as common as it deserves to be. The colour of the bracts has been described as brick red or a very fine shade of jasper red, fading to brazil red as the bracts fade. The variety follows its parent in the seasonal flowering and it only bursts into full colour after periods of drought.

Var. *Maharaja of Mysore.* It is not quite certain whether this is a hybrid between *B. spectabilis* var. *lateritia* × *B. glabra* or whether it is a chance variety of *B. spectabilis*. At all events the leaves and other parts of the plant are hairy. The bracts are described as being spinel red to rose dorée in hue, and to change to brick or brazil red with age. It is a fairly free flowering variety and can be propagated very easily.

Var. *Mrs. Frazer.* This variety which is probably a hybrid has bracts which are spinel red in colour turning to eugenia red with age. It again exhibits seasonal flowering and does not do well unless it is placed in a hot and dry position.

Var. *Rosa Catalina.* This distinct variety is known as the 'pink' Bougainvillea in our Indian gardens. It was introduced into England from the Canaries and first exhibited in London in 1910. The bracts are spinel red in colour or pale red or a bright rosy scarlet. One advantage of this plant is that the colour of the bracts is capable of harmonising with other flowers in the garden. It flowers more freely

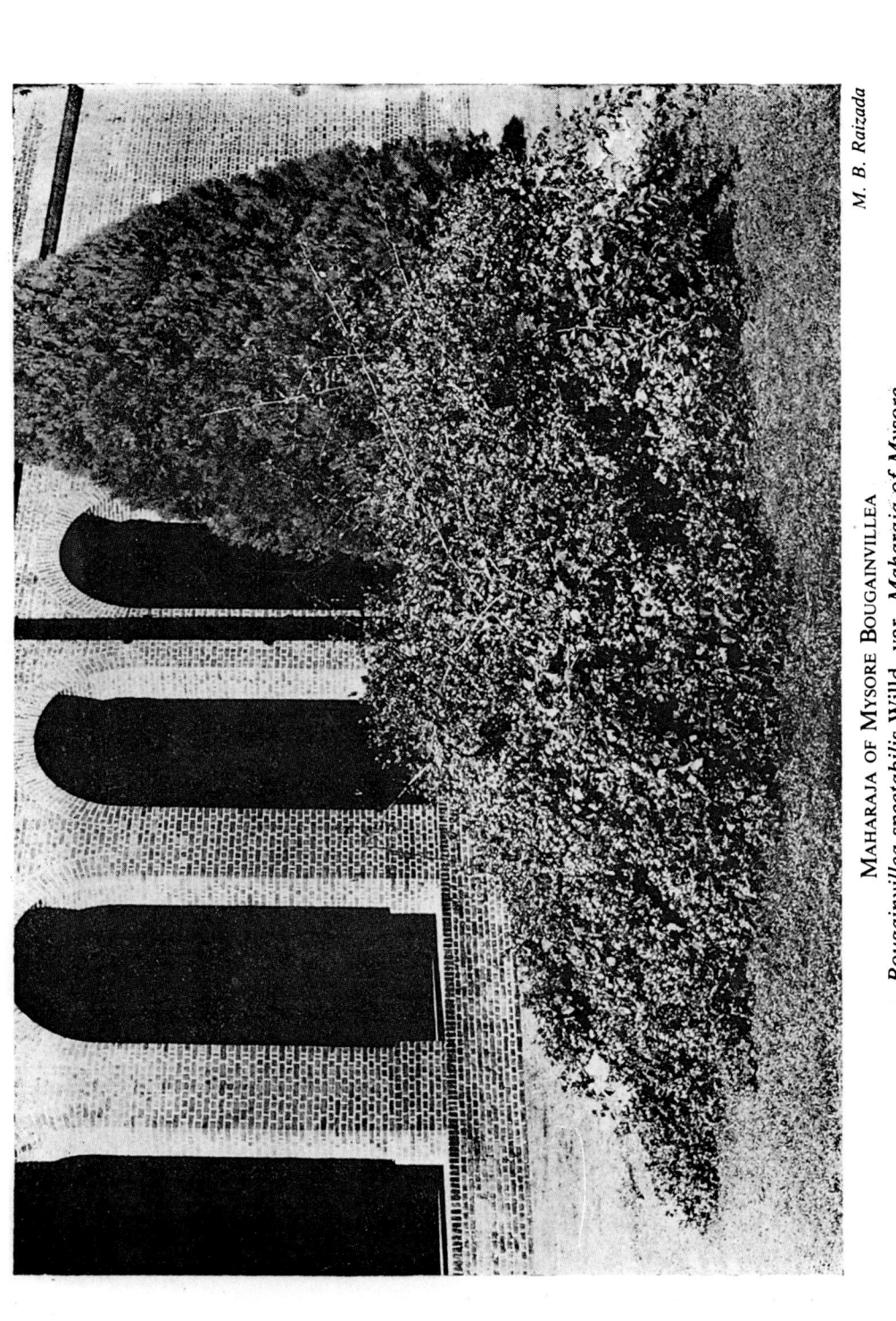

MAHARAJA OF MYSORE BOUGAINVILLEA
Bougainvillea spectabilis Willd. var. *Maharaja of Mysore*

M. B. Raizada

PLATE 97

MRS. BUTT BOUGAINVILLEA

M. B. Raizada

than some of the varieties of *B. spectabilis* and deserves to be more widely cultivated. Its propagation is rather difficult.

Bougainvillea glabra Choisy.

Description.—A climbing shrub with leafy, glabrous, spreading, spiny branches. Spines short, scarcely curved. Leaves glabrous on both sides, oblong-lanceolate or ovate-oblong in shape, acuminate, entire, cuneate or cordate at the base, seated on a petiole 0.2-0.3 in. long, 1-2 in. broad, firmly membranous in texture. Flower-clusters arranged in axillary or terminal panicles which are many- or few-flowered. Bracts elliptic or elliptic-lanceolate in shape, cordate at the base, acute at the tip, reticulately nerved, magenta in colour. Perianth tubular, less than 1 in. slightly hairy. Fruit turbinate, 5-ribbed, glabrous, 0.5 in. long.

Distribution: This plant is at home in evergreen woods on the mountainous parts of Rio de Janeiro and Sao Paulo.

This climber differs from *B. spectabilis* in being less hairy in all its parts, and more particularly in habit. *B. glabra* flowers almost continuously throughout the year and therefore is of far more value horticulturally than the former species. *B. glabra* arrived and was exhibited in London in 1860, it having come from Brazil by way of Mauritius. This species has been known in India for the past 60 or 70 years and is very common.

Varieties of B. glabra Choisy.

Var. *Sanderiana*. This variety has been exhibited over and over again in London and is commonly cultivated in Europe. It is very free flowering and can be grown as a small pot plant. One adavantage in North India is that it is able to withstand slight frost. The colour of the bracts is described as phlox purple to margenta.

Var. *Cypheri*. This variety appeared in 1897 and is quite common in the East. The bracts are paler than those of var. *Sanderiana*, but larger, and the flower panicles are produced in large quantities.

Bougainvillea buttiana Holttum & Standley.

Description.—A widely climbing spiny shrub of open growth. The stem and other parts are usually hairy when young but become almost glabrous with age. Leaves deep green in colour, ovate-rotundate or elliptic-ovate in shape, 5-6 in. long by 4 in. wide, acuminate at the tip, truncate or rotundate at the base, usually with some hairs even when old.

The inflorescence is lax and few-flowered. The flower clusters are seated on glabrous peduncles about 0.3 in. long. The bracts are rotundate-elliptic or ovate-orbicuar in shape, crimson in colour, upto

1.5 in. long by 1.5 in. broad, rotundate or broadly obtuse at the tip, broadly rotundate, or somewhat cordate at the base, glabrous or occasionally minutely hairy on the veins, reticulately nerved. Perianth about 0.75 in. long, very minutely hairy to almost glabrous.

Distribution.—The species is supposed to be Brazilian in origin.

This magnificent plant has an interesting history which has been so well summarised by Mr. Holttum in the Malayan Agri-Horticultural Association Magazine April 1941 that it is reproduced here verbation.

'How long this *Bougainvillea* has been cultivated in South America is unknown. The first recorded item of history is that Mrs. R. V. Butt, of Trinidad, visiting Colombia in 1910, found a plant in a garden near Cartagena, and brought cuttings to Trinidad. It was found easy to propagate and was soon distributed to the other West Indian islands.

'In October 1916 the U.S. Department of Agriculture recorded the introduction from British Guiana to Florida of a crimson flowered *Bougainvillea* of which sufficient information is given to make its identity with *Mrs. Butt* certain, though that name is not mentioned. But it seems that the species had been introduced into Florida at an even earlier date.

'It seems that this *Bougainvellea* did not cross the Atlantic until after the 1914-18 war, and then it was introduced to Europe both as Mrs. Butt and as Crimson Lake.

'The next item of history comes from India. Mr. Percy-Lancaster, Secretary of Royal Agri-Horticultural Society of India, informed us that a *Bougainvillea* resembling Mrs. Butt arrived in India direct from the West Indies in 1920. Mr. Percy-Lancaster gave the name 'Scarlet Queen' to this plant. He writes "The coloured bracts are slightly darker than in Mrs. Butt but the actual flower is malformed so that instead of there being three white or cream flowers in the bracts, there are merely little bunches of anthers. In growth and foliage these plants are identical. The original Scralet Queen gave rise at Madras in 1932 to a bud sport with orange bracts which was called 'Louis Wathen'. Mrs. Butt was soon distributed from Kew to the eastern tropics and to Africa. In 1931 in the garden of Mrs. McLean of Trinidad, a plant of Mrs. Butt. produced a branch with apricot-orange instead of crimson bracts. This was shown to Mr. R. O. Williams, who propagated it and called it 'Mrs. McLean'. The next year the variety Scralet Queen did the same thing in Madras, in the garden of Mrs. Wathen, and the orange variety so porduced was called 'Louis Wathen' by Mr. B. S. Nirody, Honorary Secretary of the Madras Agri-Horticultural Society. Mr. Nirody actually said that Louis Wathen was derived from Mrs. Butt, but Mr. Percy-Lancaster in his letter on Scarlet Queen informed us that the latter variety was concerned. This statement is borne out by the fact the 'Louis Wathen' can be distinguished from 'Mrs. Mc-

Lean' by its imperfect flowers. I know of no other distinction between the two. To be quite accurate. Louis Wathen does occasionally have a perfect flower, which one may see as a little cream coloured star, among the orange bracts, but most of the flowers lack the white star-like end entirely, the remainder consisting of an inconspicuous group of stamens protruding from a short tube."

'Mrs. Butt, like *Bougainvillea peruviana* is evidently native of more tropical regions than the original purple *Bougainvilleas* which were found in the south of Brazil, on the borders of the tropics. It has also a quite distinct habit, with its stronger, more open growth, very broad leaves and close branches of flowers. It is needless to describe in detail its many well-known qualities, but will point out two facts which have not, so far as I know, appeared in print.

The bracts of Mrs. Butt and its derivatives turn from crimson or orange when young to purple or mauve as they grow older but the bracts of the orange varieties (var. *lateritia* and hybrids) of the older *Bougainvilleas* start purple when they are young and pass through red to orange when they are old. The other feature of Mrs. Butt is one of which gardeners should beware. It will not stand hard pruning back so that no leafy shoots are left. The old purple *Bougainvilleas* can be cut back almost to the ground, and they rise again more vigorous than before, but Mrs. Butt will not stand rough treatment.'

The varieties, var. *Mrs. McLean* and var. *Louis Wathen*, are very common in Indian gardens and their history is given above. As stated, it is not certain that var. *Louis Wathen* is actually a variety of *B. buttiana* since the origin of Scarlet Queen is not known, but if the only difference between the two, *Louis Wathen* and *Mrs. Mclean*, is the presence of imperfect flowers in *Louis Wathen*, the two plants surely have a common origin.

Bougainvillea peruviana Humb. et Bonp.

Description.—A climbing, spreading, spiny shrub. The branches are covered with a greenish bark and bear spines which are straight when young but become curved when old. This species is said to lose its leaves entirely each year. The leaves are quite glabrous, alternate, ovate in shape, upto 4 in. long by 3 in. wide, entire, acute at the tip, cuneate at the base, thin in texture, seated on very slender petioles. Flower-clusters panicled in the axils of the leaves. Bracts ovate-obtuse, upto 1 in. long by 0.75 in. wide. delicate mauve pink in colour. Perianth tubular, much more slender than that in any other of the species, quite glabrous except for a few hairs on the subtending bracts, ending above in a 10-toothed limb, of which 5 teeth are longer, entire, while the other 5 are broader, 2-fid, yellowish in colour. Stamens 7, shorter than the perianth. Fruit oblong.

Distribution.—The home of this plant is said to be Peru, Colombia and Ecuador.

Again Trinidad seems to have been the scene of the first introduction of this species, when a Mrs. Rover in 1920 brought back a specimen to the island. It was introduced into Florida about 12 years afterwards and is a recent acquisition in the East. There are two varieties.

1. var. *Lady Hudson*
2. var. *Princess Margaret Rose*.

The two varieties are very similar, the latter has longer bracts (upto 1.25 in. long as against 1 in. in Lady Hudson) of a decidedly deep shade of pink. Both varieties do well in pots.

Cultivars

There are numerous cultivars of Bougainvillea. These have arisen as a result of bud sports, or as seedling variation as a result of chance crossing in nature. There is considerable confusion in the nomenclature of cultivars because in many cases the same cultivar has been named differently in different places, and in some cases the same name has been given to cultivars of different origin. In many cases the origin of a cultivar is not known. No complete systematic record has been maintained of all the cultivars, except those grown in Britain, which have been systematically recorded by the Royal Horticultural Society, London. Recently, the Division of Horticulture (now called the Division of Vegetable Crops and Floriculture) of the Agricultural Research Institute, New Delhi, has been appointed to be the International Registration Authority, by the International Society for Horticultural Science, for the record of names of cultivars of the Bougainvillea. The Division has recorded and standardized more than 250 cultivars, indicating the synonyms and the priorities of names to be used. The list of cultivars thus recorded is expected to be published soon.

The important cultivars belonging to different species, as well as some promising hybrids, particularly those being commonly grown in India, are listed below:

B. peruviana

'Princess Margaret Rose'
'Lady Hudson'
'Mary Palmer' (syn. 'Surprise' as patented in U.S.A.)
'Dr. B. P. Pal'

BRICK-RED BOUGAINVILLEA
Bougainvillea spectabilis Willd. var. *lateritia*

M. B. Raizada

PLATE 99

M. B. Raizada

THE SMOOTH BOUGAINVILLEA
Bougainvillea glabra Choisy

B. glabra

'Sanderiana'
'Cypheri'
'Formosa'
'Splendens'
'Snow white'

B. spectabilis

'Speciosa'
'Thomasii' (syn. 'Rosea')
'Lateritia' (syn. 'Dar-es-Salam', 'Brasiliensis')

B. × buttiana

'Louise Wathen' (syn. 'Orange Glory')
'Lady Mary Baring'
    Introduced into India by the Lal Bagh Botanic Gardens, Bangalore in 1961. A bud sport from Golden Glow. The best yellow cultivar.
'Mrs. McLean' (syn. 'Orange King')
'Alick Lancaster' (syn. 'Lilac Queen')

Hybrids and seedlings

Several cultivars of the Bougainvillea have arisen from seedling variations as a result of natural crossing. In many cases, only the seed parent is known, whereas in others the probable pollen parent is also indicated. In several cases, even the seed parent is not recorded. Hybrids of known parentage derived from artificial crossing are few. Holttum [*Malay Agri. Hort. Ass. Mag.* 14(1): 13, 1957 and 14(2): 58, 1957 and in 'Bougainvillea' (pp. 233-4) *in* Edwin A. Manninger's FLOWERING VINES OF THE WORLD. AN ENCYCLOPEDIA OF CLIMBING PLANTS (Hearthside Press Inc., Publishers, New York 1970] described several hybrid cultivars of the Bougainvillea.

Holttum described a number of hybrid cultivars derived from crosses between *B. spectabilis* and *B. glabra* and called them Spectoglabra hybrids.

Although several cultivars are self sterile, there are some in which natural self-pollination occurs, as recorded by Percy-Lancaster and others. It is therefore considered more appropriate to classify the hybrid seedlings on the basis of their seed parent species, if known, and indicate the pollen parent, wherever recorded.

Seedlings of B. peruviana

'Mrs. H. H. Buck' (syn. 'Mahatma Gandhi')
'Maharaja of Mysore'

'Mrs. Fraser'
'Mrs. Lancaster'
'Rosa Catalina'
'Tomato Red'
'Lord Willingdon'

Seedlings of B. glabra

'Happiness'
'Kalyani'
'Dream'

Seedlings of B. × buttiana

'Meera'
'Jayalakshmi'
'Dr. R. R. Pal'
'Blonde'

Seedlings of unknown parentage

'Nawab Ali Yavurjung'
'Begum Ali Yavurjung'
'Eclipse'

In addition several multibracted cultivars have been reported, e.g. 'Maharajah' (syn. 'Million Dollar'), 'Cherry Blossom', etc.

Also a number of cultivars having variegated leaves have been reported a few of which are: 'Thimma'; 'Rao'; 'Bhabha'; 'Laxminarain'; 'L. N. Birla'.

Cultivars of bracts of two colours on the same plant are also known as for example 'Sunset', 'Fantasy', 'Colour Splash', etc.

DISTINCTIONS BETWEEN SPECIES

Flower-tube very slender, little constricted in the middle, hairless; leaves with a very broad base, surface quite smooth......*B. peruviana*

Flower-tube distinctly swollen in basal part, more or less hairy; leaves ovate to elliptic, more or less hairy,

Hairs on flower-tube short (to 1/5 mm) and curved upwards; leaves widest in middle, smooth to tough but with short hairs as flower-tube; flowering when leafy in warm wet season (continuously in wet tropics) .... *B. glabra*

Hairs on flower-tube to 1 mm long, straight and spreading; leaves somewhat ovate (widest below middle), softly hairy to touch; flowering mainly in cool dry season, often when leafless .... *B. spectabilis*

B. peruviana Humboldt & Bonpland ('Ecuador Pink'; 'Lady Hudson'). Native in Ecuador, Peru, and Colombia. Inflorescences crowded near ends of branches; bracts delicate pale magenta-pink, crinkled; flowering in response to dry weather; much less vigorous than its offspring × *buttiana*, but with an elegance of its own. First recorded in cultivation at Trinidad in 1920; no distinct cultivars known.

B. glabra Choisy (syn. '*B. splendens*'). Species described in 1849 from northern Brazil; first records of cultivation in 1860 (in Britain and Mauritius), introduced to India soon afterwards. Inflorescences borne all along smaller branches; bracts not crisped, usually with green veins, changing little in colour with age. Several cultivars, of unrecorded origin exist:

cv. 'Sanderiana'. Bracts rather small (30 × 18 mm at flowering) with distinctly acute tips, medium purple; very free-flowering. Exhibited and named in London 1894 (FCC); probably the best-known cultivar of *B. glabra*.

cv. 'Cypheri'. Bracts at flowering 45 × 29 mm, apex blunt, a fresh bright medium purple, slightly redder when old. Named and exhibited at Shrewsbury in 1897; a very fine cultivar.

cv. 'Formosa' (a form of *B. glabra* var. *brachycarpa*). Bracts to 35 × 23 mm at flowering, elliptic with a short abrupt tip, light mauve; flower-tube much swollen and angled; old bracts persistent. Named and exhibited in London, 1904; very handsome when in full flower but old bracts unsightly.

cv. 'Magnifica' (also *magnifica* var. *traillii*). Bracts a little smaller than in cv. 'Cypheri' but of a deeper colour. Named in Australia in 1893, said to have been brought from the Seychelles; probably the finest cultivar of *B. glabra*.

White-bract cultivars. There have been more than one; they have been too little compared and described.

B. spectabilis Willd. Though having the habit of *B. glabra*, in most varieties it is more vigorous, attaining a very large size and flowering best when climbing high on a tree. The first *Bougainvillea* plants grown in Europe were of this species, flowering at Paris from 1835 or earlier. It is not known whether the original cultivar still exists; it had rather deep purple bracts. Among the most important cultivars and varieties of *B. spectabilis* now being grown are the following:

cv. 'Speciosa' (*B. speciosa*). Bracts at flowering 55 × 37 mm, ovate, hairy, the apex slightly rounded, a fine purple, changing little with age. Described 1849; figured from plants grown in Britain 1854.

cv. 'Thomasii'. Habit and size of bracts about as in cv. 'Speciosa', but bracts a fine bright carmine-pink. Grown by Mr. Thomas near Brisbane, Queensland, from 1905; probably the finest cultivar of

B. spectabilis.

var. *lateritia* (*B. lateritia*). Bracts when young red-purple, at flowering brick-red, fading to orange; hairs on flower-tube spreading as in typical *B. spectabilis* but (in cases observed) only 0.5 mm long. Named and exhibited at London in 1865. There are several cultivars (not clearly distinguished) having these characters, with bracts of varying size; they have been important as parents of *glabra-spectabilis* hybrids, contributing much to the colour-range now available.

DISCUSSION OF HYBRIDS

B. spectabilis × B. glabra *hybrids.* Grex name: Spectoglabra. These are intermediate between the two parents in hairiness of flower-tube; they show a large range of size and colour in bracts, and of colour-change from young to old bracts, also in relationship between climate and flowering. Probably the first of these hybrids developed by natural crossing at Teneriffe from about 1900; the first to be named and cultivated outside the Canaries was 'Rosa Catalina' (FCC, London, 1909). Colour-change is usually from purplish to less purple (to red, pink, or orange); this is the reverse of the change in × *buttiana*.

In the 1930's hybrids of this group were raised in the West Indies, at Calcutta by S. Percy-Lancaster, in Queensland by W. F. Turley, and doubtless elsewhere; they have been inadequately recorded. Some are very large and vigorous, approaching *B. spectabilis* in habit. Good cultivars are 'Turley's Special', 'Mrs. Lancaster', 'Aida', 'Jubilee', 'Mrs. Fraser'. 'Maharajah of Mysore', and 'Mrs. H. C. Buck'. 'Mary Palmer' originated in India as a bud-sport from 'Mrs. H. C. Buck' after drastic pruning of a very large plant; the young shoots bore bracts from very pale pink to creamy white, or sometimes partly purple (patented in the U.S.A. as 'Surprise'). 'Ardeni' is a recent S. African cultivar, a fine tomato-red. 'Margaret Bacon' and 'Barbara Karst' were raised in Florida by J. E. Hendry; I have not seen descriptions of any other cultivars named in Florida, some of which appear to duplicate names used elsewhere. 'Afterglow' (Florida) belongs to this group.

B. × buttiana (*B. peruviana × glabra*). First recorded at Trinidad, 1910, from a plant brought by Mrs. R. V. Butt from a garden in Cartagena, named 'Mrs. Butt' by R. O. Williams; introduced to Florida from Trinidad by E. N. Reasoner in 1913 and sold by him from 1917 as 'Crimson Lake'; a similar form was produced later by controlled cross of *B. peruviana* in Peru by S. C. Harland. Bracts in close bunches at ends of branches, crisped (not flat), crimson when young, fading and turning to mauve when old; leaves with a very broad base and slender tip, those of sucker shoots very large (to 18 × 14 cm). Flowering in response to dry weather but not leafless.

In cultivation, 'Mrs. Butt' has produced bud-sports with bracts of lighter colour; and seedlings have been raised, first in Kenya and Natal (some being back-crosses with *B. glabra*). I do not know of hybrid involving also *B. spectabilis,* but probably such have been produced in Kenya, where 'Mary Palmer' (probably also other Spectoglabra hybrids) has been grown together with × *buttiana* varieties. Cultivars in this group include these:

cv. 'Scarlet Queen' (India from 1920) differs from 'Mrs. Butt' in having flowers that appear as if their tips had been cut off.

cv. 'Mrs McLean' (syn. 'Grange King'). A bud-sport from 'Mrs. Butt', originating at Trinidad, 1931; young bracts orange, old ones pale mauve.

cv. 'Louis Wathen'. A bud-sport from 'Scarlet Queen', Madras, 1932; colour as 'Mrs. McLean', flower-tubes imperfect.

cv. 'Done Luz Magsaysay'. Origin Manila, brick-red to salmon-pink.

General Notes on Gardening

The species of *Bougainvillea* are deservedly very popular in tropical gardens for the gorgeous splashes of colour they provide and, moreover, they are in flower for the greater part of the year. The species with magenta flowers are apt to kill all other colours round about and for that reason it is unwise to make much use of them in a small garden. These species are seen at their best when used on a grand scale, say for climbing over tall trees, as arches and pergolas, and they are particularly effective rambling down a slope, but they must be kept away from the other colour effects in the garden.

All the species can be grown as standards or as isolated shrubs and can be exceedingly attractive when used in this way. All the species and varieties (except *B. buttiana*) can stand pruning and the genus is used very commonly in South India as a hedge.

Bougaincilleas thrive in almost any kind of soil and will grow with ordinary care almost anywhere. They are, however, most luxuriant between 2,000-4,000 ft. above sea level. They prefer full sunshine and thorough drainage. An occasional top dressing and periodic manuring is all that is necessary in the latter stages of growth. The majority of the varieties are propagated by cuttings or by layering or gootee, but certain varieties like *lateritia, Rosa Catalina* and *Louis Wathen* are not easy to propagate and strike root with difficulty. The usual method is to select well ripened shoots of the previous season's growth. Cuttings 9-10 inches long and as thick as a pencil are selected and inserted in beds of sandy, well drained soil. Most of these take root in 4-5 weeks from the time of insertion. Those varieties which fail by this method should be either layered or gooteed and when sufficiently rooted they are transplanted into larger containers and kept there for some time before finally planting out.

Though *Bougainvilleas* will grow in almost any kind of soil they require a certain amount of manure and humus in the early stages of growth. Later on periodic top dressing of manure and clean weeding immediately round the plant is all that is required.

GLOSSARY OF BOTANICAL TERMS USED IN THIS BOOK

abruptly acuminate, passing suddenly into a tapering point at the apex.
accrescent, continuing to grow.
achene, a dry 1-seeded carpel of an apocarpous fruit.
acicular, needle-like, long, slender and rigid.
actinomorphic, regular.
acuminate, ending in a tapering apex.
-adelphous, combined in groups; e.g. monadelphous, combined in one group.
adherent, when the members of a flower become united in the course of growth to the members in a different whorl and of a different character, e.g. when the stamens become united to the corolla.
adnate, united with a member of another series.
aestivation, the method of arrangement of parts in the bud.
alternate, the relative position of lateral members on an axis when inserted at different levels.
androecium, a collective word for all the stamens (and staminodes) in a flower.
anterior, the side remote from the parent axis.
anther, that part of the stamen which contains the pollen.
apiculate, ending abruptly in a short point.
apocarpous, carpels free from one another.
aril, an envelope which grows up from the stalk of the seed and more or less completely covers it.
aristate, with an awn or arista.
ascending, becoming erect from a prostrate or sub-prostrate base.
auricle, an ear-like appendage.
auriculate, having an ear-like lobe or appendage.
awn, a rigid, very fine or almost hair-like terminal appendage.
axil, the angle between leaf and shoot.
axile, used of ovules which are attached to a central column situated round an axis.
axillary, situated in an axil.
axis, of inflorescence, is that part of the shoot which bears individual flowers
baccate, berry-like.
bark, all the tissues alive or dead situated outside the cambium ring.
barren, of flowers which do not produce seed.
beaked, provided with a firm prolongation. (The term is not applied to leaves.)
berry, a fleshy indehiscent fruit with many seeds in the pulp.
bifid, 2-fid, divided into 2 parts about half-way down.
binate, 2-nate, 2 arising together from the same point.
bi-sexual, 2-sexual, containing both fertile stamens and carpels with ovules.
blade, the expanded part of a leaf, bract, etc., as distinct from the stalk.
bract, a reduced leaf.
bracteole, small bracts occurring on the axis of a next higher older than that on which the bract is situated.
caducous, quickly falling off.
calycinal, relating to the calyx.
calycine, resembling a calyx in texture rather than petals.
calyx, the outer floral envelope where there are two.
calyx-tube, the tube or cup formed by the cohesion of the leaves of the calyx.
canescent, see hoary.

capitate, (1) clusted together into a head or ball, (2) knob-like.

capsule, a form of fruit which becomes dry when ripe and opens by two or more valves.

carpel, the modified leaves which bear the ovules.

caruncle, a peculiar growth at the apical or micropylar end of the seed.

caudate, furnished with a long slender tail-like tip.

chartaceous, paper-like in texture.

ciliate, with a fringe of hairs along the margin.

circinate, (1) rolled up longitudinally with a growing tip inside, (2) coiled.

clavate, club-shaped.

claw, the narrow or stalk-like base found in some petals.

coherent, said of members of the same kind joined together.

compound, composed of two or more similar parts.

connate, said of members of the same kind united one to another.

connective, that part of an anther which connects its two lobes.

connivent, weakly cohering

convolute, rolled up from one or both margins.

cordate, shaped like the conventional heart (as on playing-cards), or with the base heart-shaped.

corolla, one of the envelopes of the flower and a collective name for the petals.

corona, a circle of appendages between corolla and stamens.

coriaceous, leathery.

corymb, a form of inflorescence in which the several branches or flower-stalks arising at different levels reach more or less the same level at the top.

cotyledon, a leaf present on the embryonic plant while yet in the seed.

crenate, with margin notched with regular rounded teeth.

crustaceous, firm and brittle.

cuneate, wedge-shaped.

cusp, a short hard point or tip.

cuspidate, furnished with a cusp.

cyme, a system of branching in which the main axis ceases to grow or terminates in a flower; the secondary or lateral axes from beneath the apex continue to grow beyond the parent axis and may be likewise superseded by branches or axes of a higher order. cf. raceme.

deciduous, falling off. cf. caducous.

decumbent, having the lower parts prostrate.

decurrent, prolonged downwards from the base.

decussate, in planes at right angles to one another.

deflexed, bent downwards.

dehisce, to open by the separation of the walls or valves.

dehiscent, dehiscing when ripe.

deltoid, shaped like an equilateral triangle.

dentate, with teeth projecting more or less perpendicularly from the margin.

denticulate, with little teeth, or points along the margin.

diadelphous, 2-adelphous, in two bundles.

dichotomous, method of branching in which each axis bifurcates at the tip.

didymous, consisting of two equal or similar connected halves or lobes. In the case of anthers, the term is especially applied to those with two rounded lobes without separating connective.

didynamous, in two unequal pairs.

digitate, spreading like the fingers of the hand.

dimorphic, of two forms.

dioecious, where the sexes occur on different individuals, the male flower on distinct plants from the female.

disc, a swelling of the torus inside the calyx and under or outside the pistil sometimes glandular.

disciform, disc-like.

dissected or *divided*, when the incisions between the segments just reach the midrib or petiole, but the parts or segments so divided off do not separate from the axis without tearing.

distichous, disposed alternately in two opposite rows.

divaricate, spreading in opposite directions from a common base.

dorsal, situated at the back of.

dorsifixed, fixed by the back of, in contrast to the state of being attached by the end or margin, etc.

drupaceous, more or less resembling a drupe.

drupe, a form of fruit consisting of a more or less succulent pericarp which encloses a single 1-many-celled stone, e.g. a plum.

ebracteate, without bracts.

emarginate, having a deep dent at the apex.

endosperm, the tissue formed within the embryo-sac or macroscore subsequent to fertilization (in the case of angiosperms) and destined to feed the embryo.

entire, with the margin or edges not toothed or cut but even and continuous.

epipetalous, situated on the corolla or petals.

episepalous, situated on the sepals.

exstipulate, without stipules.

extra-axillary, situated away from the axil of the leaf to which it is nearest.

extrorse, opening outwards, of anthers.

falcate, somewhat curved.

fascicled, clustered.

-fid, used in composition, divided about half-way down.

filament, the stalk of an anther.

filiform, very slender, thread-like.

fimbriate, clothed with narrow or filiform appendages.

flaccid, soft.

fleshy, thick and of somewhat firmed texture than succulent.

foliaceous, leaf-like.

-foliolate, in composition refers to the leaflets in a compound leaf, e.g. 3-foliolate means with 3 leaflets.

follicle, a pod-like structure containing seeds.

free, not united with other members.

free central placentation, where the ovules are situated on the axis of a unilocular ovary, which may be produced above the base of the ovary or not.

fruit, the ovary (in the case of an apocarpous ovary all the carpels) and its contents after the fertilization of the ovules, including in the case of inferior ovaries the accrescent hypanthium or investing part of the floral axis, e.g. apple.

fulvous, yellowish golden.

furcate, folked.

gamopetalous, with united petals.

gamosepalous, with united sepals.

glabrate, nearly glabrous.

glabrescent, with deciduous hairs, becoming glabrous.

glabrous, without any hairs.

glaucous, of a blue green colour.

gynaecium, the carpel ovary or assemblage of carpels in a flower, together with their appendages (style, stigma).

gynophore, an internode of the floral axis between the stamens and the pistil, so that the pistil is considerably separated from the stamens.

gynostemium, stamens and ovary on a column as in Aristolochia.

hairy, clothed with somewhat long, not very dense hairs. cf. pubescent, villous, etc.

hastate, shaped like an arrows-head in which the barbs, basal lobes or auricles spread more or less at right angles to the rest of the blade.

herbaceous, (1) not woody (2) green.

hermaphrodite (flower), a flower in which both stamens and ovary are present and functional.

hirsute, with a thick covering of somewhat firm, moderately long and spreading hairs.

hispid, with short scattered very stiff hairs or bristles.

hoary or *conescent,*, when the hairs are so short as not to be distinguished by the naked eye and yet give a general whitish or grey hue to the surface.

hypocrateriform, salver-shaped.

imbricate, a mode of aestivation in which one member of the whorl is outside all the others (i.e. its margins are free) and one inside all the others (i.e. both margins are overlapped); the others usually overlap by one margin only. Also used for leaves etc., where they overlap one another like the tiles of a house.

imparipinnate, pinnate with an odd terminal leaflet.

incurved, turned inwards.

indefinite, of varying number and usually numerous.

indehiscent, not opening by valves or pores.

indumentum, the clothing of hairs, scales, etc.

inferior, an inferior calyx, stamens etc., implies insertion at a level below, or near, the base of the ovary; an inferior ovary implies that the sepals, stamens, etc., are inserted on the torus at a level above or near the top of the ovary.

inflorescence, an axis or assemblage of axes especially devoted to the bearing of flowers and including the flowers and their bracts and bracteoles.

infundibuliform, funnel-shaped.

internode, the space between two leaves or metamorphosed leaves.

interpetiolar, said of stipules situated between the bases of opposite leaves.

intrapetiolar, said of stipules when each pair of a single leaf unite together within the axil of the leaf.

irregular, unsymmetrical.

keel, in a papilionaceous flower, the two anterior petals which are more or less united.

lacerate, irregularly cleft, as if torn.

laciniate, irregularly cut into very narrow lobes.

lanceolate, shaped like a lance-head.

lateral, situated to the right and left of the median plane. See anterior.

latex, milky juice.

laticiferous, possessing latex.

leaf, the flat blade or lamina, usually green and more or less horizontal, attached to the stem by a petiole.

leaflet, one of the blades of a compound leaf. A leaflet may usually be distinguished from a simple leaf from its position (one very frequently terminating the foliarnating the foliar axis), and from bearing no bud in its axil.

leguminous, resembling the peas and beans, especially in the nature of the fruit.

limb, the expanded part of a corolla, petal, etc., in contradistinction to the tube or claw.

linear, at least four or five times as long as broad.

lobed, cut less than half way down into more or less rounded segments.

-locular, used in composition to indicate the number, etc., of cells or compartments in an ovary.

loculicidal, dehiscing along the dorsal suture.

male flower, a flower which bears fertile stamens but not fertile carpels. An abortive pistil may be present in a male flower or not.

marginate, with a margin of a different character from the rest of the member.

micropyle, the canal through the integuments of an ovule at the apex of the nucellus.

monadelphous, more or less united into one bundle by the filaments.

monoecious, bearing both male and female flowers on the same individual.

mucronate, tipped with a very short, hard, usually blunt point.

muricate, covered with scattered, short, firm, thick or conical spines.

nervation, the arrangement of the fibre-vascular bundles in the leaves.

node, the plane of insertion of a leaf on the axis.

ob-, in composition means inversely. Thus an ovate leaf has the wider part towards the base, an obovate leaf is inversely ovate and has the wider part towards the apex.

oblanceolate, inversely lanceolate.

oblique, when referring to shape means with one half more developed than the other.

oblong, longer than broad and with the sides more or less parallel.

obovate, inversely ovate.

obovoid, inversely ovoid.

obsolete, not developed.

obtuse, blunt but scarcely rounded.

operculum, a lid.

opposite, on different sides of the axis with the bases on the same level.

oval, broadly elliptical.

ovate, egg-shaped in outline with the broader end towards the base-scarcely twice as long as broad.

ovate-lanceolate, ovate-oblong, etc., between ovate and lanceolate, between ovate and oblong, etc.

palmate, with the segments radiating like the spreading fingers of the hand.

palmatifid, palmate with the sinuses reaching about half-way down.

panduriform, fiddle-shaped with the base and end broader than above the base.

panicle, a repeatedly branched inflorescence.

papilionaceous, shaped somewhat like the flowers of a sweet pea or bean. A typical papilionaceous flower has a corolla with a large posterior petal (standard), two lateral petals (alae, wings) and two anterior petals more or less combined into a keel.

parallel-nerved, with numerous nerves from the base running more or less parallel and close to one another.

parasitic, drawing sustenance from the living tissues of other plants.
paripinnate, pinnate, with an even number of leaflets.
partite, divided but not quite to the base.
pedicel, a small stalk.
peduncle, the stalk of an inflorescence.
peltate, of leaves, attached to the petiole in the centre of the blade, or at least not by the margin.
penninerved, with one mid-rib and secondary nerves branching from it.
pentamerous, with 5 members in each whorl.
perianth, a general term for the floral envelopes including both calyx and corolla, but more especially when there is no differentiation into calyx and corolla.
persistent, not falling off.
petal, one of the divisions of the corolla.
petaloid, of a more or less delicate texture and white or coloured.
petiole, the stalk of a leaf.
petiolule, the stalk of a leaflet in a compound leaf.
pilose, covered with rather long, not matted nor very silky hairs.
pinna, the branches of a bi-pinnate leaf. See pinnate.
pinnate, a compound leaf with two or more leaflets springing from each side of the axis or rachis.
pinnately, in a pinnate manner.
pinnatifid, deeply lobed to about half-way down or more with the lobes pinnately arranged.
pinnatisect, pinnatifid down to the mid-rib.
pistil, a collective word for the ovary, style and stigma.
pistillode, a rudimentary pistil.
placenta, the surface to which are attached the ovules.
placentation, position of the placenta.
pod, typically a dry fruit derived from a mono-carpellary ovary, elongated in shape and dehiscing along one or both sutures, such, for instances, as a pea-pod.
pollen, the male spores which are developed in the pollen-sacs or loculi of anthers.
polygamous, bearing male, female and hermaphrodite flowers on the same plant.
prostrate, when they lie close to the ground.
puberulous, with very short soft hairs or down.
pubescence, the hairy covering of members.
pubescent, covered with close short fine hair.
punctate, marked with small dots or points.
raceme, an inflorescence in which the main axis continues to grow and the lowest flowers are the oldest and open first.
racemose, a form of branching in which the main axis continues to grow and remain stronger than the lateral axes, which successively spring from it, with the youngest nearest the apex.
regular, with all the members symmetrically disposed around the geometric centre of the flower.
reniform, kidney-shaped.
reticulation, the network of veins in a leaf.
retrorse, directed backwards.
rhomboid, shaped like a rhomboid.

rotate, a corolla with a very short tube and a horizontally spreading limb, or without a tube.
rotund, roundish.
rugose, with numerous minute elevations and depressions.
sagittate, allow-shaped with the basal lobes directed backward.
salver-shaped, with a long time tube and horizontally spreading limb.
scabrid, covered with small hard hairs or points so as to feel rough to the touch.
scandent, climbing.
scape, a peduncle which rises direct from the root.
scorpioid, with the (apparently) lateral axes forming a double now on one side of the usually curved (apparent) main axis or sympodium. Hole's definition differs. He says the lateral branch develops alternately on opposite sides.
seed, the ovule after fertilization and development of the embryo.
sepal, one of the divisions of the calyx.
sepaloid, green and resembling a sepal in texture rather than a petal.
septum, an interior wall.
serrate, toothed like a saw with the teeth inclined forwards.
serrulate, serrate but with the teeth very minute.
sessile, without a stalk.
silky, sericeous, covered with very fine adpressed silky hairs.
simple, not composed of a number of similar parts, opposed to compound.
sinuate, somewhat deeply waved.
spadix, a spike with an enlarged fleshy axis and usually enclosed when young in a spathe.
spathaceous, sheath.
spathulate, *spatulate*, spoon-shaped.
spicate, spiked, with the flowers in a spike.
spiciform, resembling a spike in appearance.
spike, a form of racemose inflorescence in which the flowers are sessile on the axis.
stamen, a modified leaf or sporophyll in the flowering plants which bears the microsporangia or anthers.
staminodes, imperfect or reduced or rudimentary stamens which do not bear fertile pollen.
stellate, spreading in a star-shaped manner.
stigma, the part of a carpel especially adapted by means of papillae, viscosity, etc., to receive the pollen-grains.
stipulate, having stipules.
stipule, stipules are a pair of processes (often absent), which spring from each side of the leaf-base (i.e. where the stalk of a leaf or the base of a sessile leaf leaves the stem).
subsessile, almost sessile.
subulate, awl-shaped.
succulent, soft and juicy.
sulcate, grooved.
syncarpous, with united carpels.
tendril, a filiform sensitive organ which winds round supports to enable weak stems to reach the light.
ternate, in groups of 3.
tomentose, with exceedingly close matted short pubescence.
tomentum, a covering of tomentose hairs.

torus, the portion of the floral axis from which spring the perianth, stamens, carpels or any portion of the flower.

trichotomous, divided with the division in threes.

triple-nerved, 3-nerved, with 3 nerves from base.

triquetrous, with 3 sharp corners.

truncate, as though cut off at the end.

turbinate, top-shaped.

umbel, an inflorescence in which the branches all radiate from the top of the peduncle. If these branches each terminate in a flower the umbel is simple; if they are again umbellately branched, the umbel is compound.

unilocular, applied to an ovary not divided up by partitions into separate compartments.

urceolate, flask-shaped and broadest below the middle.

valvate, said of sepals, etc., when they are only connate in bud by their edges, which do not overlap.

ventral, relating or attached to the front or inner angle of a carpel.

ventricose, suddenly bulged.

versatile, said of an anther which is attached above its base to the attenuated tip of the filament on which it swings.

verticillate, whorled.

vestigial, remaining as a trace.

villose, villous, covered with long fine soft hairs.

zygomorphic, symmetrical right and left of the median plane only, as in many lipped flowers. Sometimes equivalent to irregular.

INDEX

(Italic figures denote text-figures; CP after a figure denotes coloured plate; MP denotes monochrome plate).

Abelmoschus esculentus (Linn.) Moench, 254
Abutilon, 252
Acanthaceae, 110
Acocanthera G. Don., 222
 spectabilis Hook. f., 222, *121*
Allemanda Linn., 213
 cathartica Linn., 214, *117*
  v. *nobilis*, 216
  v. *schottii*, 216, 217
 hendersonii, 216
 neriifolia Hook., 214
 nobilis T. Moore, 216
 schottii Pohl, 216
 violacea Gardn., 216
Allemanda, Key to species, 213
Alstonia R. Br., 209
 scholaris R. Br., 209
Althaea officinalis, 253
Althaea, Shrubby, 260, *138*
Ambari, 253
Ambari Hemp, 253
American Sumach, 70, *42*
Angel's Trumpet, 143, *83*
Anogeissus, 196
Antigonon Endl., 289
 guatemalense Meissn., 290
 leptopus Hook. & Arn., 289, 290, 30 (CP), 95 (MP)
 v. *albus* Hort., 290
Antirrhinum, 248
Apocynaceae, 208,, 210
 Key to genera, 211
Arachis, 50
Aristolochia, 14, 15, 16, 17
 bracteata Retz., 18, 29, *21*
 bracteolata Lamk., 18, 29, *21*
 brasiliensis Mart. et Zucc., 21
 ciliata Hook., 27, *11*, *19*
 elegans Mast., 30, *22*
 fimbriata Cham. 27
 grandiflora Sw., 18, *12*
  v. *sturtevantii*, 20
 indica Linn., 23, *16*
 littoralis Parodi, 30, *22*
 macroura Gomez, 20, *13*, *14*
 ornithocephala Hook., 21, 2 (CP),

Aristolochia (cont.)
 4 (MP), 5 (MP)
 pallida, 14
 ridicula N. E. Brown, 26, *18*
 ringens Vahl., 25, *17*
 roxburghiana Klotz., 22, *15*
 tagala Cham., 22
 tomentosa Sims, 28, *20*
 Key to the species, 18
 Fringe-flowered, 27, *28*
Aristolochiaceae, 14
Arribidaea magnifica Sprague, 39
Asclepiadaceae, 205, 208, 209
Atropa Linn., 124
 belladona Linn., 124
Atropos, 124

Banisteria Linn., 200, 201
 laevifolia Juss., 202, *112*
Barbados Pride, 62
Bauhinia Linn., 72
 acuminata Linn., 76, *45*, 22 (MP)
 anguina Roxb., 79, *46*, *47*
 candicans Benth., 73, *43*
 corymbosa Roxb., 82, 83, *49*
 galpini N. E. Brown, 8 (CP), 23 (MP)
 purpurea, 72
 tomentosa Linn., 74, *44*
 vahlii, 73, 81, *48*
 variegata, 72
Bauhinia, Key to the species, 73
 Corymbose, 82, *49*
 Galpin's, 78, 8 (CP), 23 (MP)
 Sharp-leaved, 76, *45*
Bean Family, 50
Beaumontia Wall., 224
 grandiflora (Roxb.) Wall., 209, 225, *115*, 25 (CP)
Bela, 239
Belladona, 124
Bellyache Bush, 192, *108*, 64 (MP), 65 (MP)
Bhendi, 254
Bignonia Linn., 32, 33
 capreolata Linn., 36, *24*
 magnifica Bull., 39, *27*

Bignonia (cont.)
  speciosa R. Grah., 38, *26*, 7 (MP) 8 (MP)
  unguis-cati Linn., 37, *25*
  venusta Ker., 33, 34,, 35, *23*, 3 (CP), 6 (MP)
Bignonia, Key to the species, 34
Bignoniaceae, 32, 33, 199
  Key to the genera, 33
Bird of Paradise, 61, *36*
Birthwort, Birdshead, 21, 2 (CP), 4 (MP), 5 (MP)
  Bracteated, 29, *21*
  Hairy, 28,, *20*
  Livid-flowered, 20, *13, 14*
Bitter-sweet, 124
Blue Acacia, 52, 5 (CP), 13 (MP)
Blue Dawn-flower, 6, *3*
Blue Rain, 52, 2 (CP), 13 (MP)
Bougainvillea Comm., 291, 292, 293, 294, 296, 301, 303
  buttiana Holtt. & St., 292, 295, 303
    cv. Alick Lancaster, 299
      Blond, 300
      Done Luz Magsaysay, 303
      Dr R. R. Pal, 300
      Jayalakshmi, 300
      Lady Mary Baring, 299
      Lilac Queen, 299
      Louis Wathen, 292, 297, 299, 303
      Meera, 300
      Mrs McLean, 297, 303
      Orange Glory, 299
      Orange King, 299
      Scarlet Queen, 303
  glabra Choisy, 292, 294, 295, 299, 300, 301, 302, 303, 31 (CP), 99 (MP)
    v. *brachycarpa,* 301
      Cypheri, 295, 299
    cv. Cypheri, 301
      Dream, 300
      Formosa, 301
    v. *Formosa,* 299
    cv. Happiness, 300
      Kalyani, 300
      Magnifica, 301
      Sanderiana, 301
    v. *Sanderiana,* 295, 299
      Snow White, 299
      Splendens, 299
lateritia, 302
magnifica, 301
  v. *traillii,* 301

Bougainvillea Comm. (cont.)
  peruviana Humb. & Bonpl., 297, 298, 300, 301, 302
    cv. H. H. Buck, 299
      Lady Hudson, 298
      Lord Willingdon, 300
      Maharaja of Mysore, 299, 96 (MP)
      Mahatma Gandhi, 299
      Mary Palmer, 298
      Mrs Fraser, 300
      Mrs Lancaster, 300
      Princess Margaret Rose, 298
      Rosa Catalina, 300
      Surprise, 298
      Tomato Red, 300
  speciosa, 301
  spectabilis Willd., 292, 293, 294, 295, 299, 300, 301, 302, 303
    cv. Brasiliensis, 299
      Dar-es-Salam, 299
      Mrs Fraser, 294
    v. *lateritia* 294, 297, 299, 302, 303, 98 (MP)
    cv. *Rosa Catalina,* 294, 303
      Rosea, 299
      Speciosa, 299, 301
      Thomasii, 299, 301
  splendens, 301
Bougainvillea, discussion of hybrids, 302
  Distinctions between species, 300
  Gardening, 303
  Key to the Species, 293
Bower Plant of Australia, 47, *31*
Browallia, 124
Brugmansia suaveolens (Humb. & Bonpl. ex Willd.) Bercht. & Presl., 143, *83*
Brunfelsia Sw., 124, 144
  americana Linn., 147, *86*
  latifolia Benth., 145, *84*
  undulata Sw., 146, *85*
Brunfelsia, key to the Species, 145

Cactaceae, 182
Caesalpinia Linn., 60
  boduc (Linn.) emend. Dandy & Exell., 64
  boducella (Linn.) Flem., 64, *37*
  coriaria Willd., 70, *42*
  decapetala (Roth) Alston, 68
  digyna Rottl., 69, *41*
  gilliesii Wall., 61, *36*

Caesalpinia (cont.)
  nuga Ait., 66, *39*
  pulcherrima (Linn.) Swartz, 62, *7* (CP), 21 (MP)
    v. *flava* Hort., 63
  sappan Linn., 65, *38*, 20 (MP)
  sepiaria Roxb., 68, *40*, 18 (MP), 19 (MP)
Caesalpinia, key to the Species, 60
Caesalpiniaceae, 51, 60, 72
Calonyction, 1
Campsis grandiflora (Thunb.) K. Schum., 46
  radicans Seem., 45
Calceolaria, 248
Calico Flower, 30, *22*
Cape Honeysuckle, 44, *29*
Cape Jasmine, 91, *52*
Caprifoliaceae, 171
Caprifolium, 171
Carissa carandas Linn., 208
Catesbaea Linn., 108
  spinosa Linn., 108, *63*
Catharanthus roseus (L.) G. Don, 219
Cat's Claw, 37, *25*
Cephaelis (*Psychotria*) *ipecacuanha* Rich., 86
Cereus, 182
Cestrum Linn., 127
  aurantiacum Lindl., 129, 13 (CP), 41 (MP), 42 (MP)
  diurnum Linn., 130, *75*, 43 (MP), 44 (MP)
  elegans Schl., 128, *74*
  nocturnum Linn., 131, *76*, 45 (MP), 46 (MP)
  parqui L'Her., 132, *77*, 47 (MP), 48 (MP)
Cestrum, key to the Species, 128
  Golden, 129, 13 (CP), 41 (MP), 42 (MP)
Chandnee, 224
Changeable Rose, 256, *135*, 28 (CP), 86 (MP)
Chenopodiaceae, 199
Chinese Rose, 259, *137*
  Shoe-flower, 259, *137*
  Trumpet Creeper, 46
  Wisteria, 52
Chonemorpha G. Don., 226
  fragrans (Moon) Alst., 226
  macrophylla G. Don, 226, *122*
Christmas Flower, 188, 19 (CP), 60

(MP), 61 (MP)
Cinchona, 84, 85, 86
  ledgeriana, 84
  officinalis Linn., 85
Clerodendrum Linn., 149, 150, 157
  fragrans (Vent.) Willd., 164, *94*
  indicum O. Ktze, 158
  inerme Gaertn., 161, *92*
  infortunatum Linn., 165, *95*
  japonicum (Thunb.) Sweet, 162
  nutans Wall. ex Don., 167, *96*
  phlomidis Linn. f., 169, *98*
  phlomoides Willd., 169
  siphonanthus R. Br., 157, 158, *90*
  squamatum Vahl., 162, *93*
  thomsonae Balf., 160, *91*
  trichotomum Thunb., 168, *97*
  wallichii Merr., 167
Clerodendrum, key to the Species, 158
Clytostegia callistegioides (Cham.) Bur., 38
Coffea Linn., 102
  arabica, 86, 102
  bengalensis Roxb., 102, *60*
Combretaceae, 196
Combretum, 196
Common Garden Hibiscus, 259, 137
  Morning Glory, 11, *9*
Compositae, 50
Congea Roxb., 149, 153
  tomentosa Roxb., 153, *89*
Convolvulaceae, 1
Coral Creeper, 289, 30 (CP), 95 (MP)
Coral Plant, 194, *109*; 249, 27 (CP), 82 (MP), 83 (MP)
Cortex quebracho, 209
Cream-fruit, 229, *124*
Cross Vine, 36, *24*
Crown of Thorns, 186, *104*, *106*
Cup-and-Saucer Plant, 156, 16 (CP)
Cynanchum erectum, 208
Cypress-Vine, 4

Datura Linn., 124, 125, 141
  fastuosa Linn., 141, *50* (MP)
  metel Linn., 141, *82*
  stramonium Linn., 141
  suaveolens Humb. & Bonpl., 143, *83*, 51 (MP)
Datura, key to the Species, 141
Day Jasmine, 130, *75*
Deadly Nightshade family, 124

Deccan Hemp, 253
Delima, 210
Desmodium, 50
Dhoby's Tree, 89, *9* (CP), *26* (MP), *27* (MP)
   Yellowish. 87, *50, 25* (MP)
Diervilla, 171
Digitalis, 248
Dipelta, 171
Divi-divi Plant, 70, *42*
Dogbane family, 208
Durantia plumieri, 149
Dwarf Poinciana, 62

Emblica officinalis, 184
Ervatamia divaricata (L.) Burkill, 224
Euphorbia Linn., 184, 185
   antiquorum, 184
   bojeri Hook. 186, *104, 106*
   officinarum, 182
   pulcherrima Willd. ex Klotzsch., 188
      v. *alba* Hort., 189
Euphorbia, key to the Species, 186
Euphorbiaceae, 182, *105,* 185
Excoecaria agallocha, 185

Fever-nut, 64, *37*
Fire-cracker Plant, 249
Flos Passionis, 264 -
Forsteronia, 209
Fountain Plant, 249
Four-o'clock Plant, 291
Franciscea, 144
Fraxinus, 236

Galphimia Cav., 200, 206
   gracilis Bartl., 206, *23* (CP), *69* (MP), *70* (MP)
Gardenia Linn., 91
   florida Linn., 91, *52*
   jasminoides Ellis., 91
Giant Granadilla, 268, *142*
Giant Potato Vine, 137, *80*
Gmelina arborea, 149
Granadilla, Giant, 268, *142*
   Purple, 277, *148*
Greek Periwinkle, 221 *120*
Griffith's Sophora, 55, *14* (MP), *15* (MP)

Hairy Birthwort, 28, *20*
Hamelia Jacq., 104

Hamelia (cont.)
   patens Jacq., 105, 61, *32* (MP), *33* (MP)
Hamelia, Spreading, 105 *61, 32* (MP), *33* (MP)
Hamiltonia Roxb., 106
   suaveolens Roxb., 106, *62, 34* (MP), *35* (MP)
Hamiltonia, Sweet-scented, 106, *62, 34* (MP), *35* (MP)
Hancornia, 209
Harsingar, 236
Heavenly Blue, 3
Hemp, Deccan, 253
Henbane, 124
Hevea brasiliensis, 184
Hibiscus Linn., 252, 253, 254, 255
   abelmoschus Linn., 253
   cannabinus Linn., 253
   esculentus Linn., 254
   mutabilis Linn., 256, *135, 28* (CP), 86, (MP)
   rosa-sinensis Linn., 255, 259, *137*
   sabdariffa Linn., 254
   schizopetalus (Mast.) Hook. f., 257, *136,* 87 (MP)
   syriacus Linn., 260, *138,* 90 (MP), 91 (MP)
Hibiscus, Coral, 257, *136,* 87 (MP)
   Common Garden, 259
   In the garden, 254
   Key to the Species, 256
Hippomane mancinella, 185
Hiptage Gaertn., 200
   benghalensis (L.) Kurz., 200
   madablota Gaertn., 200, *111*
Holmskioldia Retz., 149, 156
   sanguinea Retz., 156, *16* (CP)
Honeysuckle, Japanese, 175 *101*
   Trumpet, 173
Hyoscyamus niger Linn., 124

Indian Pink, 4
Ipecacuanha, 86
Ipomoea Linn., 1
   batatas, 2
   cairica (Linn.) Sweet, 7
   carnea Jacq., 7, 5
   coccinea Linn., 9, 7
   congesta R. Br., 6
   fistulosa Mart. ex Choisy, 7
   hederacea, 11
   hederifolia House, 10

Ipomoea (cont.)
  horsfalliae Hook., 5, *2*
    v. *briggsi* Hort., 6
  learii Pax, 6, *6*
  lobata Thell, 10, *8*
  (*Mina*) *lobata*, 2
  palmata Forsk., 7, 4, 1 (MP), 2 (MP)
  pes-caprae, 2
  purga, 2
  purpurea (Linn.) Roth 11, *9*
  quamoclit Linn., 3, *1*
    v. *alba* Hort., *5*
  rubro-caerulea Hook., 3, 1 (CP)
    v. *alba* Hort., 4
  sinuata Ortega, 12, *10*
  tricolor Cav., 3
  turpethum, 2
  vitifolia Sw., 8, *6*
Ipomoea, key to the Species, 2
  Star, 9
Ixora Linn., 86, 93
  arborea Roxb. ex Sm., 96
  barbata Roxb., 97, *55*
  chinensis Lamk., 98, *56*
  coccinea Linn., 102, *10*, 28 (MP)
  fulgens Roxb., 100, *58*
  lutea Hutch., 101, *59*
  parviflora Vahl, 96, 54, 29 (MP)
  rosea Wall., 94, *53*
  undulata Roxb., 99, *57*
Ixora, key to the Species, 93
  Bearded, 97, *55*
  Chinese, 98, *56*
  Pink, 94, *53*
  Scarlet, 93, *10*, 28 (MP)
  Small-flowered, 96, 54, 29 (MP)
  Wavy-leafed, 99, *57*
  Yellow, 101, *59*

Japanese Honeysuckle, 175, *101*
Jasmin, 237
Jasmine, 237
  Arabian, 238, *128*
  Cape, 91, *52*
  Day, 130, *75*, 43 (MP), 44 (MP)
  Hairy, 239, *129*, 78 (MP)
  Hill, 231, *125*, 76 (MP), 77 (MP)
  Primrose, 241, *130*, 79 (MP)
  Spanish, 245, *133*, 81 (MP)
  White, 243, *132*
  Willow-leaved, 132, 77, 47 (MP), 48 (MP)
  Yellow, 242, *131*, 26 (CP), 80 (MP)

Jasminum Linn., 236, 237
  grandiflorum Linn., 244, 245, *133*, 81 (MP)
  humile Linn., 242, *131*, 26 (CP), 80 (MP)
  multiflorum (Burm. f.) Andr. 239
  officinale Linn., 243, 246, *132*
    v. *grandiflorum* (L.) Kobuski, 245
  primulinum, 241, *130*, 79 (MP)
  pubescens Willd., 239, *129*, 78 (MP)
  sambac (Linn.) Ait., 238, *128*
Jasminum, key to the Species, 238
Jatropha Linn., 182 189
  curcas Linn., 190
  gossypifolia Linn., 190, *108*, 64 (MP), 65 (MP)
  hastata Jacq., 192
  multifida Linn., 194, *109*
  panduraefolia Andr., 192, 20 (CP), 62 (MP), 63 (MP)
  podagrica Hook., 190, *107*
Jatropha, key to the Species, 190
  Fiddle-leaved, 192, 20 (CP), 62 (MP), 63 (MP)
Jerusalem Cherry, 135, *78*
Jessamin, 237
Jessamine, 131
  Night, 131, *76*, 45 (MP), 46 (MP)
  Star, 232, *126*

Karaunda, 208
Kickxia, 209

Ladies' Fingers, 254
Lady of the Night, 131, *76*, 45 (MP), 46 (MP)
Landolphia, 209
Lantana, 149, 150
Lasiobema scandens (Linn.) De Wit, 79
Leadwort, Cape, 179, 58 (MP), 59 (MP)
  Ceylon, 178
  Rosecoloured, 180, *103*
Leguminosae, 50, 51, 60, 72
  Key to the Genera, 51
Leycesteria, 171
Ligustrum, 237
Limonium, 177
Linaria, 248
Livid-flowered Birthwort, 20, *13*, *14*
Lonicera Linn., 171
  confusa DC., 174, *100*

Lonicera (cont.)
　japonica Thunb., 175, *101*, 57 (MP)
　　v. *chinensis*, 176
　periclymenum Linn., 172, *99*
　sempervirens Linn., 173, 17 (CP), 56 (MP)
Lonicera, key to the Species, 173

Madagascar Periwinkle, 219, *119*
Mallow, Musk, 253
Malpighia, 206
Malpighiaceae, 199, 200, 203
　Key to the Genera, 200
Malva, 251, 252
Malvaceae, 251, 252
Malvastrum, 252
Malvaviscus, 252
Mandragora, 125
Mandrake, 125, *73*
Manihot glaziovii, 184
　utilissima, 184
Marvel of Peru, 291
Maypop, 275, *147*
Melodinus Forst., 234
　monogynus Roxb., 234, *127*
Merremia, 1
Mimosaceae, 51, 60, 72
Mimulus, 248
Mina, 1
　lobata Llav. & Lex., 10
Mirabilis jalapa, 291
Mogra, 239
Moonbeam, 224, 24 (CP)
Morinda, 86
Morning-glory, 3
Motiya, 239
Musk Mallow, 253
Mussaenda Linn., 87
　frondosa Linn., 89, 9 (CP), 26 (MP), 27 (MP)
　luteola Del., 87, *50*, 25 (MP)
Mussaenda, key to the Species, 87
Mysore-thorn, 68, *40*, 18 (MP), 19 (MP)

Nerium Linn., 217
　indicum Mill., 217
　odorum Sol., 210, 217, *118*
Nicotiana tabacum Linn., 124
Nyctaginaceae, 291
Nyctanthes arbor-tristis, 236

Olea, 236
　europaea Linn., 236
　fragrans, 247
Oleaceae, 236
　Key to the Genera, 237
Oleander, 217, *118*
　Yellow, 212, *116*
Oilve family, 236
Osmanthus Lour., 246
　fragrans Lour., 246, *134*

Paederia foetida, 86
Pandorea jasminoides (Lindl.) K. Schum., 47
　pandorana (Andrews), V. Steen., 48
Paperchase Tree, 89, 9 (CP), 26 (MP), 27 (MP)
Papilionaceae, 51, 53, 60, 72
Passiflora Linn., 262, 265, 278
　alata, 266, 274
　biflora Lamk., 270
　caerulea Linn., 263, 264, 272, *140, 145*
　caerulea-racemosa, 274
　calcarata Mast., 286, *153*
　ciliata Dryand., 278, *149*
　edulis Sims., 277, *148*, 93 (MP)
　foetida Linn., 281, *151*
　　v. *ciliata*, 278
　gracilis Jacq., 287, *154*
　holosericea Linn., 280, *150*, 94 (MP)
　incarnata Linn., 275, *147* 92 (MP)
　leschenaultii DC., 271 *144*
　lunata Sm., 270, *143*
　minima Linn., 267, *141*
　morifolia Mast., 283, *152*
　quandrangularis Linn., 268, 278, *142*
　　v. *macrocarpa*, 269
　racemosa, 274, 284, 29 (CP)
　raddina, 274
　suberosa Linn., 267, 274, *146*
Passiflora, key to the Species, 266
Passifloraceae, 262
Passion-flower, Crescent-leaved, 270, *143*
　Edible, 277, *148*
　Fringe-leafed, 278, *149*
　Madagascar, 286, *153*
　Princess Charlotte's, 284, 29 (CP)
　Silky-leafed, 280, *150*
　Small, 267, *141*
　Stinking, 281, *151*
　Wild, 275, *147*

Patwa, 254
Paulownia, 248
Pavonia, 252
Peacock flower, 62, *7* (CP), 21 (MP)
Pelican Flower, 18, *12*
Pentstemon, 248
Periploca graeca Linn., 205
Periwinkle, 219
   Greek, 221, *120*
   Madagascar, 219, *119*
Petrea Houst., 149, 155
   volubilis Linn., 155, 15 (CP), 54 (MP), 55 (MP)
Petunia, 124
Phanera vahlii (W. & A.) Benth., 81
Physic Nut, 64, *37*, 194, *109*
Plumbaginaceae, 177
Plumbago Linn., 177, 178
   capensis Thunb., 179, 18 (CP), 58 (MP), 59 (MP)
   europaea, 177
   rosea Linn., 180, *103*
     v. *coccinea,* 181
   zeylanica Linn., 178, *102*, 181
Plumbago, key to Species, 178
   Cape, 179, 18 (CP), 58 (MP), 59 (MP)
Poinciana, 62
   gilliesii Hook., 62
Poinsettia pulcherrima (Willd.) Grah., 188, 19 (CP), 60 (MP), 61 (MP)
Poison Hog-meat, 18, *12*
Polygonaceae, 289
Polygonum, 289
Potato Creeper, 139, 14 (CP)
   Vine, 139, *81*
Prickly-apple, 108, *63*
Psychotria, 86
Purple Granadilla, 277, *148*
Purple Wreath, 155, 15 (CP), 54 (MP), 55 (MP)
Pyrostegia venusta (Ker.-Gawl.) Miers, 33, 34

Quamoclit, 1
   coccinea Moench., 9
   lobata House, 10
   pinnata (Linn.) Boj., 4
Quarter Vine, 36, *24*
Quisqualis Linn., 196
   indica Linn., 196, 21 (CP), 66 (MP)
Railway Creeper, 7, *4*, 1 (MP), 2 (MP)

Rangoon Creeper, 196, 21 (CP), 66 (MP)
Rat-ki-rani, 131, *76*
Rauwolfia serpentina, 178
Red Sorrel, 254
Ricinus communis Linn., 185
Rondeletia Linn., 103
   odorata Jacq., 104, 11 (CP), 30 (MP), 31 (MP)
Rondeletia, Sweet-smelling, 104, 11 (CP), 30 (MP), 31 (MP)
Rose, Changeable, 256, *135*
Rose-of-Sharon, 260, *138*
Roupellia Wall. et Hook., 229
   grata Wall. & Hook. ex Benth., 229, 231
Rozella, 254
Rubia cordifolia, 84, 86
Rubiaceae, 84, 171
   Key to the Genera, 87
Russelia Jacq., 248
   coccinea (Linn.) Wettst., 249
   equisetiformis Cham. et Schl., 249
   juncea Zucc., 249, 27 (CP), 82 (MP), 83 (MP)
   multiflora Sims., 250
   sarmentosa Jacq., 249, 27 (CP), 84 (MP), 85 (MP)
Russelia, key to the Species, 248
   Twiggy, 249, 27 (CP), 84 (MP), 85 (MP)

Salpiglossis, 124
Sandwich-Island Creeper, 289
Saponaria officinalis, 149
Sappan Wood, 65, *38*
Saritaea magnifica, 39
Schizanthus, 124
Scrophulariaceae, 248
Serissa Comm., 89
   foetida Lamk., 90, *51*
Sharon, Rose of, 260, *138*
Shrubby Althaea, 260, *138*
Small-flowered Ixora, 96, *54*
Snake Climber, 79, *46, 47*
Solamen, 134
Solanaceae, 124, 125, 134
   Key to the Genera, 126
Solanum Linn., 124, 134, 137
   dulcamara Linn., 124
   jasminoides Paxt., 139, *81*
   lycopersicum, 134

Solanum (cont.)
  melongena, 134
  nigrum Linn., 134
  pseudo-capsicum Linn., 135, *78*
  rantonnetii Carr., 136, *79*
  seaforthianum Andr., 139, 14 (CP)
  tuberosum, 134
  wendlandii Hook., 137, *80*, 49 (MP)
Solanum, key to the Species, 134
Solaticum, 134
Solatrum, 134
Sophora Linn., 53, 54
  griffithii Stocks, 55, *33*, 14 (MP), 15 (MP)
  japonica Linn., 54
  secundiflora DC., 57, *34*, 17 (MP)
  tomentosa Linn., 56, 6 (CP), 16 (MP)
  viciifolia Hance, 58, *35*
Sophora, key to the Species, 55
  Griffith's, 55, *33*, 14 (MP), 15 (MP)
  Hairy, 56
  One-sided, 57
Sorrel, Red, 254
Spanish Guava, 108, *63*
Spermadictyon suaveolens Roxb., 106
St. Thomas-Tree, 74, *44*
Stachys alopecuroides, 127
  officinalis, 127
Stachytarpheta Vahl, 149, 150
  indica (Linn.) Vahl., 151, *87*
  mutabilis (Jacq.) Vahl., 152, 52 (MP), 53 (MP), *88*
  jamaicensis Vahl., 151
Stachytarpheta, key to the Species, 151
  Variable, 152, 52 (MP), 53 (MP), *88*
Star Jessamine, 232, *126*
Statice, 177
Stenolobium stans, 42
Stigmaphyllon Juss., 200, 203
  ciliatum (Lamk.) A. Juss., 204, 113, 22 (CP), 67 (MP), 68 (MP)
  periplocifolium (Desf.) A. Juss., 205, *114*
Stigmaphyllon, key to the Species, 203
  Fringed, 204, *113*, 22 (CP), 67 (MP), 68 (MP)
Strophanthus DC., 209, 229
  gratus (Benth.) Baill., 229, *124*, 74 (MP), 75 (MP)
Symphoricarpus, 171
Syringa, 237
  vulgaris Linn., 237
Tabernaemontana Linn., 223

  coronaria R. Br., 224, 24 (CP)
Tamarind, 50
Tecoma Juss., 40
  australis R. Br., 48, *32*
  capensis Lindl., 44, 39 (MP), 40 (MP), *29*
  grandiflora Delaun., 46, 4 (CP)
  jasminoides Lindl., 47, *31*
  mollis Humb. & Bonpl., 45
  radicans Juss., 45, *30*, 11 (MP), 12 (MP)
  smithii Wats., 45
  stans H.B.K., 42, *28*, 9 (MP), 10 (MP)
    v. *incisa* Hort., 43
Tecoma, key to the Species, 42
  Large-flowered, 46, 4 (CP)
Tecomaria capensis (Thunb.) Spach., 44
Tectona grandis, 149
Terminalia arjuna, 196
  tomentosa, 196
Thespesia, 252
Thevetia Linn., 211
  neriifolia Juss., 212 *116*
  peruviana (Pers.) Schum., 212
Thunbergia Retz.; 110, 112, *64, 65*
  alata Boj., 114, *66*
    v. *alba*, 115
      aurantiaca, 115
  coccinea Wall., 122, *72*
  erecta, T. Anders., 116, *68*, 38 (MP), 39 (MP)
    v. *alba*, 117
  fragrans Roxb., 115, *67*, 36 (MP), 37 (MP)
  grandiflora Roxb., 118
  laurifolia Lindl., 119, *70*
  mysorensis T. Anders., 120 *71*
  natalensis Hook., 117, *69*
Thunbergia, key to the Species, 113
  Large-flowered, 118, 12 (CP), 40 (MP)
  Upright, 116, *68*, 38 (MP), 39 (MP)
  White, 115, 36 (MP), 37 (MP)
Torenia, 248
Trachelospermum Lem., 231
  divaricatum (Thunb.) K. Schum., 232
  fragrans Hook. f., 231, *125*, 76 (MP), 77 (MP)
  jasminoides Lem., 232, *126*
  lucidum (Don.) K. Schum., 231
Trachelospermum, key to the Species, 231

INDEX

Trumpet-Flower, 36, 24, 42, *28*, 9 (MP), 10 (MP)
Trumpet Flower, 212, *116*
Trumpet Honeysuckle, 173, 17 (CP), 56 (MP)
Trumpet Vine, 45, *30*, 11 (MP), 12 (MP)
Tube-Flower, 158, *90*
Turk's turban, 158, *90*

Uncaria, 86
　gambir Roxb., 86
Urena lobata, 252

Vallaris Burm., 227
　heynei Spr., 227, *123*
　solanacea (Linn.) O. K., 227
Verbena, 149
　officinalis, 149
Verbenaceae, 149
　Key to the Genera, 150
Veronica, 248
Viburnum, 171
Vinca Linn., 210, 219
　major Linn., 221, *120*

Vinca (cont.)
　pervinca, 219
　rosea Linn., 219, *119*, 73 (MP)
Vinca, key to the Species, 219
Vitex agnus-castus, 149
　littoralis, 149
　peduncularis, 149

Wavy-leafed Ixora, 99, *57*
Wax-flower, 224
Wightia, 248
Willow-leaved Jasmine, 132, *77*
Willughbeia, 209
Wintergreen, 222, *121*
Wisteria Nutt., 51, 52
　sinensis (Simms) DC., 52, 5 (CP), 13 (MP)
　v. *alba*, 53
Wisteria, Chinese, 52, 5 (CP), 13 (MP)
Withania Pauq., 125
　coagulans Dunal, 125
　somnifera Dunal, 125
Wonga-Wonga Vine, 48, *32*

Yellow Bells, 42, *28*
Yellow Elder, 42, *28*
Yellow Oleander, 212, *116*